SIMULATION OF WAITING-LINE SYSTEMS

SUSAN L. SOLOMON

Eastern Washington University
Cheney, Washington

PRENTICE-HALL, INC.
Englewood Cliffs, New Jersey 07632

Library of Congress Cataloging in Publication Data

SOLOMON, SUSAN, L. (date)
    Simulation of waiting-line systems.

    Bibliography: p.
    Includes index.
    1. Queuing theory.   2. Simulation methods.   I. Title.
T57.9.S64 1983   658.4'03'3   83-8647
ISBN 0-13-810044-6

Editorial/production supervision
    and interior design: *Kathryn Gollin Marshak*
Cover design: *Edsal Enterprises*
Manufacturing buyer: *Gordon Osbourne*

Printed in the United States of America

10 9 8 7 6 5 4 3 2 1

ISBN   0-13-810044-6

Prentice-Hall International, Inc., *London*
Prentice-Hall of Australia Pty. Limited, *Sydney*
Editora Prentice-Hall do Brasil, Ltda., *Rio de Janeiro*
Prentice-Hall Canada Inc., *Toronto*
Prentice-Hall of India Private Limited, *New Delhi*
Prentice-Hall of Japan, Inc., *Tokyo*
Prentice-Hall of Southeast Asia Pte. Ltd., *Singapore*
Whitehall Books Limited, *Wellington, New Zealand*

CONTENTS

Chapter 5   **INTRODUCTION TO GPSS**   **122**

Chapter 6   **INTERMEDIATE GPSS**   **148**

Contents

PREFACE

Simulation is simultaneously one of the easiest to understand and one of the most misunderstood of the management science techniques. Mathematicians, computer scientists, and engineers sometimes denigrate simulation because it is purportedly not based on elegant, theoretical, general models as is, for example, linear programming. Managers and business people sometimes have been told that simulation is a panacea that always solves any problem; these individuals may then be dismayed to find that simulation may be more expensive, more time consuming, and less accurate than they were lead to believe.

The purpose of this book is to present a balanced, realistic picture of the entire process of simulation at a level consistent with the backgrounds and interests of most managers and students of business. Real-world ''glitches'' and often-forgotten caveats are discussed and dealt with. While waiting-line systems are the main vehicle for discussion, the principles apply to many types of simulation. Anyone conversant with computer usage and introductory applied statistics should be adequately prepared for the technical material. Nonparametric statistics are used almost exclusively because they are intuitively easy to understand and avoid many assumptions. Large-sample applications of nonparametric statistics call for χ^2 or normal distributions with which most business students are familiar. The book is intended as a text in a three-semester hour or four-quarter hour survey course in simulation or as a self-study practical introduction to simulation. Numerical answers to selected end-of-chapter problems are given. A complete statistical support package (SIM-STAT) written in conversational FORTRAN is provided along with an extensive bibliography, not just of individual articles but of professional simulation organizations and their regular publications. Chapter appendices explain how the examples given in the chapter were obtained by using the programs in the SIMSTAT package. A few simple illustrations are offered and are carried through the narrative as well as the problem sections of the book to provide clarity, continuity, and perspective on the overall modeling process.

Nearly two decades of teaching and research in simulation have convinced the author that it is possible and desirable to provide beginners with a theoretical and practical foundation in the principles of model building, the distinction between simulation and other modeling methods, statistical analysis of model inputs and outputs, and simulation programming in sufficient detail to enable the student to formulate and implement a complete model of a simple, real-world waiting-line system within a semester or quarter. This author prefers to acquaint the student with all aspects of simulation modeling in the context of a simple situation rather than immerse him or her in one aspect, such as simulation programming, in great detail. In my classes, each student or pair of students must submit such a project contributing about 50 percent of the total course grade. The projects must be completed in segments synchronized with chapters in the book to ensure that students do not fall irretrievably behind. Class members brainstorm unforeseen difficulties encountered by their colleagues. It takes discipline and perseverance to accomplish these objectives within a limited time span, but it can be done. The SIMSTAT statistical support package is expressly designed to enable students to tackle end-of-chapter and project problems without spending scarce time on review of statistical methods, table lookups, data transcription, formula interpretation, and repetitious arithmetic. Each student is provided with statement of computer costs incurred in the simulation effort at the end of the course.

Chapter 1 is an introduction to modeling and model types. Chapter 2 discusses goodness-of-fit tests. Chapter 3 introduces analytical waiting-line models along with a methodology for evaluating the cost components of service systems that can be extended to simulation models. Chapter 4 examines random variate generation and criteria for their suitability. Chapters 5, 6, and 7 give more depth of coverage of GPSS than do most other simulation survey books but without the detail of compiler operation and options characteristic of books dedicated to the GPSS language. Chapter 8 encompasses tests for transients and serial correlation in simulation output data and methods for coping with them. Chapter 9 enables the novice modeler finally to use "sanitized" simulation output data to make estimates of system performance or to test hypotheses about differences in performance of alternative configurations. Chapter 10 offers a selection of simulation studies which were chosen according to the following criteria: readability, relevance of subject area, recentness, practicality, quality and completeness of modeling effort, and apparent benefits or cost savings. This chapter includes some waiting-line simulations as well as some other types of simulation, some simulations in the private sector, and some in the public sector. Chapter 11 investigates the management of the modeling process as well as the future of simulation. Techniques employed throughout the book were chosen for their simplicity of comprehension and execution rather than for the efficiency or state-of-the-art aspects. It is assumed that a reader who decides to become a simulation professional will undertake further continuing study.

I would like to thank Geoffrey Gordon of IBM for his reviews of the entire manuscript, and Eastern Washington University for providing me with professional leave and computer support without which this work would have been impossible.

Susan L. Solomon
Cheney, Washington

chapter one

THE NATURE AND PURPOSE OF MODELS

THE RATIONALE FOR MODELING

Our world is complex and is growing ever more so. The complexity results from both a rapidly increasing store of facts and the intricate interrelationships among these facts. A decision maker, no matter how bright and how well trained, finds it difficult to sift the relevant facts from those which are irrelevant and to fathom the complexities of the system that the decision will affect. Whether the problem is to modify a currently existing system or to design a brand-new system, the development of a model of the system on which to experiment is often desirable. First, the decision maker can try out alternatives (and probably make some mistakes) using the model rather than the real world as the laboratory. This is not only less embarrassing but is usually less costly and difficult than experimentation with the real system. Also, when the real system is not yet in existence, waiting for it to be implemented before experimentation can impose a long delay and incur the possibility that the implemented system will be unsatisfactory.

WHAT IS A SYSTEM?

A system may be defined as a collection of inputs which pass through certain processing phases to produce outputs. A manufacturing system may use crude oil as an input and a cracking plant for processing and produce various types of oil and gasoline as outputs.

The performance of the system might be evaluated by criteria such as how many gallons of output result from each gallon of input, how long it takes to

1

transform the inputs into outputs, how expensive it is to transform the inputs into outputs, and how much space the processing phases occupy.

In many systems there is a mechanism for evaluating the performance of the system and altering the nature of the process depending on the degree to which performance criteria have been satisfactorily met. If the oil refinery's ratio of inputs used to ouputs produced is considered too low by some predefined criterion, this may be reported to a manager or an operator who can alter the mix of inputs, change the nature of processing, or vary the composition of outputs in an effort to improve the performance of the system. When the performance of the system is monitored and evaluated, with the evaluation affecting the future state of the system, the system is said to incorporate feedback.

A service system, likewise, has inputs, processing phases, and outputs, as well as measures of system performance. The inputs are customers and any items the customer wishes to obtain; the processing components are servers, equipment, and space; and the outputs are satisfied customers. Performance measures for a service-oriented system might include the amount of time a customer spends waiting for service or waiting for processing to be completed, the number of customers waiting in line, or the proportion of time the servers are idle. By changing the number or quality of servers or equipment, the manager can bring the performance of the system into better alignment with predefined criteria.

Not all variables which are relevant to system performance are under the manager's control. A downturn in the economy or the emergence of a strong new competitor can influence system performance in a negative manner, but they cannot be wished away by the manager. However, the manager may evaluate their probable effect on the system and manipulate those variables which he or she can alter in an attempt to compensate. Thus, the manager might elect to improve the quality of merchandise or service or to cut costs in an endeavor to offset the competitor's challenge. Variables which affect a system but cannot be manipulated directly by the manager are called *exogenous variables*. Variables which can be controlled by the manager are called *endogenous variables*. Variables which are exogenous to one system may be endogenous to another. For example, the new competitor is exogenous to the service system of the manager just described, but they are both endogenous to the system which is the economic structure of the United States.

STANDARDIZED VERSUS CUSTOM MODELS

Management science, decision science, and operations research are all names for a set of quantitative tools which have assisted decision makers with their task. Many of these tools specify which variables, or sets of facts, are to be examined, what the nature of their interrelationships is presumed to be, and what results or performance measures may be observed as output of the model. To the extent that these tools are computer-based, the decision maker reaps the additional advantage of speed and accuracy of computation.

For many well-known decision situations, these restrictive, "packaged" models yield good results. Production and operations managers are familiar with models of inventory systems in which the relevant variables—volume of sales per unit of time, cost of ordering new merchandise, cost of carrying unsold inventory, purchase price per unit incorporating possible discounts, and cost of running out of stock—are related in an intuitively plausible mathematical expression to provide the manager with answers to inevitable questions—how many units to order at a time, how often to place an order, how often stockouts will occur, and how much this inventory management policy will cost per unit time.

Other tools are more general in their applicability, but by convention, they have tended to be used more frequently in some decision situations than in others. An example is linear programming, a technique which enables the decision maker to optimize some objective such as finding the combination of possible products or services that minimizes cost or maximizes profits. Unlimited optimization is generally precluded by some restrictions or constraints, such as meeting specified levels of demand, confining labor-intensive activities to a 40-hour week, limiting usage of materials to the quantities on hand, or ensuring that the proportion of a certain input or output is not more than, not less than, or exactly equal to some fraction of the total input or output.

A bit of imagination or exposure to more advanced treatment of the subject may lead the decision maker to employ linear programming in a decision situation other than the traditional product mix determination. For example, linear programming may be used to route deliveries from warehouses to retail stores, to assign people to jobs, to budget capital and operating expenditures, or to schedule and monitor the progress of large-scale projects. Many more potential but less obvious applications exist. Sensitivity analysis methods permit the decision maker to ask and answer "what if" questions about the effects of variations of the stated objective function and the constraints.

Nevertheless, any general-purpose packaged model has limitations which may not be ignored on pain of inaccurate results and false conclusions. The standard formulation of the inventory model assumes that the cost of holding unsold inventory is proportional to the average level of that inventory. This assumption is reasonable for the segment of holding costs, which comprises costs of spoilage, pilferage, obsolescence, and insurance and which may be assumed to increase and decrease with the level of inventory over the order cycle. However, one of the largest components of holding cost is the cost of warehousing; a firm usually needs and pays for sufficient warehouse space to hold the maximum as opposed to the average amount of inventory during the order cycle, since that maximum is the quantity initially delivered with each new order.

Fortunately, a well-trained, experienced, clever, and aware decision maker can modify the standard inventory model to accommodate some variations. Similarly, an informed user of linear programming who encounters a crucial nonlinearity in the situation to be modeled may select a nonlinear optimization technique as an alternative. Eventually, however, the multitude of idiosyncrasies in a particular situation may render all known packaged models ineffective. At that point, the decision maker must assess whether the deviations of the particular situation from the general formulation of a packaged model are sufficiently important to warrant the cost of constructing a complex, customized model. Sometimes the decision maker will tolerate modest discrepancies in order to use a packaged model which is convenient, accessible, and inexpensive. The construction of a fairly simple customized model may be no more expensive and considerably more accurate than the use of a package. This possibility should be explored before choosing the vehicle for modeling.

ATTRIBUTES OF MODELS

A model may or may not be a faithful replication of reality or proposed reality. For reasons of complexity and cost previously mentioned, most decision makers prefer models which extract the salient characteristics of reality while omitting the "noise" or irrelevancies. In some cases all characteristics are relevant, and identical or scale models are constructed, often at great expense. The aerospace industry exemplifies instances in which faithful replication is an essential prerequisite to a successful model in which the decision maker can have confidence. While the cost of model construction and testing is high, the decision maker can rest assured that most design defects will be exposed, discovered, and rectified as a result of experimentation with the model rather than after a real-system catastrophe.

Models may include characteristics which are vastly different from the real system as long as those characteristics have no impact on the performance outputs which are relevant to the decision maker. No one would take the trouble to include irrelevancies in a customized model, but they may be present in a packaged model. A human fashion model may display attributes of attractiveness, proportion, style, and grace which are desirable for the apparel marketing system, yet the same model may possess or lack intelligence, knowledge, lucidity, and other characteristics which are generally construed as valuable but which are not germane to the task at hand.

In general, simplicity is a virtue in modeling. Simplicity clarifies the system in the eyes of the decision maker, illuminating alternatives which might be profitable. A simple model is also less expensive to construct and manipulate. However, simplicity should not be pursued at the expense of relevant elements.

The decision maker must assess whether the model is to describe an optimum condition or reality, which is generally suboptimal. The fashion model may represent an ideal but may be very different from reality as characterized by an average customer for the garments which the model displays. In designing a new system, the search for optimality may weigh more heavily than in modifying an existing system in which few variables are left to the decision maker to manipulate. The manager may wish that the number of customers desiring service while the employees are having lunch would be few, or zero, but may have no control over this variable and should plan the service system to accommodate rather than ignore this unpleasant reality.

Some models, particularly packages such as inventory control and linear programming, lend themselves well to the goal of optimization as well as experimentation with selected alternatives. However, many models are not amenable to mathematical optimization. In these instances, the decision maker must be content with the facility to test a set of possible model configurations and select the best of the sample tested, realizing that it is unlikely to be a universal optimum but the cost of infinite search is prohibitive.

The question of optimization requires the decision maker to focus on objectives—what is to be maximized or minimized and how it is to be measured. In a service system, the manager may wish to minimize the cost of staffing subject to the constraint that the average waiting time per customer should not exceed 5 minutes. This statement facilitates development and evaluation of models incorporating various numbers and qualities of servers. However, it makes an implicit assumption about the value of customer waiting time and the probability that a disgruntled customer will elect to do business elsewhere after enduring a long wait.

TYPES OF MODELS

According to one dimension of classification, the most common types of models are physical, schematic, mathematical, and heuristic. A physical model may be an identical replication of the real system, such as an experimental aircraft or a fashion model, or it may be scaled down, such as the wind tunnel version of the same aircraft or a doll analogous to the fashion model. A schematic model is a pictorial representation of the system, such as a blueprint or a graph. A mathematical model consists of expressions containing variables, constants, and operators which describe the process of interest. An heuristic model is a collection of descriptors and decision rules, usually computer-based, which is not limited by the physical, diagrammatic, or mathematical bounds of the other types of models. While mathematical models may be implemented on a computer, they are restricted to purely mathematical operations such as arithmetic, algebra, and calculus. Heuristic models may be programmed to search data sets and perform logical comparisons as well as mathematics.

Inventory and linear programming models are analytic in nature. That is, they offer a general formulation of a common class of problems for which the user needs only to supply specific parameters. Synthetic models, on the other hand, allow the modeler to extrapolate from a particular situation to the more general case which has not been previously described and evaluated.

Another distinction in model types is static versus dynamic. A static model is a snapshot of a system at a particular point in time; a dynamic model conveys the essence of changes in the system over time. The contrast is akin to the accountant's balance sheet, which reflects the stock of certain variables of the firm at the end of an accounting period, as compared with the profit and loss or cash flow statements, which itemize movements among categories during the accounting period. A dynamic model can produce static summaries in addition to dynamic flow information, if desired. (It might be noted that an accounting system is a model or mathematical presentation of the ongoing activities of a business firm or other organization.)

One final contrast in model types compares models which incorporate probabilities—stochastic models—with those which do not—deterministic models. Linear programming and simple inventory models are deterministic; more sophisticated inventory and mathematical programming models are stochastic.

SIMULATION MODELS

Models which permit the decision maker to observe the status of a system over time as well as at particular points in time are often called simulation models. Physical simulation models are called analog simulations. An analog cockpit simulator surrounds a pilot in training like an airplane cockpit itself, presents computer-generated views of what the pilot would see while flying, and alters the picture based on the pilot's manipulation of the simulated controls. While a digital computer usually does the calculation in an analog simulation, analog/digital conversion devices are needed to translate the turn of a wheel or the reading on an instrument dial into finite, discrete, digital form for computer processing. Nonphysical models are purely digital in nature.

Many physical events can be represented by mathematical models using difference and differential equations to express performance relationships during and at the end of selected time periods. This type of simulation, known as continuous simulation, eliminates the necessity of analog devices but still requires approximation of continuous random variables by finite, discrete, digital computers. Fluid dynamics problems of dams and waste disposal systems are common subjects of continuous simulations. Likewise, econometric models which postulate the levels of GNP (gross national product), retail sales, inventories, and other business and economic variables may be continuous simulations. The flow of money, like the flow of water or oil, can be treated as a continuum rather than as separable, recognizable particles.

By far the most frequent kind of simulation in the business environment is discrete-event digital simulation. Within this category, the most popular subjects of study are waiting-line or queuing situations. As compared with water flow, customers desiring service are integral, discrete entities, and so are the employees and devices that serve them. Models which combine certain characteristics of both continuous and discrete digital simulation are called hybrid simulations.* A model of an elevator system, for instance, comprises discrete elements (e.g., passengers and floors) as well as continuous elements (e.g., the rate of travel from one floor to another).

---

*Occasionally the term ''hybrid simulation'' is applied to models which combine analog and digital simulation, such as the cockpit simulator.

SIMULATION LANGUAGES

Since most realistic simulations must be done by computer because of the number of calculations required, the decision maker must choose a computer programming language in which to communicate the essence of the model to the computer. It is possible to write simulation models in common computer languages such as FORTRAN, BASIC, COBOL, or Assembler, and indeed, this is sometimes done when the analyst has only a small computer system available or is thoroughly conversant with the language. Nevertheless, analysts who want to minimize the portion of model construction time devoted to programming and debugging the model tend to prefer languages which have been expressly designed for simulation. These languages imbed in the compiler certain functions necessary for every simulation: establishing and updating a time clock; generating random occurrences; and initializing, incrementing, and printing system performance statistics such as utilization of service facilities and waiting time. Dozens of simulation languages currently exist, but the most popular are GPSS (General-Purpose Systems Simulator) and SIMSCRIPT for discrete-event simulation, CSMP (Continuous Simulation Modeling Package) and DYNAMO for continuous simulation, and GASP and SLAM for hybrid simulation. GPSS is considered the simplest simulation language to learn, but it is relatively inflexible in terms of the situations which it can model conveniently and in the nature of its output.

Simulation analysts are increasingly attempting to interface formal statistical routines with simulation models. The structure of GPSS and SIMSCRIPT permits the user to call a FORTRAN, PL/1, or Assembler subprogram for this purpose, although, in practice, discovering the correct way in which to implement this feature through the operating system of a particular computer can be tedious. SIMSCRIPT incorporates most of the arithmetic capability of FORTRAN in its own grammar, allowing the user to write statistical routines in SIMSCRIPT itself. The GASP ''language'' is actually a collection of FORTRAN subprograms which perform the common housekeeping operations of simulation, so it is a simple matter to include additional FORTRAN statements for statistical analysis within a GASP program, to be compiled simultaneously with the simulation routines.

At a higher level are special-purpose simulation languages. These languages feature macros or subprograms which describe features common to the particular type of situation being modeled. For instance, the CSS language is designed for the simulation of computer systems. As such, its compiler ''understands'' the properties of computer hardware and software types, such as disk and tape access methods, without their being described in detail in the simulation program itself. Likewise, the SIMPLAN language facilitates the simulation of financial and economic systems. These languages have been written by software vendors who sell or lease the compilers to users.

Finally, there are special-purpose simulation packages, which require the user to input only fixed-format alphanumeric information rather than conventional programming statements. Input to these packages looks much like program code in the RPG language. One such package is SCERT, which enables an analyst to model a computer's workload easily on many different prospective computer systems whose properties, such as tape transfer speeds, are stored in a factor library maintained by the software vendor and are immediately accessible to the SCERT user. Thus, the user can expeditiously decide which computer system should be procured by examining the output of the SCERT package. As with the special-purpose simulation languages, simulation packages are available for purchase or lease through software vendors.

SIMULATION APPLICATIONS

There is virtually no limit to the potential for simulation models to help management evaluate alternative solutions to a problem or predict what would result if various conditions should occur.

Fluid flow simulations allow engineers to assess the rate of flow of oil, water, sewage, or other fluid media through pipelines, dams, and other impediments, to decide how these facilities should be built or modified to meet anticipated demand.

Financial simulations can enable the manager to see the profits or losses resulting from changing investments from stocks to bonds, from short-term investments to long-term investments, or from financing capital improvements internally rather than externally.

Economic simulations can help government officials to predict the effect on the U.S. economy if the Federal Reserve raises the discount rate for banks or if another country alters its dollar exchange rate.

Production and operations simulations can help a shop floor supervisor to decide the sequence in which orders for the coming week should be processed, given their due dates and resources required.

Health care simulations can assist in the scheduling of patients, doctors, and rooms.

Police and fire department simulations can help to position vehicles, equipment, and personnel at locations which will minimize travel time to sites needing emergency assistance.

A data processing manager may use simulation to help decide which of a considerable number of competing, complex computer systems would handle the anticipated stream of data processing jobs best.

Perhaps most common of all is the simulation of a queuing or waiting-line situation. Inputs are customers at a bank, gas station, post office, grocery, or any of innumerable similar service situations. Processing involves clerks and possibly materials. Outputs are customers whose service has been completed. Performance of the system may be measured by cost of servers, cost of waiting, the opportunity cost of future sales lost because customers refuse to tolerate slow service, time spent in the waiting line and in service, and percentage of the time the servers are idle. Indeed, many of the preceding applications can be redefined to fit this input-processing-output waiting-line scenario. The police, fire, health care, production, and data processing applications are particularly amenable to treatment as waiting-line systems. This book focuses principally on the simulation of waiting-line systems. The reasons for emphasizing waiting-line simulation are the ubiquity of the situations to be modeled, the ease with which a novice can model such systems, and the generalizations which can be made to other model types. Every phase of modeling necessary to support managerial decision making is covered. At each stage there is a possibility that errors may occur, forcing the analyst to return to prior stages. The detection of these errors and how to cope with the lack of neatness in real-world data and systems is stressed. The orderliness and common sense behind the methodology should convince management that simulation is a reliable, if potentially expensive, decision tool.

STEPS IN THE CONSTRUCTION OF A SIMULATION MODEL

Simulation is directly analogous to laboratory experimentation in that tests may be replicated in a controlled environment on demand. The phases of simulation modeling and analysis are as follows:

1. Recognize the problem and define its limits.
2. Collect data to set values for system parameters.
3. Write a program to describe the system.
4. Verify and validate the output of the simulation.
5. Change the model to test alternative system configurations.
6. Apply statistical analysis to prepare the outputs for formal evaluation.
7. Apply statistical analysis to evaluate the alternative systems tested.
8. Select the best system configuration for implementation.

Often the result at the conclusion of one of the intermediate stages impels the analyst to backtrack to a previous stage. Suppose that the output of the simulation model is found to be inconsistent with the observed performance of the real system. It would then be necessary to isolate whether:

1. The parameter settings in the model were incorrect.
2. The programming of the model was faulty.
3. The output of the real system was atypical or observed inaccurately.
4. Some combination of these problems existed.

Depending on the disposition of these issues, the analyst might have to return to phases 1, 2, or 3 of the modeling procedure.

PITFALLS AND PROBLEMS IN SIMULATION

Perhaps the most pervasive problem in simulation modeling is the failure to define and circumscribe the limits and level of detail of the system to be simulated. Defining the limits too narrowly can lead to suboptimization of a larger system. Defining the limits too broadly or including excessive detail can make the model unwieldy and uneconomic to manipulate. For example, in simulating a computer system, the analyst might consider manual operations such as accounting controls and delivery of output to be part of the system. The internal functions of the computer are accomplished in milliseconds or less, while the external activities cited may take hours. The simulation model timer unit must be the lesser of these. To simulate the beginning and end of a clerical activity might require billions of simulated milliseconds and tremendous costs to run the model.

In addition to the failure to define the scope and detail of the model appropriately, some other common pitfalls of modeling are:

1. Underestimating the time and expense involved in the modeling process.
2. Failure to seek the simplest and most economical type of model for the task.
3. Absence or misuse of statistical methodology.
4. Inaccurately approximating the attributes of a system which does not yet exist.
5. Superficial understanding of the system to be modeled.
6. Poor skills for communications with managers and staff who will finance and use the model.

QUESTIONS

1. Define and contrast
   (a) Analog versus digital
   (b) Continuous versus discrete
   (c) Deterministic versus stochastic
   (d) Analytic versus synthetic
2. For each model type in Question 1, give a real-world example.
3. Under what conditions is it preferable to use a simulation model rather than some other model types? What are the disadvantages of simulation compared with other model types?
4. Choose two articles on applications of simulation from the bibliography and summarize them. Indicate the potential benefits of each application in terms of system performance measures and suggest why other types of models might not be appropriate.
5. Select a waiting-line system at your university or place of employment and specify
   (a) System inputs
   (b) System processing
   (c) System outputs
   (d) System performance measures
6. What are the steps in the execution of a simulation project?
7. What are the most common problems hampering the satisfactory completion of simulation projects?

chapter two

DATA COLLECTION, REDUCTION, AND ANALYSIS

THE IMPORTANCE OF DATA COLLECTION

Probably the least glamorous and most essential task in model building is gathering the data which will permit the analyst to estimate model parameters. In a model of an economic system, the principal parameters might be the level of gross national product; the rates of saving, consumption and investment; and the rate of interest. In a model of a waiting-line system, the principal parameters are the arrival and service patterns for each work station. If the model is to be accurate, dependable, and worthwhile, it is important to know not just the average rates at which customers arrive and are served, but also the statistical distributions which characterize arrival and service.

In some lucky circumstances, data on arrivals and service may already be recorded in convenient form. For example, meticulous records are kept on actual (as well as scheduled) arrivals and departures of aircraft and trains. At the other end of the spectrum, there are systems that an analyst may desire to model that do not yet exist in any form. As an illustration, imagine that a manufacturer of computer systems may want to evaluate several possible disk drives which could be built and offered as peripherals on its next-generation systems. The drives do not yet exist, and indeed never will exist if their simulated performance is unsatisfactory, so the service parameters are provided from engineering design specifications rather than from observation. In the absence of contrary information, the general nature of the statistical distribution may be assumed to be the same as for other disk drives which are currently offered.

TYPES OF DATA TO BE GATHERED

Usually, the manager is interested in evaluating possible modifications to an exist-ing system which seems unsatisfactory or inappropriate in some respects but whose basic framework may be acceptable. It is impossible to specify exactly how much performance data will be needed to build a model of the system, since the data requirement will vary with the nature of the still-undetermined statistical distribu-tions which characterize the process. Someone must observe the time at which each successive customer arrives and the time at which the customer begins and ends service. From these three pieces of data for each customer, the analyst can compute certain system performance attributes:

1. Interarrival time (the time between successive arrivals)
2. Service time (the difference between the time service ended and the time serv-ice began)
3. Queue length (the number of customers in the waiting line just before the cus-tomer arrived)
4. Waiting time (the time spent in the queue, not including service time)
5. Transit time (the sum of waiting time plus service time)
6. Server idle time (the time when the server is free between successive customer arrivals)

Often the distribution of interarrival time is considered to be exogenously determined,* while the other system performance measures may vary according to the nature of the system which is designed to process the arriving customers. Thus, with all other factors held constant, a system with two servers should offer shorter waiting time, shorter transit time, and shorter queues, while server idle time should be higher than for a single-server system. Since each of these time components has a cost, the manager must attempt to balance the costs to optimize the system. The longer the time spent by customers being processed, the higher the probability that they will take their business elsewhere in the future; this creates costs of ill will and lost sales for the manager. The greater the amount of server idle time, the greater the proportion of employee salaries spent for nonproductive activity. Since most systems are stochastic in nature, they will exhibit some occurrences of customers having to wait as well as some occurrences of servers being idle in the course of nor-mal system operation.

---

*This is not necessarily the case. Eventually a system which offers expeditious service should attract more customers, changing the mean and possibly also the distribution of interarrival time. Like-wise, a system which imposes long wating times will ultimately lose patrons, unless the serving organi-zation is a monopoly.

RECORDING OBSERVATIONS

Record keeping for simulation model input data can be burdensome and prone to error. To minimize difficulties, at least two people should share the record-keeping duties—one to observe and describe the system activities, the other to operate a stopwatch and record system performance data on paper. It is helpful if a form for data collection is prepared in advance of observing the system and if the base time unit (e.g., seconds, hours, days) appropriate to the system has been previously established. If the system to be modeled is more complex than a one-server facility, accuracy might be best achieved by breaking the system down into components which can be observed separately, by other teams or on other occasions. If data collection is to take place on different occasions, the analyst should endeavor to make sure that the occasions are comparable. For example, if it is desired to model system performance at a peak traffic hour, additional data which may be needed for the model should not be collected at a normal or slow time, unless these conditions are to be modeled separately.

TYPES OF WAITING-LINE SITUATIONS

Waiting-line systems are often classified according to three attributes:

1. Number of phases (how many work stations a customer may visit before leaving the system)
2. Number of channels (how many servers staff each work station)
3. Queue discipline (the sequencing rules which determine which customer in the waiting line will be served next)

A system with one work station staffed by a single server is called a single-channel, single-phase system. An example is a one-lane bank drive-up window. A system with one work station staffed by multiple servers is called a multiple-channel, single-phase system. An example is a set of parallel bank drive-up lanes staffed by a group of tellers. A system with several successive work stations, each staffed by a single server, is called a single-channel, multiple-phase system. An example might be an assembly line where each successive work station is staffed by only one employee. Finally, a system with several successive work stations, at least some of which are staffed by more than one server, is called a multiple-channel, multiple-phase system. An example is an office where each new arrival is greeted by a receptionist who then routes the visitor to the appropriate employees.

Some common queue disciplines are first-come-first-served (FIFO), last-come-first-served (LIFO), and random or prioritized according to some attribute of the customer, job, or time of day. For example, if the president of a large corporation and a file clerk for the same corporation were to appear simultaneously at the desk of the president's receptionist, the receptionist would undoubtedly serve the president first. In fact, even if the receptionist were in the process of serving the file clerk when the president arrived for service, the receptionist might interrupt service of the file clerk to deal with the president's needs. When service of a lower-priority customer can be interrupted, deferred, or terminated in favor of a higher-priority customer, queue discipline is said to be based on preemptive priority. Other familiar priority queue disciplines might include shortest jobs first, longest jobs first, jobs with earliest due dates first, most attractive or most obnoxious customers first, and most lucrative jobs first.

Queue disciplines may be hierarchical or nested. Thus, any members of the queue with priority 2 (higher) status will be served before those members of the queue with priority 1 (lower) status. Within the set of priority 2 customers, however, the server will process members in first-come, first-served order.

The number of channels and the queue discipline are not always obvious to the analyst observing a system. At a distance, all that may be clear is the arrival of a new customer, the commencement of service for that customer, and the eventual departure of the customer from the service area. It is these three bits of data which are of greatest importance to capture, since all the other performance measures previously mentioned can be derived from them. Eventually, however, queue discipline may be an aspect of the model which the analyst may wish to vary. If it is suspected that customer attributes affect current or possible queue disciplines, the analyst should attempt to note any attributes of customers which might entitle them to preferential treatment. Interviews with the management and staff of the subject system prior to data collection can assist the analyst in determining and recording possible queue disciplines.

For some systems, it is feasible to reduce the analyst's responsibility for data collection and to enhance the accuracy of data by automating the capture of some data or by having the customers or the servers participate in the data gathering endeavor. For example, the customer might be asked to complete a card with demographic information and to record a time-clock stamp on the card each time the customer arrives at or departs from a system checkpoint. In a busy medical clinic, for instance, patients might be given a time-stamped card on arrival; asked to fill in age and nature of complaint while they wait; requested to time-stamp the card before and after speaking to the receptionist, the nurse, and the doctor; and told to deposit the card in a box when leaving the clinic. The analyst can retrieve the cards at the end of the day and process the data in a leisurely, careful manner.

AN EXAMPLE: THE TICKET SELLER

Suppose that we have been asked to observe a ticket seller, to model the current arrival and service patterns, and to modify the model to report performance characteristics of possible alternative systems. We arrive one morning at 9:59:20 A.M. to record arrivals and service. At that time a customer is in the process of being served, but since we do not know the time at which he arrived or commenced being served, we cannot compute interarrival time or service time for him, and we do not bother to record any information about him.

When we arrived, we noticed two customers in the queue. We cannot determine when they arrived, but we can observe and record the facts that the first began service at 10:00:00, 40 time units (seconds) after the observation period began and completed service at 10:00:47, and that the second began service at 10:00:47 and completed service at 10:01:57. If the server had been interrupted between customers, the service start time of the second customer might not have followed immediately upon the departure of the first customer, and the gap would have been recorded as server idle time unless the reason for the hiatus was known and important to the analyst. Likewise, interruption of the server while a customer is being served is often included in service time unless the reason for the interruption is known and important to the analyst.

At 10:01:07 we record our first customer arrival. Since the system is busy, he is forced to wait until the prior customer finishes at 10:01:57. We might estimate interarrival time for this first observed arrival as the difference between the current time (10:01:07) and the time observation began (9:59:20), or 87 seconds, but this is only an estimate because we do not know exactly when the prior customer arrived. We will discard this value and begin interarrival statistics with the second arrival actually observed.

In Table 2-1 we record the time that service began and ended for the two customers who were in the queue when the observation period began and the time arrived as well for 51 customers* who followed them during the observation period. Clock time for each of the three data items for each customer is given as hours (two digits), minutes (two digits), and seconds (two digits).

In Table 2-2 the raw data on time arrived, time service began, and time service ended for each customer observed are used to compute six system performance measures:

1. Interarrival time†
2. Queue length
3. Service time
4. Transit time
5. Waiting time
6. Server idle time

---

*There is nothing magical about the number 50. It may or may not prove to be an adequate sample size for determining the statistical distributions of interarrival time and service time. The initial sample size limit may be determined by the availability of preexisting data, the availability of the system for observation, and/or the endurance of the data collectors.

†Interarrival time for the first arrival observed is listed as −99999.9 in Table 2-2 to indicate that it is not possible to calculate interarrival time for the first customer. The arrival time of the first customer is used only as a benchmark for subsequent arrivals.

TABLE 2-1   TALLY SHEET FOR RECORDING
ARRIVAL TIME, SERVICE START TIME, AND SERVICE
END TIME

100000,	100047	
100047,	100157	
100107,	100157,	100628
100124,	100628,	100740
100133,	100740,	100858
100227,	100858,	101038
100332,	101038,	101058
100427,	101058,	101246
100635,	101246,	101450
101144,	101450,	101542
101326,	101542,	101820
101818,	101820,	101951
101837,	101951,	102050
102125,	102125,	102252
102231,	102252,	102552
102353,	102552,	102636
102551,	102636,	102942
102608,	102942,	103017
102811,	103017,	103150
103205,	103205,	103418
103215,	103418,	103422
103249,	103422,	103847
103335,	103847,	103946
103740,	103946,	104008
103821,	104008,	104144
104105,	104144,	104217
104109,	104217,	104232
104231,	104232,	104319
104337,	104337,	104343
104459,	104459,	104502
104644,	104644,	104929
104752,	104929,	105111
104832,	105111,	105157
104952,	105157,	105214
105043,	105214,	105254
105143,	105254,	105254
105411,	105411,	105557
105423,	105557,	105641
105429,	105641,	105755
105537,	105755,	105813
110054,	110054,	110118
110219,	110219,	110400
110512,	110512,	110527
110553,	110553,	110710

TABLE 2-1 (cont.)

110828, 110828, 110844
111039, 111039, 111145
111315, 111315, 111721
111614, 111721, 112223
111735, 112223, 112523
111826, 112523, 112716
112128, 112716, 112802
112347, 112802, 112846
112531, 112846, 112913

Table 2-2 also gives the mean or average values of these system performance measures, where the averages are computed from the raw observations of performance measures for each individual transaction.

Average number in queue,* average waiting time, and average transit time are computed on the basis of 51 observations, excluding the two customers in the initial queue. Average service time is computed using 53 observations, including those in the initial queue. Average interarrival time is computed using 50 observations, excluding those in the initial queue as well as the first one of the 51 regular arrivals, whose arrival time is used as a benchmark to begin comparisons. Average server idle time as a percentage is calculated as the time the server was busy, divided by the time the system was under observation, times 100. The system was observed from 09:59:20 A.M. to 11:29:13 A.M., a total of 5393 seconds. The server was idle for a total of 926 seconds. Idle time percent, then, is $(926/5393) \times 100 = 17.17\%$.

EVALUATING INTERARRIVAL TIME AND SERVICE TIME DATA

We are now in a position to perform some detailed statistical analysis of the distribution of interarrival time and service time. We already know the mean values from the summary statistics in Table 2-2. However, it is desirable to establish whether the data tend to fit some common pattern so that it is possible to extrapolate to other values in the distribution which did not happen to occur in our limited sample.

Some common statistical distributions which often characterize interarrival time and service time are

1. The uniform or rectangular distribution, in which all values on an interval are equally likely.

---

*Average number in queue is estimated as the number in the queue prior to each arrival, summed over all arrivals and divided by the total number of arrivals. A more accurate but much more time-consuming estimate would be given by weighting each observed queue length by the total number of time units for which that queue length appeared during the observation period.

TABLE 2-2 RAW DATA AND PERFORMANCE MEASURES COMPUTED FROM SAMPLE OBSERVATIONS

OBSERVATION NUMBER	TIME ARRIVED	INTERARRIVAL TIME	QUEUE LENGTH	SERVICE START TIME	SERVICE END TIME	SERVICE TIME	TRANSIT TIME	WAITING TIME	SERVER IDLE TIME
1				40.0	87.0	47.0			0.0
2				87.0	157.0	70.0			0.0
1	107.0	-99999.9	1	157.0	428.0	271.0	321.0	50.0	0.0
2	124.0	17.0	2	428.0	500.0	72.0	376.0	304.0	0.0
3	133.0	9.0	3	500.0	578.0	78.0	445.0	367.0	0.0
4	187.0	54.0	3	578.0	678.0	100.0	491.0	391.0	0.0
5	252.0	65.0	4	678.0	698.0	20.0	446.0	426.0	0.0
6	307.0	55.0	5	698.0	806.0	108.0	499.0	391.0	0.0
7	435.0	128.0	5	806.0	930.0	124.0	495.0	371.0	0.0
8	744.0	309.0	2	930.0	982.0	52.0	238.0	186.0	0.0
9	846.0	102.0	2	982.0	1140.0	158.0	294.0	136.0	0.0
10	1138.0	292.0	1	1140.0	1231.0	91.0	93.0	2.0	0.0
11	1157.0	19.0	1	1231.0	1290.0	59.0	133.0	74.0	0.0
12	1325.0	168.0	0	1325.0	1412.0	87.0	87.0	0.0	35.0
13	1391.0	66.0	1	1412.0	1592.0	180.0	201.0	21.0	0.0
14	1473.0	82.0	1	1592.0	1636.0	44.0	163.0	119.0	0.0
15	1591.0	118.0	2	1636.0	1822.0	186.0	231.0	45.0	0.0
16	1608.0	17.0	2	1822.0	1857.0	35.0	249.0	214.0	0.0
17	1731.0	123.0	2	1857.0	1950.0	93.0	219.0	126.0	0.0
18	1965.0	234.0	0	1965.0	2098.0	133.0	133.0	0.0	15.0
19	1975.0	10.0	1	2098.0	2102.0	4.0	127.0	123.0	0.0
20	2009.0	34.0	2	2102.0	2367.0	265.0	358.0	93.0	0.0
21	2055.0	46.0	3	2367.0	2426.0	59.0	371.0	312.0	0.0
22	2300.0	245.0	2	2426.0	2448.0	22.0	148.0	126.0	0.0
23	2341.0	41.0	3	2448.0	2544.0	96.0	203.0	107.0	0.0
24	2505.0	164.0	1	2544.0	2577.0	33.0	72.0	39.0	0.0
25	2509.0	4.0	2	2577.0	2592.0	15.0	83.0	68.0	0.0
26	2591.0	82.0	1	2592.0	2639.0	47.0	48.0	1.0	0.0

TABLE 2-2 (cont.)

27	2657.0	66.0	0	2657.0	2663.0	6.0	6.0	0.0	18.0
28	2739.0	82.0	0	2739.0	2742.0	3.0	3.0	0.0	76.0
29	2844.0	105.0	0	2844.0	3009.0	165.0	165.0	0.0	102.0
30	2912.0	68.0	1	3009.0	3111.0	102.0	199.0	97.0	0.0
31	2952.0	40.0	2	3111.0	3157.0	46.0	205.0	159.0	0.0
32	3032.0	80.0	2	3157.0	3174.0	17.0	142.0	125.0	0.0
33	3083.0	51.0	3	3174.0	3214.0	40.0	131.0	91.0	0.0
34	3143.0	60.0	3	3214.0	3214.0	0.0	71.0	71.0	0.0
35	3291.0	148.0	0	3291.0	3397.0	106.0	106.0	0.0	77.0
36	3303.0	12.0	1	3397.0	3441.0	44.0	138.0	94.0	0.0
37	3309.0	6.0	2	3441.0	3515.0	74.0	206.0	132.0	0.0
38	3377.0	68.0	3	3515.0	3533.0	18.0	156.0	138.0	0.0
39	3694.0	317.0	0	3694.0	3718.0	24.0	24.0	0.0	161.0
40	3779.0	85.0	0	3779.0	3880.0	101.0	101.0	0.0	61.0
41	3952.0	173.0	0	3952.0	3967.0	15.0	15.0	0.0	72.0
42	3993.0	41.0	0	3993.0	4070.0	77.0	77.0	0.0	26.0
43	4148.0	155.0	0	4148.0	4164.0	16.0	16.0	0.0	78.0
44	4279.0	131.0	0	4279.0	4345.0	66.0	66.0	0.0	115.0
45	4435.0	156.0	0	4435.0	4681.0	246.0	246.0	0.0	90.0
46	4614.0	179.0	1	4681.0	4983.0	302.0	369.0	67.0	0.0
47	4695.0	81.0	1	4983.0	5163.0	180.0	468.0	288.0	0.0
48	4746.0	51.0	2	5163.0	5276.0	113.0	530.0	417.0	0.0
49	4928.0	182.0	3	5276.0	5322.0	46.0	394.0	348.0	0.0
50	5067.0	139.0	3	5322.0	5366.0	44.0	299.0	255.0	0.0
51	5171.0	104.0	3	5366.0	5393.0	27.0	222.0	195.0	0.0

AVERAGE INTERARRIVAL TIME   = 101.280
AVERAGE QUEUE LENGTH   =   1.608
AVERAGE SERVICE TIME   = 83.528
AVERAGE TRANSIT TIME   = 213.314
AVERAGE WAITING TIME   = 128.804
AVERAGE SERVER IDLE TIME (%)   = 17.170

2. The normal distribution, in which values follow a bell-shaped curve with a particular mean and standard deviation.*

3. The negative exponential distribution (sometimes erroneously called the exponential distribution), in which all nonnegative values are possible, with decreasing probability as the value increases.

Another common distribution, the Poisson distribution, sometimes characterizes arrival and service *rates*. Indeed, when the interarrival time distribution is negative exponential, the arrival rate distribution is, by definition, Poisson. The same relationship holds between the service time and service rate distributions. When the analyst is forced to use precompiled data in tabular form, it may be presented as a rate distribution. In this case, the Poisson test rather than the negative exponential test would be appropriate. When there is a choice because the analyst collects the raw data, it is better to collect time data rather than rate data because time data are measured on a continuous scale whereas rate data are integer in nature. If rate data are gathered and a customer is observed to begin service during one rate interval (e.g., the first 100-second time period) but finishes service during a subsequent rate interval (e.g., the second 100-second time period), the analyst must establish a rule for deciding in which interval to count the service as occurring. Double counting renders the data useless. It is desirable to avoid this complication by gathering time data rather than rate data when possible.

Sometimes it seems that no common theoretical distribution fits the observed arrival and service patterns. In this case, we might want to test whether the patterns were similar to those found in sample data acquired previously, from this or some other system. It is possible to build a model in which the arrival and service distributions do not conform to any theoretical model but are based solely on limited empirical sample evidence.

It may not be possible to establish conclusively which of several possible theoretical distributions fit the data best. Then it is appropriate to gather additional data until the most appropriate distribution becomes statistically obvious. If more data cannot be gathered, the analyst must select the distribution which seems to be the most likely candidate.

GOODNESS-OF-FIT TESTS

Goodness-of-fit tests, sometimes called one-sample tests, are designed to compare a set of data with some predefined standard and establish whether or not the set of data differs from the standard more than might be explained by chance variation. The standard, given in the null hypothesis (H_0), specifies the form of the statistical distribution and all population parameters needed to define the specific distribution

---

*Sometimes it is necessary to truncate the negative range of values of the normal distribution to represent accurately the fact that interarrival times and service times can never be negative. Other distributions, such as lognormal, gamma, and Erlang, also appear frequently.

completely. These population parameters might include the range (for a uniform distribution), the mean (for a Poisson, negative exponential, or normal distribution), and the standard deviation (for a normal distribution). The alternative hypothesis (H_1) is that some distribution other than the one detailed in the null hypothesis is the true descriptor of the population from which the sample data came. The level of significance, α, is the risk the analyst is willing to tolerate that the null hypothesis might be erroneously rejected when it does, in fact, accurately describe the population from which the sample data were drawn; suitable levels of significance are 0.10, 0.05, and 0.01.

There are two common types of goodness-of-fit tests: the Kolmogorov-Smirnov test and the χ^2 test. We will use the χ^2 test for establishing arrival and service distributions because generally we have no independent information about the population parameters which must be specified in the null hypothesis. The K-S goodness-of-fit procedure assumes that these population parameters can be estimated from some source other than the sample data themselves. The χ^2 test uses the mechanism of degrees of freedom to compensate for a potential bias when the sample statistics themselves (mean and standard deviation) are substituted for unknown population parameters in the null hypothesis.

A disadvantage of the χ^2 test is that it is affected by sample size and by the definition of class boundaries which alter expected class frequencies. Many statisticians feel that classes with expected frequencies smaller than 5 should not be used in a χ^2 test. We will honor this convention and undertake to define class boundaries, when possible, to take maximum advantage of the available sample size.

AN INTERARRIVAL TIME DISTRIBUTION

From Table 2-2 we can extract 50 observations of interarrival time (see Table 2-3). Some preliminary descriptive statistics—confirming the mean, computing the standard deviation and variance, and ranking the observations in ascending order—can offer clues about the possible nature of the interarrival time distribution for the population of ticket customers.

Of the four distributions (uniform, normal, negative exponential, and Poisson) which we will consider as possible theoretical models, only the normal distribution allows negative data values. Since there are no negative data values in our interarrival time data, we are unable to exclude any possible distributions on this basis. In a Poisson distribution, the data values are strictly integer and the mean and the variance are equal. All our data happen to be recorded as integers, although for time data this is a matter of convenience of record keeping rather than of accuracy. However, for our data there is an enormous difference in the order of magnitude of the mean and the variance, so we rule out the Poisson distribution as a possibility. In a negative exponential distribution, the mean and the standard deviation are equal; we might wonder whether the difference between the sample mean and sample standard deviation is small enough to be ascribed to chance.

TABLE 2-3 DESCRIPTIVE STATISTICS FROM TICKET SELLER
INTERARRIVAL TIMES

INTERARRIVAL TIME RAW DATA

17.0	9.0	54.0	65.0	55.0	128.0	309.0	102.0	292.0	19.0
168.0	66.0	82.0	118.0	17.0	123.0	234.0	10.0	34.0	46.0
245.0	41.0	164.0	4.0	82.0	66.0	82.0	105.0	68.0	40.0
80.0	51.0	60.0	148.0	12.0	6.0	68.0	317.0	85.0	173.0
41.0	155.0	131.0	156.0	179.0	81.0	51.0	182.0	139.0	104.0

INTERARRIVAL TIME SUMMARY STATISTICS

THE MEAN IS 101.2800.
THE VARIANCE IS 6216.5390.
THE STANDARD DEVIATION IS 78.8450.
THE SAMPLE SIZE IS 50.

INTERARRIVAL TIME SORTED DATA

4.0	6.0	9.0	10.0	12.0	17.0	17.0	19.0	34.0	40.0
41.0	41.0	46.0	51.0	51.0	54.0	55.0	60.0	65.0	66.0
66.0	68.0	68.0	80.0	81.0	82.0	82.0	82.0	85.0	102.0
104.0	105.0	118.0	123.0	128.0	131.0	139.0	148.0	155.0	156.0
164.0	168.0	173.0	179.0	182.0	234.0	245.0	292.0	309.0	317.0

UNIFORM GOODNESS-OF-FIT TEST

We now perform goodness-of-fit tests of the remaining theoretical models using the ranked, sorted interarrival time data to determine observed class frequencies. First, we test the null hypothesis that the data are uniformly distributed on the interval from 4 to 317, respectively, the lowest and highest observed values. The alternative hypothesis is that some other distribution is the true population descriptor. Since we

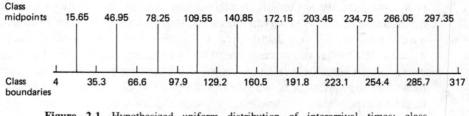

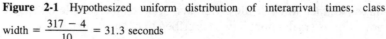

Figure 2-1 Hypothesized uniform distribution of interarrival times; class width $= \dfrac{317 - 4}{10} = 31.3$ seconds

want to ensure that expected class frequencies are not less than 5, since we can specify the distribution with more precision when we define more classes, and since our sample size is 50, we define 10 classes, each with probability 0.1 and expected class frequency of 5.0. These classes are shown in Figure 2-1. The values of the observations range from a potential low of zero to an observed high of 317.

The goodness-of-fit test of this uniform distribution is shown in Table 2-4.

Since the null hypothesis for the uniform distribution does not include an estimate of the mean or standard deviation, the only sample statistic used to compute expected class frequencies is the inevitable sample size. Degrees of freedom, then, is the number of classes with expected frequencies of at least 5 (10 classes) minus 1 for use of the sample size in the expected class frequency calculation. We refer to the χ^2 table (Appendix Table II), for 9 degrees of freedom and a level of significance of 0.05 and find that the computed value of χ^2 (27.60) far exceeds the critical level (16.919) which marks the most extreme value such that we might still cling to the belief that the null hypothesis might be true. We conclusively reject the null hypothesis that the population of interarrival times is uniformly distributed on the interval from 4 to 317. Figure 2-2 illustrates this graphically, while Table 2-4 gives the class midpoints, the observed and expected class frequencies, and the value of the χ^2 component. χ^2 is computed as the (observed class frequency − expected class frequency) squared and then divided by the expected class frequency, for each class. The χ^2 components are summed over all classes to give a total. If this total is close to zero, the null hypothesis is likely to be correct and the distribution specified in it can be assumed to characterize the population. If the χ^2 total is very different from zero, it is unlikely that the null hypothesis is correct, and the analyst will have to look elsewhere to describe the data accurately. The degree to which computed χ^2 may differ from zero without suggesting rejection of H_0 depends on the number of classes with expected frequencies of 5 or more and the number of sample statistics

TABLE 2-4 GOODNESS-OF-FIT TEST OF THE UNIFORM DISTRIBUTION FOR INTERARRIVAL TIMES

CLASS	CLASS IDENTIFIER	OBSERVED FREQUENCY	RELATIVE FREQUENCY	EXPECTED FREQUENCY	χ^2
1	15.6	9.00	0.100	5.00	3.20
2	46.9	12.00	0.100	5.00	9.80
3	78.3	8.00	0.100	5.00	1.80
4	109.6	6.00	0.100	5.00	0.20
5	140.9	5.00	0.100	5.00	0.00
6	172.1	5.00	0.100	5.00	0.00
7	203.4	0.00	0.100	5.00	5.00
8	234.8	2.00	0.100	5.00	1.80
9	266.1	0.00	0.100	5.00	5.00
10	297.4	3.00	0.100	5.00	0.80
TOTALS		50.0	1.000	50.00	27.60

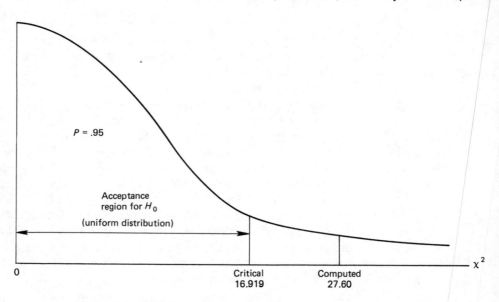

Figure 2-2 χ^2 distribution showing computed and critical values for hypothesized uniform distribution

used in the null hypothesis as well as to calculate expected class frequencies. Figure 2-2 shows the hypothesized χ^2 distribution, with acceptance and rejection regions for the null hypothesis indicated.

NORMAL GOODNESS-OF-FIT TEST

Suppose that we reformulate our null hypothesis to consider a normal distribution. We must indicate the mean and standard deviation of the particular normal distribution we postulate, and our only sense of these parameters comes from the sample mean, 101.28 seconds, and the sample standard deviation, 78.845 seconds. The null hypothesis (H_0) is that the interarrival time distribution is normal with a mean of 101.28 seconds and a standard deviation of 78.845 seconds. The alternative hypothesis (H_1) is that some other pattern actually characterizes the data.

We will follow the same general procedure as before, attempting to define 10 classes, each with an expected class frequency of 5. To define class boundaries for these classes, each of which will have a probability of 0.1, we must refer to a table of the standard normal distribution (the "z" table) found in Appendix Table I. Figure 2-3 illustrates the definition of boundaries given the sample mean, sample standard deviation, and the tabulated values of z.

We will use the class probabilities to find class boundaries in terms of z from the standard normal table. Then, we will convert the boundaries in terms of z to boundaries in seconds by inserting our sample mean and standard deviation into the equation:

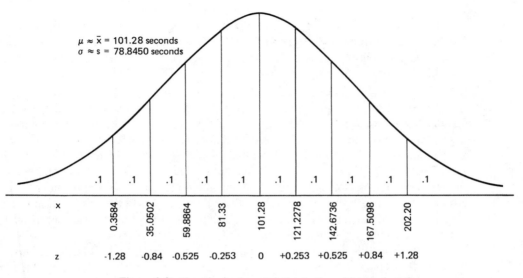

.1	.1	.1	.1	.1	.1	.1	.1	.1	.1

x

| 0.3584 | 35.0502 | 59.8864 | 81.33 | 101.28 | 121.2278 | 142.6736 | 167.5098 | 202.20 |

z   -1.28   -0.84   -0.525   -0.253   0   +0.253   +0.525   +0.84   +1.28

Figure 2-3   Hypothesized normal distribution of interarrival times

$$\text{REAL BOUNDARY} = \text{SAMPLE MEAN} + (z \text{ VALUE} \\ \times \text{SAMPLE STANDARD DEVIATION})$$

As an example, we find that a z value of -1.28 cuts off the lowest 10 percent of any normal distribution, just as a z value of $+1.28$ cuts off the highest 10 percent of any normal distribution. For the normal distribution mentioned in our null hypothesis, then, the lowest class boundary would be $101.28 + (-1.28 \times 78.845) = 0.3584$. Likewise, the highest class boundary would be $101.28 + (+1.28 \times 78.845) = 202.20$. Table 2-5 shows the class boundaries (including 999.0 as a designator for

TABLE 2-5   GOODNESS-OF-FIT TEST OF THE NORMAL DISTRIBUTION FOR INTERARRIVAL TIMES

CLASS	CLASS IDENTIFIER	OBSERVED FREQUENCY	RELATIVE FREQUENCY	EXPECTED FREQUENCY	χ^2
1	0.4	0.00	0.100	5.01	5.01
2	35.1	9.00	0.100	5.01	3.18
3	59.9	8.00	0.099	4.97	1.85
4	81.3	8.00	0.100	5.02	1.77
5	101.3	4.00	0.100	4.99	0.20
6	121.2	4.00	0.100	4.99	0.20
7	142.7	4.00	0.100	5.02	0.21
8	167.5	4.00	0.099	4.97	0.19
9	202.2	4.00	0.100	5.01	0.20
10	999.0	5.00	0.100	5.02	0.00
TOTALS		50.00	1.000	50.00	12.81

+ infinity, the theoretical upper end of the normal curve). It also give the observed and expected frequencies for each of the 10 classes and the value of χ^2. Theoretical relative frequencies differ from 0.1, and expected class frequencies differ from 5.0 only because of computer rounding errors.

There are 10 classes with expected frequencies of 5 or more (except for insignificant rounding errors). We have used three sample statistics (sample mean, sample standard deviation, and sample size) in defining the null hypothesis and in computing expected class frequencies. Degrees of freedom is, therefore, $10 - 3 = 7$. From the χ^2 table in Appendix Table II, we find that critical χ^2 at the 0.05 level of significance with 7 degrees of freedom is 14.067. We are unable to reject the null hypothesis that the interarrival time data are normally distributed with the specified mean and standard deviation, although the computed value of χ^2 is closer to the critical value than it is to zero. We defer making a decision to accept the normal distribution until we test the two remaining theoretical distributions for a possible better fit.

The probability density function for the normal distribution cannot be integrated by analytic means, and so the values in the z table which are produced by computerized numerical approximations are the only reasonable way of dealing with normal probabilities and boundaries. By contrast, Figure 2-4 shows the formulas for the probability density function and the cumulative distribution function of the negative exponential distribution with mean μ.

$$P(\text{time} < x) = 1 - e^{(-x/\text{mean time})}$$

$$P(x_1 < \text{time} < x_2) = e^{(-x_1/\text{mean time})} - e^{(-x_2/\text{mean time})}$$
$$\text{for all } x \geq 0$$

Figure 2-4 Negative exponential formulas

The negative exponential distribution has a probability density function which is solvable analytically, but the formulas include the mathematical constant e which requires a table, a calculator, or a computer for easy, accurate production of probabilities and class boundaries. Since tabled values are quite limited (see Appendix Table V), a calculator or a computer is recommended.

NEGATIVE EXPONENTIAL TEST

We will now test the null hypothesis that interarrival times are negative exponentially distributed with a mean of 101.28 seconds. Since the negative exponential distribution is a one-parameter distribution—its mean and standard deviation are equal—we do not need to specify the standard deviation separately, as we did in the case of the normal test. As before, the alternative hypothesis is that the null

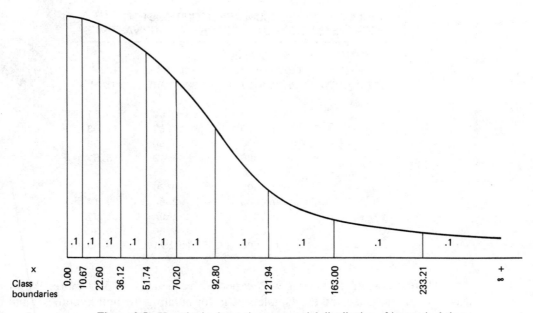

Figure 2-5 Hypothesized negative exponential distribution of interarrival time

hypothesis is false and that some other unspecified distribution really characterizes the population. Again, we will attempt to set up 10 classes, each with a probability of 0.1 and expected class frequency of 5, because our sample size is 50. We use the cumulative distribution function for the negative exponential distribution to solve for the class boundaries which cut off, respectively, 0.1, 0.2, . . . , 0.9 of the area under the curve. Figure 2-5 shows the class boundaries for the hypothesized negative exponential distribution.

Tables 2-6 and 2-7 show the same class boundaries, observed and expected class frequencies, and χ^2.

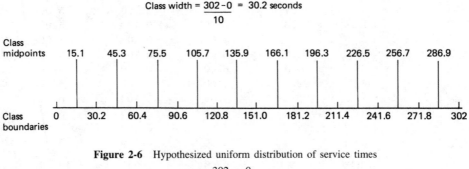

Figure 2-6 Hypothesized uniform distribution of service times

$$\text{class width} = \frac{302 - 0}{10} = 30.2 \text{ seconds}$$

TABLE 2-6   CLASS BOUNDARY DEFINITIONS FOR
NEGATIVE EXPONENTIAL TEST FOR INTERARRIVAL TIMES

CLASS IDENTIFIED	CUMULATIVE GREATER-THAN PROBABILITY	CLASS BOUNDARY	CLASS WIDTH
1	1.0000	0.00	10.67
2	0.9000	10.67	11.93
3	0.8000	22.60	13.52
4	0.7000	36.12	15.61
5	0.6000	51.74	18.47
6	0.5000	70.20	22.60
7	0.4000	92.80	29.14
8	0.3000	121.94	41.07
9	0.2000	163.00	70.20
10	0.1000	233.21	999.99

There are again 10 classes with expected frequencies of 5 or more. We have used the sample mean and the sample size in formulating the null hypothesis and in calculating expected class frequencies. Thus, degrees of freedom $= 10 - 2 = 8$. Critical χ^2 with 8 degrees of freedom, at the 0.05 level of significance is 15.507, while computed χ^2 is 6.80. Again, as in the normal test, we are unable to reject the null hypothesis that the sample data were drawn from a population which is negative exponentially distributed with a mean of 101.28 seconds. If it were feasible to collect additional data and re-perform the goodness-of-fit tests, this would be the best

TABLE 2-7   GOODNESS-OF-FIT TEST OF THE NEGATIVE EXPONENTIAL
DISTRIBUTION FOR INTERARRIVAL TIMES

CLASS	CLASS IDENTIFIER	OBSERVED FREQUENCY	RELATIVE FREQUENCY	EXPECTED FREQUENCY	χ^2
1	10.7	4.00	0.100	5.00	0.20
2	22.6	4.00	0.100	5.00	0.20
3	36.1	1.00	0.100	5.00	3.20
4	51.7	6.00	0.100	5.00	0.20
5	70.2	8.00	0.100	5.00	1.80
6	92.8	6.00	0.100	5.00	0.20
7	121.9	4.00	0.100	5.00	0.20
8	163.0	7.00	0.100	5.00	0.80
9	233.2	5.00	0.100	5.00	0.00
10	999.0	5.00	0.100	5.00	0.00
TOTALS		50.00	1.000	50.00	6.80

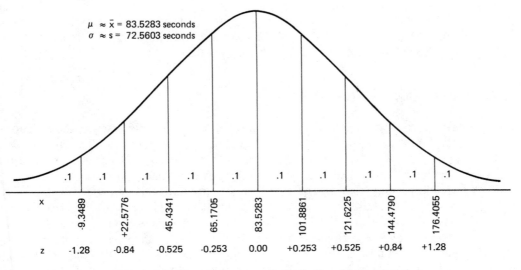

$\mu \approx \bar{x} = 83.5283$ seconds
$\sigma \approx s = 72.5603$ seconds

| | .1 | .1 | .1 | .1 | .1 | .1 | .1 | .1 | .1 | .1 |

| x | -9.3489 | +22.5776 | 45.4341 | 65.1705 | 83.5283 | 101.8861 | 121.6225 | 144.4790 | 176.4055 |

| z | -1.28 | -0.84 | -0.525 | -0.253 | 0.00 | +0.253 | +0.525 | +0.84 | +1.28 |

Figure 2-7   Hypothesized normal distribution of interarrival times

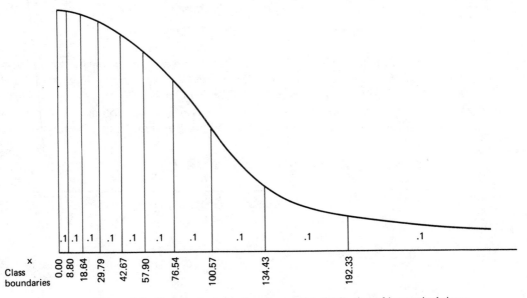

| | .1 | .1 | .1 | .1 | .1 | .1 | .1 | .1 | .1 | .1 |

x
Class
boundaries   0.00 8.80 18.64 29.79 42.67 57.90 76.54 100.57 134.43 192.33

Figure 2-8   Hypothesized negative exponential distribution of interarrival times

way to decide which distribution fits best. Eventually, however, the cost of sampling becomes prohibitive, and the analyst might simply decide to accept that null hypothesis for which the value of χ^2 was smallest.

POISSON GOODNESS-OF-FIT TEST

To corroborate the negative exponential test and to illustrate the relationship between the Poisson distribution and the negative exponential distribution, we will convert the data from interarrival time format to arrival rate format. Suppose that we aggregate the sequential arrival data in Table 2-2 into 100-second periods. In the second period, from time 101 to time 200, four customers arrived, respectively, at times 107, 124, 133, and 187. In the next 100-second period, from time 201 to 300, only one customer arrived, at time 252. So far, we have one time period with an arrival rate of four customers and one time period with an arrival rate of one customer per 100 seconds. If we were to continue this for the 52 100-second periods which elapsed during the data collection period, we would find that there were 16 100-second periods in which there were no customer arrivals, 25 100-second time periods in which there was one arrival, 8 100-second time periods in which there were two arrivals, 2 100-second time periods in which there were three arrivals, and 1 100-second time period in which there were four arrivals. The mean arrival rate per 100 seconds is

$$(0 \times 16) + (1 \times 25) + (2 \times 8) + (3 \times 2) + (4 \times 1)$$

= approximately 1 customer per 100 seconds

The null hypothesis for this test is that the arrival rate of customers is Poisson distributed with a mean of 1 per 100 seconds. The alternative hypothesis is that some other distribution is correct. To ensure that classes are collectively exhaustive, we redefine the last class as "four or more arrivals per 100 seconds." Note that the number of arrivals in any 100-second period must be an integer, although the mean

TABLE 2-8   GOODNESS-OF-FIT TEST OF THE POISSON DISTRIBUTION FOR ARRIVAL RATES

CLASS	CLASS IDENTIFIER	OBSERVED FREQUENCY	RELATIVE FREQUENCY	EXPECTED FREQUENCY	χ^2
1	0.0	16.00	0.368	19.13	0.51
2	1.0	25.00	0.368	19.13	1.80
3	2.0	8.00	0.184	9.56	0.26
4	3.0	2.00	0.061	3.19	0.44
5	99.0	1.00	0.019	0.99	0.00
TOTALS		52.00	1.000	52.00	3.01

number of arrivals averaged over all observation periods may be a decimal number. Values of the Poisson distribution for small means are given in Appendix Table IV.

Poisson classes are defined to include only single integer values, except for the uppermost class, which is open ended. The "Class Identifier" columns in Tables 2-8 and 2-9 indicate the integers included in each class. For Table 2-8 the fifth class includes "four or more" and for Table 2-9 the third class includes "two or more."

Some of the expected class frequencies are substantially less than five because we were not able to pinpoint class boundaries at our discretion, as in the previous cases. It will be necessary to combine the classes representing two, three, and four or more arrivals per 100 seconds to meet the conventional expected frequency criterion. The revised Poisson test appears in Table 2-9.

TABLE 2-9   REVISED GOODNESS-OF-FIT TEST OF THE POISSON DISTRIBUTION

CLASS	CLASS IDENTIFIER	OBSERVED FREQUENCY	RELATIVE FREQUENCY	EXPECTED FREQUENCY	χ^2
1	0.0	16.00	0.368	19.13	0.51
2	1.0	25.00	0.368	19.13	1.80
3	99.0	11.00	0.264	13.74	0.55
TOTALS		52.00	1.000	52.00	2.86

Expected class frequencies now exceed the acceptable minimum. Total observed frequency is 52 *time periods*, not customers. With three classes remaining and the sample mean and sample size used in the calculation of expected class frequencies, degrees of freedom is $3 - 2 = 1$. Critical χ^2 at the 0.05 level of significance, with 1 degree of freedom, is 3.841 compared with computed χ^2 of 2.86. Thus, we are unable to reject the null hypothesis of Poisson-distributed arrival rates just as we were unable to reject the null hypothesis of negative exponentially distributed interarrival times.

TESTING FIT OF OTHER DISTRIBUTIONS

Many other goodness-of-fit tests might be necessary or desirable. If none of the theoretical distributions tested proves to be acceptable, the analyst might alter the mean and/or standard deviation and re-perform the test. For example, a negative exponential distribution with a mean of 90 might be a logical candidate to test using our current interarrival time data. The rationale is that for a perfect negative exponential distribution, the mean and the standard deviation are equal, so why not average the sample mean and the sample standard deviation to come up with a best guess for the parameter setting in the null hypothesis?

It is also possible to compare the data with some prior data set. Suppose last year's experience in selling tickets was that interarrival time was twice as likely to be 100 seconds or less as it was to be more than 100 seconds. We specify this ratio as the null hypothesis and construct an empirical distribution with two classes, 0 to 100 seconds and over 100 seconds, having respective probabilities of 0.67 and 0.33 (see Table 2-10).

TABLE 2-10   GOODNESS-OF-FIT TEST OF AN EMPIRICAL DISTRIBUTION FOR INTERARRIVAL TIMES

CLASS	CLASS IDENTIFIER	OBSERVED FREQUENCY	RELATIVE FREQUENCY	EXPECTED FREQUENCY	χ^2
1	1.0	29.00	0.670	33.50	0.60
2	2.0	21.00	0.330	16.50	1.23
TOTALS		50.00	1.000	50.00	1.83

The data pass this goodness-of-fit test also, with only 1 degree of freedom lost for use of the sample size in the expected frequency calculation. In the absence of other plausible theoretical distributions, this empirical distribution might be used to describe interarrival time in a simulation model. However, a more detailed set of classes is always preferable when possible to achieve.

TESTING SERVICE TIME DISTRIBUTIONS

We have tentatively determined that the most likely distribution of interarrival time for the ticket seller is negative exponential with a mean of 101.28 seconds. What can we conclude about the distribution of service time? A glance back to Table 2-2 reveals 53 observations of service time. We now restate those, along with their mean, variance, standard deviation, and rank ordering, in Table 2-11.

If the sample size were 55 instead of 53, it would be legitimate to define 11 classes, each with an expected class frequency of 5. However, we are a bit short of that magic number, so we continue to use 10 classes; this time each class will have an expected frequency of 5.3 instead of 5.0.

The sorted data list shows 0 as the smallest value of service time. Zero may not seem to be a reasonable value for service time, but this could occur in two ways. First, the measuring instrument might have been too imprecise to record fractions of a second, so that zero was the nearest whole second to the true service time. Another possibility is that the customer decided not to buy a ticket at the last instant and simply bypassed the server.

The sorted data list suggests a possible null hypothesis of a uniform distribution of service time on the interval from 0 to 302. Table 2-12 shows the results of the test.

TABLE 2-11   DESCRIPTIVE STATISTICS FROM TICKET SELLER SERVICE TIMES

SERVICE TIME RAW DATA

47.0	70.0	271.0	72.0	78.0	100.0	20.0	108.0	124.0	52.0
158.0	91.0	59.0	87.0	180.0	44.0	186.0	35.0	93.0	133.0
4.0	265.0	59.0	22.0	96.0	33.0	15.0	47.0	6.0	3.0
165.0	102.0	46.0	17.0	40.0	0.0	106.0	44.0	74.0	18.0
24.0	101.0	15.0	77.0	16.0	66.0	246.0	302.0	180.0	113.0
46.0	44.0	27.0							

SERVICE TIME SUMMARY STATISTICS

THE MEAN IS 83.5283.
THE VARIANCE IS 5265.0000.
THE STANDARD DEVIATION IS 72.5603.
THE SAMPLE SIZE IS 53.

SERVICE TIME SORTED DATA

0.0	3.0	4.0	6.0	15.0	15.0	16.0	17.0	18.0	20.0
22.0	24.0	27.0	33.0	35.0	40.0	44.0	44.0	44.0	46.0
46.0	47.0	47.0	52.0	59.0	59.0	66.0	70.0	72.0	74.0
77.0	78.0	87.0	91.0	93.0	96.0	100.0	101.0	102.0	106.0
108.0	113.0	124.0	133.0	158.0	165.0	180.0	180.0	186.0	246.0
265.0	271.0	302.0							

TABLE 2-12   GOODNESS-OF-FIT TEST OF THE UNIFORM DISTRIBUTION FOR SERVICE TIMES

CLASS	CLASS IDENTIFIER	OBSERVED FREQUENCY	RELATIVE FREQUENCY	EXPECTED FREQUENCY	χ^2
1	15.1	13.00	0.100	5.30	11.19
2	45.3	13.00	0.100	5.30	11.19
3	75.5	7.00	0.100	5.30	0.55
4	105.7	9.00	0.100	5.30	2.58
5	135.9	2.00	0.100	5.30	2.05
6	166.1	4.00	0.100	5.30	0.32
7	196.3	1.00	0.100	5.30	3.49
8	226.5	0.00	0.100	5.30	5.30
9	256.7	3.00	0.100	5.30	1.00
10	286.9	1.00	0.100	5.30	3.49
TOTALS		53.00	1.000	53.00	41.15

Clearly, this uniform distribution does not fit well at all.

What about a normal distribution with a mean of 83.5283 and a standard deviation of 72.5603? It appears that the normal distribution does not fit the service time data well either. The negative exponential distribution remains a possibility (see Tables 2-14 and 2-15).

TABLE 2-13   GOODNESS-OF-FIT TEST OF THE NORMAL DISTRIBUTION FOR SERVICE TIMES

CLASS	CLASS IDENTIFIER	OBSERVED FREQUENCY	RELATIVE FREQUENCY	EXPECTED FREQUENCY	χ^2
1	−9.3	0.00	0.100	5.31	5.31
2	22.6	11.00	0.100	5.31	6.10
3	45.4	8.00	0.099	5.26	1.42
4	65.2	7.00	0.100	5.32	0.53
5	83.5	6.00	0.100	5.29	0.09
6	101.9	6.00	0.100	5.29	0.09
7	121.6	4.00	0.100	5.32	0.33
8	144.5	2.00	0.099	5.26	2.02
9	176.4	2.00	0.100	5.31	2.06
10	999.0	7.00	0.100	5.32	0.53
TOTALS		53.00	1.000	53.00	18.50

TABLE 2-14   CLASS BOUNDARY DEFINITIONS FOR NEGATIVE EXPONENTIAL TEST OF SERVICE TIMES

CLASS IDENTITY	CUMULATIVE GREATER-THAN PROBABILITY	CLASS BOUNDARY	CLASS WIDTH
1	1.0000	0.00	8.80
2	0.9000	8.80	9.84
3	0.8000	18.64	11.15
4	0.7000	29.79	12.88
5	0.6000	42.67	15.23
6	0.5000	57.90	18.64
7	0.4000	76.54	24.03
8	0.3000	100.57	33.87
9	0.2000	134.43	57.90
10	0.1000	192.33	999.99

TABLE 2-15   GOODNESS-OF-FIT TEST OF THE NEGATIVE EXPONENTIAL
DISTRIBUTION FOR SERVICE TIME

CLASS	CLASS IDENTIFIER	OBSERVED FREQUENCY	RELATIVE FREQUENCY	EXPECTED FREQUENCY	χ^2
1	8.8	4.00	0.100	5.30	0.32
2	18.6	5.00	0.100	5.30	0.02
3	29.8	4.00	0.100	5.30	0.32
4	42.7	3.00	0.100	5.30	1.00
5	57.9	8.00	0.100	5.30	1.38
6	76.5	6.00	0.100	5.30	0.09
7	100.6	7.00	0.100	5.30	0.55
8	134.4	7.00	0.100	5.30	0.55
9	192.3	5.00	0.100	5.30	0.02
10	999.0	4.00	0.100	5.30	0.32
TOTALS		53.00	1.000	53.00	4.55

A computed χ^2 value of 4.55 compares with a critical χ^2 value of 15.507 at the 0.05 level of significance with 8 degrees of freedom. We are unable to reject the null hypothesis that the service time data were drawn from a population which is negative exponentially distributed with a mean of 83.5283 seconds.

SUMMARY

The collection of data which will serve as input to a model is a crucially important activity. The data will help to define the interarrival and service time distributions for the current system. They will also suggest possible modifications to the current system. Data to be collected are arrival time, service start time, and service completion time for a sample of customers.

From these three data items, the following system performance measures are calculated: interarrival time, service time, number in queue, waiting time, transit time, and server idle time. A percentage value for server idle time and averages for the other performance measures give point estimates of these aspects of system performance.

Using the individual values of interarrival time and service time that have been computed from the raw data, estimates of the standard deviation, variance and range of interarrival, and service times in the general population can be prepared. In combination with a rank-ordered list of interarrival and service time values, these estimates of dispersion allow the analyst to conduct formal statistical tests of the goodness of fit of theoretical and empirical distributions.

The χ^2 goodness-of-fit procedure enables the analyst to decide whether or not a general distribution type—such as the uniform, normal, or negative exponential—with specific parameter values drawn from sample data seems to characterize the process observed. If no theoretical distribution seems to fit, more data may be gathered and the analysis redone, or an empirical distribution based only on the sample evidence may be used.

QUESTIONS

1. What customer characteristics should be observed and recorded when gathering data for a waiting-line system?
2. What performance measures are derived from observations of a waiting-line system?
3. Define "queue discipline." What are some common queue disciplines?
4. Define "channel" and "phase" in the context of waiting-line systems. Give examples of systems with various channel and phase properties.
5. In a goodness-of-fit test, what is the null hypothesis? the alternative hypothesis?
6. What is the meaning of "level of significance" in a statistical test?
7. Briefly explain how a χ^2 test enables the analyst to evaluate the likelihood that the null hypothesis is true.
8. What parameters are needed to specify the null hypothesis in a goodness-of-fit test of
    (a) a uniform distribution?
    (b) a normal distribution?
    (c) a negative exponential distribution?
    (d) a Poisson distribution?
    (e) an empirical distribution?
9. Why is it preferable to categorize interarrival and service times according to a theoretical pattern rather than using the sample data to define an empirical pattern?
10. Why is it preferable to gather time data instead of rate data?
11. What should be done if
    (a) more than one theoretical model seems to fit the data?
    (b) no theoretical model seems to fit the data?
12. Why is a χ^2 test suggested instead of a Kolmogorov-Smirnov test in establishing the fit of arrival and service patterns?
13. Why is it impossible to specify exactly how much data need to be collected to define arrival and service patterns?
14. What is the nature of the relationship between the Poisson and the negative exponential distributions?

PROBLEMS

You should use the following set of observations of arrival time, service start time, and service completion time in solving the problems that follow. You should assume that observation began at 4 P.M. (time 160000) and that one person was waiting at the time observation began.

160500, 160614
160140, 160614, 160850
160508, 160850, 161031
160724, 161031, 161104
160808, 161104, 161211
160937, 161211, 161301
161044, 161301, 161959
162001, 162001, 162727
162956, 162956, 164526
165036, 165036, 170044
170406, 170406, 170940
171131, 171131, 171821
172038, 172038, 172301
172348, 172348, 172734
172850, 172850, 172943
173001, 173001, 173026
173035, 173035, 173809
174040, 174040, 174532
174709, 174709, 175006
175105, 175105, 180049
180404, 180404, 182009
182530, 182530, 183145
183350, 183350, 183903
184047, 184047, 184209
184237, 184237, 184826
185022, 185022, 185953

1. Calculate interarrival time and service time for each customer. How many observations of interarrival time are there? How many observations of service time are there? Why might there be a difference in these two numbers?

   Optional Supplement: If you used the DATCOL program to perform the calculations in Problem 1, examine the queue-length statistics and speculate on the probable accuracy of the average queue-length calculation.

*2. Find the mean, standard deviation, and variance of the interarrival and service time data sets produced in Problem 1. Can you make any tentative assessments about the likelihood that certain theoretical distributions will fit? On what basis?

3. Arrange the interarrival time and service time data in ascending rank order. Can you exclude any theoretical distributions from further consideration by examining the data values? Explain.

*4. Give the null hypothesis, alternative hypothesis and level of significance to test the goodness of fit of the following distributions for the interarrival time and service time data:
   (a) uniform
   (b) normal
   (c) negative exponential

5. Perform the χ^2 goodness-of-fit tests suggested in your answer to Problem 4. State the number of degrees of freedom, the computed and critical levels of χ^2, and the resulting conclusion as to whether or not to reject the null hypothesis in each case.

   *Answers to problems preceded by asterisks can be found in the Selected Numerical Answers.

*6. Rearrange the original arrival data so that a Poisson test comparable to the negative exponential test can be performed. Give the null hypothesis and alternative hypothesis, and then perform the test. Compare the results with those in Problem 4c.

7. Test the empirical hypothesis that service time has a 20 percent chance of falling below 100 seconds and has a 30 percent chance of exceeding 300 seconds. State your results and indicate under what circumstances such a hypothesis might be appropriate to test.

APPENDIX

Three programs in the SIMSTAT package are used to generate the output shown in Chapter 2. These programs are DATCOL (data collection), DESCST (descriptive statistics), and GOFFIT (goodness-of-fit tests). DATCOL requests as input the values of service start time and service completion time for all customers observed, as well as arrival time for customers other than those in the initial queue. Table 2A-1 replicates the data in Table 2-1 and provides this input to the DATCOL program. If disk file space is not available or if the user prefers to enter the data directly from the terminal, DATCOL will accept the data as well as the answers to the interactive housekeeping questions from the terminal.

Table 2A-2 displays the "conversation" with DATCOL. DATCOL first asks whether the data are on disk or are to be entered from the terminal. The file name containing the data shown in Table 2A-1 was previously defined as FORTRAN DATASET 20 through a system-dependent job control language (JCL) instruction. If this file definition has not previously been provided, program execution is temporarily interrupted to request it. All SIMSTAT programs expect input disk files to be defined as dataset 20 and output files to be defined as dataset 30.

DATCOL permits data to be entered in 24-hour military clock style, with two digits for hours, two digits for minutes, and two digits for seconds. Times for consecutive observations must be in ascending order, so this type of coding is not suitable when the observation period overlaps midnight (i.e., times progress from 23 hours to 0 hours). In that case, or whenever the analyst prefers, times may be recorded in standard units (seconds) beginning with zero, stopwatch style.

DATCOL asks at what time observation began. This time should be given in either the hour-minute-second format or in standard units, depending on the answer to the previous question about data format.

Finally, DATCOL asks how many customers were in the queue when observation began. For those customers it expects to receive only data on service start and completion times, not data on the unknown arrival time. It admonishes the user *not* to count as part of the initial queue any customer who may have been in the process of being served when observation began, since that customer's service start time would be unknown.

After the data on service start and completion times for any customers in the initial queue have been entered, DATCOL seeks arrival time, service start time, and service completion time for the customers who arrived after the observation period began. If data are entered directly from a terminal, DATCOL identifies the end of the sample data set by a dummy value of 999999, 0, 0. If the data are on disk, the end of the disk file is sensed automatically and no dummy end-of-file data should be used.

DATCOL converts the three input data elements for each customer to standard units (seconds) if they are not already in that form and echoes the values in standard units back to the analyst as part of a table with one line per customer, sequenced by observation number. For customers in the initial queue, only service start time, service completion time, and the resulting difference, service time, are shown. A separate tally is made for customers who arrived after the observation period began. For these customers, interarrival time, queue length, transit time, waiting time, and server idle time are also calculated and displayed. The information in this table is automatically written on disk as dataset 30 (which must be defined through system JCL), so that it may be easily extracted for further processing. The contents of the output disk file are shown in Table 2A-3.

The most probable items from the disk file to be processed further are interarrival time and service time. Through the use of a text-editing package, these values may be placed in individual files. Interarrival time may be found in columns 14 through 21, and service time may be found in columns 47 through 53 of the output disk file created by DATCOL. Table 2A-4 shows the list of interarrival times after extraction from the DATCOL output disk file, and Table 2A-5 shows the comparable list of service times.

It will probably be desired to compute descriptive statistics for the interarrival time and service time data sets and to print them in ascending rank order to facilitate later goodness-of-fit tests. The DESCST program accepts the data from a disk file or directly from the terminal. Tables 2A-6 and 2A-7, respectively, show the DESCST ''conversations'' which produce the desired descriptive statistics for interarrival time and service time. DESCST has two options, of which only the first is used in this chapter. DESCST will display the raw data on request to verify that the correct file is being processed. It then prints the mean, variance, standard deviation, and sample size. If desired, a list of the data sorted in ascending rank order may be printed.

The last SIMSTAT program used in this chapter is GOFFIT. GOFFIT does not read or write any individual data observations, but asks the analyst questions about how classes or groupings of the data have been arranged. GOFFIT can perform either χ^2 or Kolmogorov-Smirnov goodness-of-fit tests, but only the χ^2 type is used in Chapter 2. It is possible to test the goodness of fit of the negative exponential, Poisson, normal, uniform, and empirically defined distributions. GOFFIT will ask how many classes have been defined. Then it will ask either for the class midpoint or the upper class boundary for each class. Since the formula for the negative exponential distribution is cumbersome to use for hand calculation but simple to program, GOFFIT allows the analyst to request the calculation of class boundaries for a particular negative exponential distribution which will ensure expected class frequencies of 5 for each class; then GOFFIT will proceed to perform the goodness-of-fit test given these boundaries when requested.

For each class defined, GOFFIT will ask for the observed frequency in that class. Reference to the sorted data list prepared by DESCST facilitates provision of this information to GOFFIT.

GOFFIT stores critical values of χ^2 at the 0.05 level of significance for all levels of degrees of freedom up to 20. After printing a table showing observed and expected frequencies for each class, resulting χ^2, and the total value of χ^2, it will report the critical level of χ^2 for two types of degrees of freedom. One type assumes that the parameters in the null hypothesis have been estimated from sample data, as has been the case in Chapter 2.

There are several safeguards incorporated into GOFFIT to protect the analyst from simple, but deadly, errors. First, the sums of observed and expected class frequencies are printed, so the analyst can verify that all observed class frequencies have been correctly entered. Second, when a test produces expected class frequencies of less than 5, a warning message is printed so that the analyst can correct the class boundaries or combine classes, as appropriate. Finally, there is no need to use a χ^2 table unless degrees of freedom exceed 20 or the analyst desires to perform the test at a level of significance other than 0.05.

There is one instance in which the warning message regarding expected class frequencies may print when it is not strictly warranted. Relative frequencies are calculated from formulas which are rounded or use numerical approximations, so sometimes expected frequencies which differ from 5.0 only by very small amounts occur and trigger the warning message. The analyst should use discretion in deciding which situations merit modification of classes.

The goodness-of-fit tests which appear in the chapter are shown, along with the preliminary ''conversations,'' in Tables 2A-8 through 2A-15.

TABLE 2A-1   SOURCE DATE ON ARRIVAL TIME, SERVICE START TIME, AND SERVICE END TIME

1.0000	100000, 100047	15.0000	102231, 102252, 102552
2.0000	100047, 100157	16.0000	102353, 102552, 102636
3.0000	100107, 100157, 100628	17.0000	102551, 102636, 102942
4.0000	100124, 100628, 100740	18.0000	102608, 102942, 103017
5.0000	100133, 100740, 100858	19.0000	102811, 103017, 103150
6.0000	100227, 100858, 101038	20.0000	103205, 103205, 103418
7.0000	100332, 101038, 101058	21.0000	103215, 103418, 103422
8.0000	100427, 101058, 101246	22.0000	103249, 103422, 103847
9.0000	100635, 101246, 101450	23.0000	103335, 103847, 103946
10.0000	101144, 101450, 101542	24.0000	103740, 103946, 104008
11.0000	101326, 101542, 101820	25.0000	103821, 104008, 104144
12.0000	101818, 101820, 101951	26.0000	104105, 104144, 104217
13.0000	101837, 101951, 102050	27.0000	104109, 104217, 104232
14.0000	102125, 102125, 102252	28.0000	104231, 104232, 104319

TABLE 2A-1 (cont.)

29.0000	104337, 104337, 104343	42.0000	110219, 110219, 110400
30.0000	104459, 104459, 104502	43.0000	110512, 110512, 110527
31.0000	104644, 104644, 104929	44.0000	110553, 110553, 110710
32.0000	104752, 104929, 105111	45.0000	110828, 110828, 110844
33.0000	104832, 105111, 105157	46.0000	111039, 111039, 111145
34.0000	104952, 105157, 105214	47.0000	111315, 111315, 111721
35.0000	105043, 105214, 105254	48.0000	111614, 111721, 112223
36.0000	105143, 105254, 105254	49.0000	111735, 112223, 112523
37.0000	105411, 105411, 105557	50.0000	111826, 112523, 112716
38.0000	105423, 105557, 105641	51.0000	112128, 112716, 112802
39.0000	105429, 105641, 105755	52.0000	112347, 112802, 112846
40.0000	105537, 105755, 105813	53.0000	112531, 112846, 112913
41.0000	110054, 110054, 110118		

TABLE 2A-2   DATCOL PRINTED OUTPUT

THIS PROGRAM COMPUTES PERFORMANCE MEASURES FROM SIMULATION INPUT DATA
AND WRITES THE INFORMATION ON DISK. TABLE OUTPUT IS GIVEN IN SECONDS
OR IN STANDARD UNITS. IF THE DATA ARE ON DISK, TYPE A 1.
OTHERWISE, TYPE A 0.

• 1

IF DATA ARE IN HRS, MIN AND SEC, TYPE A 1.
IF DATA ARE IN STANDARD UNITS STARTING WITH 0, TYPE A 0.

• 1

AT WHAT TIME DID OBSERVATION BEGIN?
• 095920

HOW MANY CUSTOMERS WERE IN THE QUEUE INITIALLY? DO NOT COUNT
THE CUSTOMER WHO MAY HAVE BEEN BEING SERVED.

• 2

TABLE 2A-2 (cont.)

OBS. NO.	TIME ARRIVED	I/A TIME	QUEUE LENGTH	SERVICE START TIME	SERVICE END TIME	SERVICE TIME	TRANSIT TIME	WAITING TIME	SERVER IDLE TIME
1				40.0	87.0	47.0			0.0
2				87.0	157.0	70.0			0.0
1	107.0	-99999.9	1	157.0	428.0	271.0	321.0	50.0	0.0
2	124.0	17.0	2	428.0	500.0	72.0	376.0	304.0	0.0
3	133.0	9.0	3	500.0	578.0	78.0	445.0	367.0	0.0
4	187.0	54.0	3	578.0	678.0	100.0	491.0	391.0	0.0
5	252.0	65.0	4	678.0	698.0	20.0	446.0	426.0	0.0
6	307.0	55.0	5	698.0	806.0	108.0	499.0	391.0	0.0
7	435.0	128.0	5	806.0	930.0	124.0	495.0	371.0	0.0
8	744.0	309.0	2	930.0	982.0	52.0	238.0	186.0	0.0
9	846.0	102.0	2	982.0	1140.0	158.0	294.0	136.0	0.0
10	1138.0	292.0	1	1140.0	1231.0	91.0	93.0	2.0	0.0
11	1157.0	19.0	1	1231.0	1290.0	59.0	133.0	74.0	0.0
12	1325.0	168.0	0	1325.0	1412.0	87.0	87.0	0.0	35.0
13	1391.0	66.0	1	1412.0	1592.0	180.0	201.0	21.0	0.0
14	1473.0	82.0	1	1592.0	1636.0	44.0	163.0	119.0	0.0
15	1591.0	118.0	2	1636.0	1822.0	186.0	231.0	45.0	0.0
16	1608.0	17.0	2	1822.0	1857.0	35.0	249.0	214.0	0.0
17	1731.0	123.0	2	1857.0	1950.0	93.0	219.0	126.0	0.0
18	1965.0	234.0	0	1965.0	2098.0	133.0	133.0	0.0	15.0
19	1975.0	10.0	1	2098.0	2102.0	4.0	127.0	123.0	0.0
20	2009.0	34.0	2	2102.0	2367.0	265.0	358.0	93.0	0.0
21	2055.0	46.0	3	2367.0	2426.0	59.0	371.0	312.0	0.0
22	2300.0	245.0	2	2426.0	2448.0	22.0	148.0	126.0	0.0
23	2341.0	41.0	3	2448.0	2544.0	96.0	203.0	107.0	0.0
24	2505.0	164.0	1	2544.0	2577.0	33.0	72.0	39.0	0.0
25	2509.0	4.0	2	2577.0	2592.0	15.0	83.0	68.0	0.0
26	2591.0	82.0	1	2592.0	2639.0	47.0	48.0	1.0	0.0
27	2657.0	66.0	0	2657.0	2663.0	6.0	6.0	0.0	18.0

TABLE 2A-2 (cont.)

28	2739.0	82.0	0	2739.0	2742.0	3.0	3.0	0.0	76.0
29	2844.0	105.0	0	2844.0	3009.0	165.0	165.0	0.0	102.0
30	2912.0	68.0	1	3009.0	3111.0	102.0	199.0	97.0	0.0
31	2952.0	40.0	2	3111.0	3157.0	46.0	205.0	159.0	0.0
32	3032.0	80.0	2	3157.0	3174.0	17.0	142.0	125.0	0.0
33	3083.0	51.0	3	3174.0	3214.0	40.0	131.0	91.0	0.0
34	3143.0	60.0	3	3214.0	3214.0	0.0	71.0	71.0	0.0
35	3291.0	148.0	0	3291.0	3397.0	106.0	106.0	0.0	77.0
36	3303.0	12.0	1	3397.0	3441.0	44.0	138.0	94.0	0.0
37	3309.0	6.0	2	3441.0	3515.0	74.0	206.0	132.0	0.0
38	3377.0	68.0	3	3515.0	3533.0	18.0	156.0	138.0	0.0
39	3694.0	317.0	0	3694.0	3718.0	24.0	24.0	0.0	161.0
40	3779.0	85.0	0	3779.0	3880.0	101.0	101.0	0.0	61.0
41	3952.0	173.0	0	3952.0	3967.0	15.0	15.0	0.0	72.0
42	3993.0	41.0	0	3993.0	4070.0	77.0	77.0	0.0	26.0
43	4148.0	155.0	0	4148.0	4164.0	16.0	16.0	0.0	78.0
44	4279.0	131.0	0	4279.0	4345.0	66.0	66.0	0.0	115.0
45	4435.0	156.0	0	4435.0	4681.0	246.0	246.0	0.0	90.0
46	4614.0	179.0	1	4681.0	4983.0	302.0	246.0	67.0	0.0
47	4695.0	81.0	1	4983.0	5163.0	180.0	369.0	288.0	0.0
48	4746.0	51.0	2	5163.0	5276.0	113.0	468.0	417.0	0.0
49	4928.0	182.0	3	5276.0	5322.0	46.0	530.0	348.0	0.0
50	5067.0	139.0	3	5322.0	5366.0	44.0	394.0	255.0	0.0
51	5171.0	104.0	3	5366.0	5393.0	27.0	222.0	195.0	0.0

```
AVERAGE INTERARRIVAL TIME        = 101.280
AVERAGE QUEUE LENGTH             =   1.608
AVERAGE SERVICE TIME             =  83.528
AVERAGE TRANSIT TIME             = 213.314
AVERAGE WAITING TIME             = 128.804
AVERAGE SERVER IDLE TIME PERCENT =  17.170
```

TABLE 2A-3   DATCOL DISK OUTPUT

1				40.0	87.0	47.0			
2				87.0	157.0	70.0			
1	107.0	-99999.9	1	157.0	428.0	271.0	321.0	50.0	0.0
2	124.0	17.0	2	428.0	500.0	72.0	376.0	304.0	0.0
3	133.0	9.0	3	500.0	578.0	78.0	445.0	367.0	0.0
4	187.0	54.0	3	578.0	678.0	100.0	491.0	391.0	0.0
5	252.0	65.0	4	678.0	698.0	20.0	446.0	426.0	0.0
6	307.0	55.0	5	698.0	806.0	108.0	499.0	391.0	0.0
7	435.0	128.0	5	806.0	930.0	124.0	495.0	371.0	0.0
8	744.0	309.0	2	930.0	982.0	52.0	238.0	186.0	0.0
9	846.0	102.0	2	982.0	1140.0	158.0	294.0	136.0	0.0
10	1138.0	292.0	1	1140.0	1231.0	91.0	93.0	2.0	0.0
11	1157.0	19.0	1	1231.0	1290.0	59.0	133.0	74.0	0.0
12	1325.0	168.0	0	1325.0	1412.0	87.0	87.0	0.0	35.0
13	1391.0	66.0	1	1412.0	1592.0	180.0	201.0	21.0	0.0
14	1473.0	82.0	1	1592.0	1636.0	44.0	163.0	119.0	0.0
15	1591.0	118.0	2	1636.0	1822.0	186.0	231.0	45.0	0.0
16	1608.0	17.0	2	1822.0	1857.0	35.0	249.0	214.0	0.0
17	1731.0	123.0	2	1857.0	1950.0	93.0	219.0	126.0	0.0
18	1965.0	234.0	0	1965.0	2098.0	133.0	133.0	0.0	15.0
19	1975.0	10.0	1	2098.0	2102.0	4.0	127.0	123.0	0.0
20	2009.0	34.0	2	2102.0	2367.0	265.0	358.0	93.0	0.0
21	2055.0	46.0	3	2367.0	2426.0	59.0	371.0	312.0	0.0
22	2300.0	245.0	2	2426.0	2448.0	22.0	148.0	126.0	0.0
23	2341.0	41.0	3	2448.0	2544.0	96.0	203.0	107.0	0.0
24	2505.0	164.0	1	2544.0	2577.0	33.0	72.0	39.0	0.0
25	2509.0	4.0	2	2577.0	2592.0	15.0	83.0	68.0	0.0
26	2591.0	82.0	1	2592.0	2639.0	47.0	48.0	1.0	0.0
27	2657.0	66.0	0	2657.0	2663.0	6.0	6.0	0.0	18.0
28	2739.0	82.0	0	2739.0	2742.0	3.0	3.0	0.0	76.0
29	2844.0	105.0	0	2844.0	3009.0	165.0	165.0	0.0	102.0
30	2912.0	68.0	1	3009.0	3111.0	102.0	199.0	97.0	0.0
31	2952.0	40.0	2	3111.0	3157.0	46.0	205.0	159.0	0.0
32	3032.0	80.0	2	3157.0	3174.0	17.0	142.0	125.0	0.0
33	3083.0	51.0	3	3174.0	3214.0	40.0	131.0	91.0	0.0
34	3143.0	60.0	3	3214.0	3214.0	0.0	71.0	71.0	0.0
35	3291.0	148.0	0	3291.0	3397.0	106.0	106.0	0.0	77.0
36	3303.0	12.0	1	3397.0	3441.0	44.0	138.0	94.0	0.0
37	3309.0	6.0	2	3441.0	3515.0	74.0	206.0	132.0	0.0
38	3377.0	68.0	3	3515.0	3533.0	18.0	156.0	138.0	0.0
39	3694.0	317.0	0	3694.0	3718.0	24.0	24.0	0.0	161.0
40	3779.0	85.0	0	3779.0	3880.0	101.0	101.0	0.0	61.0
41	3952.0	173.0	0	3952.0	3967.0	15.0	15.0	0.0	72.0
42	3993.0	41.0	0	3993.0	4070.0	77.0	77.0	0.0	26.0
43	4148.0	155.0	0	4148.0	4164.0	16.0	16.0	0.0	78.0
44	4279.0	131.0	0	4279.0	4345.0	66.0	66.0	0.0	115.0

TABLE 2A-3 (cont.)

45	4435.0	156.0	0	4435.0	4681.0	246.0	246.0	0.0	90.0
46	4614.0	179.0	1	4681.0	4983.0	302.0	369.0	67.0	0.0
47	4695.0	81.0	1	4983.0	5163.0	180.0	468.0	288.0	0.0
48	4746.0	51.0	2	5163.0	5276.0	113.0	530.0	417.0	0.0
49	4928.0	182.0	3	5276.0	5322.0	46.0	394.0	348.0	0.0
50	5067.0	139.0	3	5322.0	5366.0	44.0	299.0	255.0	0.0
51	5171.0	104.0	3	5366.0	5393.0	27.0	222.0	195.0	0.0

TABLE 2A-4   INTERARRIVAL TIMES
EXTRACTED FROM DISK

1.0000	17.0	26.0000	66.0
2.0000	9.0	27.0000	82.0
3.0000	54.0	28.0000	105.0
4.0000	65.0	29.0000	68.0
5.0000	55.0	30.0000	40.0
6.0000	128.0	31.0000	80.0
7.0000	309.0	32.0000	51.0
8.0000	102.0	33.0000	60.0
9.0000	292.0	34.0000	148.0
10.0000	19.0	35.0000	12.0
11.0000	168.0	36.0000	6.0
12.0000	66.0	37.0000	68.0
13.0000	82.0	38.0000	317.0
14.0000	118.0	39.0000	85.0
15.0000	17.0	40.0000	173.0
16.0000	123.0	41.0000	41.0
17.0000	234.0	42.0000	155.0
18.0000	10.0	43.0000	131.0
19.0000	34.0	44.0000	156.0
20.0000	46.0	45.0000	179.0
21.0000	245.0	46.0000	81.0
22.0000	41.0	47.0000	51.0
23.0000	164.0	48.0000	182.0
24.0000	4.0	49.0000	139.0
25.0000	82.0	50.0000	104.0

TABLE 2A-5   SERVICE TIMES
EXTRACTED FROM DISK

1.0000	47.0	6.0000	100.0
2.0000	70.0	7.0000	20.0
3.0000	271.0	8.0000	108.0
4.0000	72.0	9.0000	124.0
5.0000	78.0	10.0000	52.0

TABLE 2A-5 (cont.)

11.0000	158.0	33.0000	46.0
12.0000	91.0	34.0000	17.0
13.0000	59.0	35.0000	40.0
14.0000	87.0	36.0000	0.0
15.0000	180.0	37.0000	106.0
16.0000	44.0	38.0000	44.0
17.0000	186.0	39.0000	74.0
18.0000	35.0	40.0000	18.0
19.0000	93.0	41.0000	24.0
20.0000	133.0	42.0000	101.0
21.0000	4.0	43.0000	15.0
22.0000	265.0	44.0000	77.0
23.0000	59.0	45.0000	16.0
24.0000	22.0	46.0000	66.0
25.0000	96.0	47.0000	246.0
26.0000	33.0	48.0000	302.0
27.0000	15.0	49.0000	180.0
28.0000	47.0	50.0000	113.0
29.0000	6.0	51.0000	46.0
30.0000	3.0	52.0000	44.0
31.0000	165.0	53.0000	27.0
32.0000	102.0		

TABLE 2A-6   DESCST OUTPUT FOR INTERARRIVAL TIMES

DESCRIPTIVE STATISTICS

DO YOU WANT INSTRUCTIONS? TYPE 1 FOR YES, 0 FOR NO.
• 1

UNDER OPTION 1 THIS PROGRAM COMPUTES DESCRIPTIVE
STATISTICS, INCLUDING THE MEAN, VARIANCE AND STANDARD
DEVIATION OF A DATA SET, AND CREATES A SORTED LIST OF
INPUTS. UNDER OPTION 2 THE PROGRAM COMPUTES THE
MEANS OF EACH M ADJACENT OBSERVATIONS OR SELECTS EVERY
N-TH OBSERVATION.

WHICH OPTION? TYPE 1 OR 2?
• 1

IF INPUT IS TO BE PROVIDED FROM A DISK
FILE, TYPE A 1. OTHERWISE TYPE A 0.
• 1

DO YOU WANT TO SEE THE RAW DATA?
IF YES, TYPE A 1. IF NO, TYPE A 0.
• 1

TABLE 2A-6 (cont.)

17.0	9.0	54.0	65.0	55.0	128.0	309.0	102.0	292.0	19.0
168.0	66.0	82.0	118.0	17.0	123.0	234.0	10.0	34.0	46.0
245.0	41.0	164.0	4.0	82.0	66.0	82.0	105.0	68.0	40.0
80.0	51.0	60.0	148.0	12.0	6.0	68.0	317.0	85.0	173.0
41.0	155.0	131.0	156.0	179.0	81.0	51.0	182.0	139.0	104.0

THE MEAN IS 101.2800
THE VARIANCE IS 6216.5390
THE STANDARD DEVIATION IS 78.8450
THE SAMPLE SIZE IS 50

WOULD YOU LIKE TO SEE A SORTED DATA LIST?
IF YES, TYPE 1. IF NO, TYPE 0.
• 1

4.0	6.0	9.0	10.0	12.0	17.0	17.0	19.0	34.0	40.0
41.0	41.0	46.0	51.0	51.0	54.0	55.0	60.0	65.0	66.0
66.0	68.0	68.0	80.0	81.0	82.0	82.0	82.0	85.0	102.0
104.0	105.0	118.0	123.0	128.0	131.0	139.0	148.0	155.0	156.0
164.0	168.0	173.0	179.0	182.0	234.0	245.0	292.0	309.0	317.0

TABLE 2A-7 DESCST OUTPUT FOR SERVICE TIMES

DESCRIPTIVE STATISTICS

DO YOU WANT INSTRUCTIONS? TYPE 1 FOR YES, 0 FOR NO.
• 0

WHICH OPTION? TYPE 1 OR 2?
• 1

IF INPUT IS TO BE PROVIDED FROM A DISK
FILE, TYPE A 1. OTHERWISE TYPE A 0.
• 1

DO YOU WANT TO SEE THE RAW DATA?
IF YES, TYPE A 1. IF NO, TYPE A 0.
• 1

47.0	70.0	271.0	72.0	78.0	100.0	20.0	108.0	124.0	52.0
158.0	91.0	59.0	87.0	180.0	44.0	186.0	35.0	93.0	133.0
4.0	265.0	59.0	22.0	96.0	33.0	15.0	47.0	6.0	3.0
165.0	102.0	46.0	17.0	40.0	0.0	106.0	44.0	74.0	18.0
24.0	101.0	15.0	77.0	16.0	66.0	246.0	302.0	180.0	113.0
46.0	44.0	27.0							

TABLE 2A-7 (cont.)

THE MEAN IS 83.5283
THE VARIANCE IS 5265.0000
THE STANDARD DEVIATION IS 72.5603
THE SAMPLE SIZE IS 53

WOULD YOU LIKE TO SEE A SORTED DATA LIST?
IF YES, TYPE 1. IF NO, TYPE 0.
• 1

0.0	3.0	4.0	6.0	15.0	15.0	16.0	17.0	18.0	20.0
22.0	24.0	27.0	33.0	35.0	40.0	44.0	44.0	44.0	46.0
46.0	47.0	47.0	52.0	59.0	59.0	66.0	70.0	72.0	74.0
77.0	78.0	87.0	91.0	93.0	96.0	100.0	101.0	102.0	106.0
108.0	113.0	124.0	133.0	158.0	165.0	180.0	180.0	186.0	246.0
265.0	271.0	302.0							

TABLE 2A-8   INTERARRIVAL TIME UNIFORM GOFFIT TEST

GOODNESS OF FIT TESTS

DO YOU WANT INSTRUCTIONS? TYPE 1 FOR YES, 0 FOR NO.
• 0

YOU MAY CHOOSE EITHER A CHI SQUARE OR A
KOLMOGOROV–SMIRNOV GOODNESS OF FIT TEST.

WHICH DISTRIBUTION DO YOU WANT TO TEST
FOR GOODNESS OF FIT?
YOUR ALTERNATIVES ARE:
1. NEGATIVE EXPONENTIAL DISTRIBUTION
2. POISSON DISTRIBUTION
3. NORMAL DISTRIBUTION
4. UNIFORM DISTRIBUTION
5. EMPIRICALLY DEFINED DISTRIBUTION

• • • WARNING • • • BE SURE THAT THE CLASSES YOU
DEFINED INCLUDE A MUTUALLY EXCLUSIVE, COLLECTIVELY
EXHAUSTIVE SET OF POSSIBLE OUTCOMES, SUCH AS
"PLUS OR MINUS INFINITY", IF APPROPRIATE!

TYPE THE NUMBER OF THE DISTRIBUTION YOU WANT TO TEST
• 4

YOU HAVE CHOSEN THE GOODNESS OF FIT TEST FOR:
4. UNIFORM DISTRIBUTION.

IF YOU WANT TO DO A CHI SQUARE TEST, TYPE A 1.

TABLE 2A-8 (cont.)

IF YOU WANT TO DO A KOLMOGOROV–SMIRNOV TEST, TYPE A 0.
• 1

YOU HAVE CHOSEN A CHI SQUARE TEST.

ENTER THE NUMBER OF CLASSES
• 10

ENTER THE CLASS MIDPOINT FOR CLASS   1
• 15. 65

ENTER THE OBSERVED FREQUENCY FOR CLASS NUMBER   1
• 9

ENTER THE CLASS MIDPOINT FOR CLASS   2
• 46. 95

ENTER THE OBSERVED FREQUENCY FOR CLASS NUMBER   2
• 12

ENTER THE CLASS MIDPOINT FOR CLASS   3
• 78. 25

ENTER THE OBSERVED FREQUENCY FOR CLASS NUMBER   3
• 8

ENTER THE CLASS MIDPOINT FOR CLASS   4
• 109. 55

ENTER THE OBSERVED FREQUENCY FOR CLASS NUMBER   4
• 6

ENTER THE CLASS MIDPOINT FOR CLASS   5
• 140. 85

ENTER THE OBSERVED FREQUENCY FOR CLASS NUMBER   5
• 5

ENTER THE CLASS MIDPOINT FOR CLASS   6
• 172. 15

ENTER THE OBSERVED FREQUENCY FOR CLASS NUMBER   6
• 5

ENTER THE CLASS MIDPOINT FOR CLASS   7
• 203. 45

TABLE 2A-8 (cont.)

ENTER THE OBSERVED FREQUENCY FOR CLASS NUMBER   7
• 0

ENTER THE CLASS MIDPOINT FOR CLASS   8
• 234.75

ENTER THE OBSERVED FREQUENCY FOR CLASS NUMBER   8
• 2

ENTER THE CLASS MIDPOINT FOR CLASS   9
• 266.05

ENTER THE OBSERVED FREQUENCY FOR CLASS NUMBER   9
• 0

ENTER THE CLASS MIDPOINT FOR CLASS   10
• 297.35

ENTER THE OBSERVED FREQUENCY FOR CLASS NUMBER   10
• 3

CLASS	CLASS IDENTITY	OBSERVED FREQUENCY	RELATIVE FREQUENCY	EXPECTED FREQUENCY	χ^2
1	15.6	9.00	0.100	5.00	3.20
2	46.9	12.00	0.100	5.00	9.80
3	78.3	8.00	0.100	5.00	1.80
4	109.6	6.00	0.100	5.00	0.20
5	140.9	5.00	0.100	5.00	0.00
6	172.1	5.00	0.100	5.00	0.00
7	203.4	0.00	0.100	5.00	5.00
8	234.8	2.00	0.100	5.00	1.80
9	266.1	0.00	0.100	5.00	5.00
10	297.4	3.00	0.100	5.00	0.80
TOTALS		50.00	1.000	50.00	27.60

DEGREES OF FREEDOM EQUALS NUMBER OF CLASSES
MINUS ONE, OR 9.

CRITICAL CHI SQUARE VALUE AT THE .05 LEVEL OF SIGNIFICANCE IS
    16.919 FOR 9 DEGREES OF FREEDOM.

IF YOU WANT TO SOLVE ANOTHER PROBLEM, TYPE A 1. OTHERWISE, TYPE A 0.
• 1

TABLE 2A-9   INTERARRIVAL TIME NORMAL GOFFIT TEST

GOODNESS OF FIT TESTS

DO YOU WANT INSTRUCTIONS? TYPE 1 FOR YES, 0 FOR NO.
• 0

TYPE THE NUMBER OF THE DISTRIBUTION YOU WANT TO TEST.
• 3

YOU HAVE CHOSEN THE GOODNESS OF FIT TEST FOR:
3. NORMAL DISTRIBUTION.

IF YOU WANT TO DO A CHI SQUARE TEST, TYPE A 1.
IF YOU WANT TO DO A KOLMOGOROV-SMIRNOV TEST, TYPE A 0.
• 1

ENTER THE MEAN
• 101.28

ENTER THE STANDARD DEVIATION
• 78.845

    YOU HAVE CHOSEN A CHI SQUARE TEST.

ENTER THE NUMBER OF CLASSES
• 10

ENTER THE UPPER CLASS BOUNDARY FOR CLASS   1
• .3584

ENTER THE OBSERVED FREQUENCY FOR CLASS   1
• 0

ENTER THE UPPER CLASS BOUNDARY FOR CLASS   2
• 35.0502

ENTER THE OBSERVED FREQUENCY FOR CLASS   2
• 9

ENTER THE UPPER CLASS BOUNDARY FOR CLASS   3
• 59.8864

ENTER THE OBSERVED FREQUENCY FOR CLASS   3
• 8

ENTER THE UPPER CLASS BOUNDARY FOR CLASS   4
• 81.33

TABLE 2A-9 (cont.)

ENTER THE OBSERVED FREQUENCY FOR CLASS   4
• 8

ENTER THE UPPER CLASS BOUNDARY FOR CLASS   5
• 101.28

ENTER THE OBSERVED FREQUENCY FOR CLASS   5
• 4

ENTER THE UPPER CLASS BOUNDARY FOR CLASS   6
• 121.2278

ENTER THE OBSERVED FREQUENCY FOR CLASS   6
• 4

ENTER THE UPPER CLASS BOUNDARY FOR CLASS   7
• 142.6736

ENTER THE OBSERVED FREQUENCY FOR CLASS   7
• 4

ENTER THE UPPER CLASS BOUNDARY FOR CLASS   8
• 167.5098

ENTER THE OBSERVED FREQUENCY FOR CLASS   8
• 4

ENTER THE UPPER CLASS BOUNDARY FOR CLASS   9
• 202.20

ENTER THE OBSERVED FREQUENCY FOR CLASS   9
• 4

ENTER • • • ONLY • • • THE OBSERVED FREQUENCY FOR CLASS   10
• 5

CLASS	CLASS IDENTITY	OBSERVED FREQUENCY	RELATIVE FREQUENCY	EXPECTED FREQUENCY	χ^2
1	0.4	0.00	0.100	5.01	5.01
2	35.1	9.00	0.100	5.01	3.18
3	59.9	8.00	0.099	4.97	1.85
4	81.3	8.00	0.100	5.02	1.77
5	101.3	4.00	0.100	4.99	0.20
6	121.2	4.00	0.100	4.99	0.20

TABLE 2A-9 (cont.)

7	142.7	4.00	0.100	5.02	0.21
8	167.2	4.00	0.099	4.97	0.19
9	202.2	4.00	0.100	5.01	0.20
10	999.0	5.00	0.100	5.02	0.00
TOTALS		50.00	1.000	50.00	12.81

IF YOU USED THESE OBSERVED DATA TO ESTIMATE HYPOTHESIS PARAMETERS,
DEGREES OF FREEDOM EQUALS NUMBER OF CLASSES MINUS 3 OR 7.

IF YOU USED AN INDEPENDENT ESTIMATE OF THE HYPOTHESIS PARAMETERS,
DEGREES OF FREEDOM EQUALS NUMBER OF CLASSES MINUS 1 OR 9.

CRITICAL CHI-SQUARE VALUES AT THE .05 LEVEL OF SIGNIFICANCE ARE
   14.067 FOR 7 DEGREES OF FREEDOM AND 16.919 for 9 DEGREES OF FREEDOM.
   AT LEAST ONE EXPECTED FREQUENCY IS LESS THAN 5. YOU MIGHT COMBINE CLASSES.

IF YOU WANT TO SOLVE ANOTHER PROBLEM, TYPE A 1. OTHERWISE, TYPE A 0.
● 1

TABLE 2A-10   INTERARRIVAL TIME NEGATIVE EXPONENTIAL GOFFIT TEST

GOODNESS OF FIT TESTS

DO YOU WANT INSTRUCTIONS? TYPE 1 FOR YES, 0 FOR NO.
● 1

TYPE THE NUMBER OF THE DISTRIBUTION YOU WANT TO TEST
● 1

YOU HAVE CHOSEN THE GOODNESS OF FIT TEST FOR:
1. NEGATIVE EXPONENTIAL DISTRIBUTION.

IF YOU WANT TO DO A CHI SQUARE TEST, TYPE A 1.
IF YOU WANT TO DO A KOLMOGOROV-SMIRNOV TEST, TYPE A 0.
● 1

ENTER THE MEAN TIME
● 101.28

YOU HAVE CHOSEN A CHI SQUARE TEST.

TABLE 2A-10 (cont.)

ENTER THE NUMBER OF CLASSES
• 10

IF YOU WANT TO DO A GOODNESS OF FIT TEST, TYPE A 1.
IF YOU WANT TO COMPUTE BOUNDARIES, TYPE A 0.
• 0

CLASS IDENTITY	CUMULATIVE GT PROBABILITY	CLASS BOUNDARY	CLASS WIDTH
1	1.0000	0.00	10.67
2	0.9000	10.67	11.93
3	0.8000	22.60	13.52
4	0.7000	36.12	15.61
5	0.6000	51.74	18.47
6	0.5000	70.20	22.60
7	0.4000	92.80	29.14
8	0.3000	121.94	41.07
9	0.2000	163.00	70.20
10	0.1000	233.21	999.99

IF YOU WANT TO PROCEED TO A CHI SQUARE GOODNESS
OF FIT TEST OF THIS NEGATIVE EXPONENTIAL DISTRIBUTION,
TYPE A 1. OTHERWISE, TYPE A 0.
• 1

ENTER THE OBSERVED FREQUENCY FOR CLASS   1
• 4

ENTER THE OBSERVED FREQUENCY FOR CLASS   2
• 4

ENTER THE OBSERVED FREQUENCY FOR CLASS   3
• 1

ENTER THE OBSERVED FREQUENCY FOR CLASS   4
• 6

ENTER THE OBSERVED FREQUENCY FOR CLASS   5
• 8

ENTER THE OBSERVED FREQUENCY FOR CLASS   6
• 6

ENTER THE OBSERVED FREQUENCY FOR CLASS   7
• 4

TABLE 2A-10 (cont.)

ENTER THE OBSERVED FREQUENCY FOR CLASS   8
● 7

ENTER THE OBSERVED FREQUENCY FOR CLASS   9
● 5

ENTER THE OBSERVED FREQUENCY FOR CLASS   10
● 5

CLASS	CLASS IDENTITY	OBSERVED FREQUENCY	RELATIVE FREQUENCY	EXPECTED FREQUENCY	χ^2
1	10.7	4.00	0.100	5.00	0.20
2	22.6	4.00	0.100	5.00	0.20
3	36.1	1.00	0.100	5.00	3.20
4	51.7	6.00	0.100	5.00	0.20
5	70.2	8.00	0.100	5.00	1.80
6	92.8	6.00	0.100	5.00	0.20
7	121.9	4.00	0.100	5.00	0.20
8	163.0	7.00	0.100	5.00	0.80
9	233.2	5.00	0.100	5.00	0.00
10	999.0	5.00	0.100	5.00	0.00
TOTALS		50.00	1.000	50.00	6.80

IF YOU USED THESE OBSERVED DATA TO ESTIMATE HYPOTHESIS PARAMETERS, DEGREES OF FREEDOM EQUALS NUMBER OF CLASSES MINUS 2 OR 8.

IF YOU USED AN INDEPENDENT ESTIMATE OF THE HYPOTHESIS PARAMETERS, DEGREES OF FREEDOM EQUALS NUMBER OF CLASSES MINUS 1 OR 9.

CRITICAL CHI-SQUARE VALUES AT THE .05 LEVEL OF SIGNIFICANCE ARE 15.507 FOR 8 DEGREES OF FREEDOM AND 16.919 FOR 9 DEGREES OF FREEDOM.

IF YOU WANT TO SOLVE ANOTHER PROBLEM, TYPE A 1. OTHERWISE, TYPE A 0.
● 1

TABLE 2A-11   ARRIVAL RATE POISSON GOFFIT TEST

GOODNESS OF FIT TESTS

DO YOU WANT INSTRUCTIONS? TYPE 1 FOR YES, 0 FOR NO.
● 0

TYPE THE NUMBER OF THE DISTRIBUTION YOU WANT TO TEST
● 2

TABLE 2A-11 (cont.)

YOU HAVE CHOSEN THE GOODNESS OF FIT TEST FOR:
2. POISSON DISTRIBUTION.

IF YOU WANT TO DO A CHI SQUARE TEST, TYPE A 1.
IF YOU WANT TO DO A KOLMOGOROV–SMIRNOV TEST, TYPE A 0.
• 1

ENTER THE MEAN RATE
• 1

    YOU HAVE CHOSEN A CHI SQUARE TEST.

ENTER THE NUMBER OF CLASSES
• 5

• • NOTE • • ONLY SEQUENTIAL INTEGER VALUES BEGINNING
WITH ZERO MAY BE USED.

ENTER THE CLASS MIDPOINT FOR CLASS   1
• 0

ENTER THE OBSERVED FREQUENCY FOR CLASS   1
• 16

ENTER THE CLASS MIDPOINT FOR CLASS   2
• 1

ENTER THE OBSERVED FREQUENCY FOR CLASS   2
• 25

ENTER THE CLASS MIDPOINT FOR CLASS   3
• 2

ENTER THE OBSERVED FREQUENCY FOR CLASS   3
• 8

ENTER THE CLASS MIDPOINT FOR CLASS   4
• 3

ENTER THE OBSERVED FREQUENCY FOR CLASS   4
• 2

ENTER • • ONLY • • THE OBSERVED FREQUENCY FOR CLASS   5
• 1

TABLE 2A-11 (cont.)

CLASS	CLASS IDENTITY	OBSERVED FREQUENCY	RELATIVE FREQUENCY	EXPECTED FREQUENCY	χ^2
1	0.0	16.00	0.368	19.13	0.51
2	1.0	25.00	0.368	19.13	1.80
3	2.0	8.00	0.184	9.56	0.26
4	3.0	2.00	0.061	3.19	0.44
5	99.0	1.00	0.019	0.99	0.00
TOTALS		52.00	1.000	52.00	3.01

IF YOU USED THESE OBSERVED DATA TO ESTIMATE HYPOTHESIS PARAMETERS,
DEGREES OF FREEDOM EQUALS NUMBER OF CLASSES MINUS 2 OR 3.

IF YOU USED AN INDEPENDENT ESTIMATE OF THE HYPOTHESIS PARAMETERS,
DEGREES OF FREEDOM EQUALS NUMBER OF CLASSES MINUS 1 OR 4.

CRITICAL CHI-SQUARE VALUES AT THE .05 LEVEL OF SIGNIFICANCE ARE
7.815 FOR 3 DEGREES OF FREEDOM AND 9.488 FOR 4 DEGREES OF FREEDOM.

AT LEAST ONE EXPECTED FREQUENCY IS LESS THAN 5. YOU MIGHT COMBINE CLASSES.

IF YOU WANT TO SOLVE ANOTHER PROBLEM, TYPE A 1. OTHERWISE, TYPE A 0.
• 1

GOODNESS OF FIT TESTS

DO YOU WANT INSTRUCTIONS? TYPE 1 FOR YES, 0 FOR NO.
• 0

TYPE THE NUMBER OF THE DISTRIBUTION YOU WANT TO TEST
• 2

YOU HAVE CHOSEN THE GOODNESS OF FIT TEST FOR:
2. POISSON DISTRIBUTION.

IF YOU WANT TO DO A CHI SQUARE TEST, TYPE A 1.
IF YOU WANT TO DO A KOLMOGOROV-SMIRNOV TEST, TYPE A 0.
• 1

ENTER THE MEAN RATE
• 1

YOU HAVE CHOSEN A CHI SQUARE TEST.

TABLE 2A-11 (cont.)

ENTER THE NUMBER OF CLASSES
• 3

• • NOTE • • ONLY SEQUENTIAL INTEGER VALUES BEGINNING
WITH ZERO MAY BE USED.

ENTER THE CLASS MIDPOINT FOR CLASS   1
• 0

ENTER THE OBSERVED FREQUENCY FOR CLASS   1
• 16

ENTER THE CLASS MIDPOINT FOR CLASS   2
• 1

ENTER THE OBSERVED FREQUENCY FOR CLASS   2
• 25

ENTER • • ONLY • • THE OBSERVED FREQUENCY FOR CLASS   3
• 11

CLASS	CLASS IDENTITY	OBSERVED FREQUENCY	RELATIVE FREQUENCY	EXPECTED FREQUENCY	χ^2
1	0.0	16.00	0.368	19.13	0.51
2	1.0	25.00	0.368	19.13	1.80
3	99.0	11.00	0.264	13.74	0.55
TOTALS		52.00	1.000	52.00	2.86

IF YOU USED THESE OBSERVED DATA TO ESTIMATE HYPOTHESIS PARAMETERS,
DEGREES OF FREEDOM EQUALS NUMBER OF CLASSES MINUS 2 OR 1.

IF YOU USED AN INDEPENDENT ESTIMATE OF THE HYPOTHESIS PARAMETERS,
DEGREES OF FREEDOM EQUALS NUMBER OF CLASSES MINUS 1 OR 2.

CRITICAL CHI-SQUARE VALUES AT THE .05 LEVEL OF SIGNIFICANCE ARE
3.841 FOR 1 DEGREES OF FREEDOM AND 5.991 FOR 2 DEGREES OF FREEDOM.

IF YOU WANT TO SOLVE ANOTHER PROBLEM, TYPE A 1. OTHERWISE, TYPE A 0.
• 1

TABLE 2A-12   INTERARRIVAL TIME EMPIRICAL GOFFIT TEST

GOODNESS OF FIT TESTS

DO YOU WANT INSTRUCTIONS? TYPE 1 FOR YES, 0 FOR NO.
• 0

TABLE 2A-12 (cont.)

TYPE THE NUMBER OF THE DISTRIBUTION YOU WANT TO TEST
• 5

YOU HAVE CHOSEN THE GOODNESS OF FIT TEST FOR:
5. EMPIRICALLY DEFINED DISTRIBUTION.

IF YOU WANT TO DO A CHI SQUARE TEST, TYPE A 1.
IF YOU WANT TO DO A KOLMOGOROV-SMIRNOV TEST, TYPE A 0.
• 1

   YOU HAVE CHOSEN A CHI SQUARE TEST.

ENTER THE NUMBER OF CLASSES
• 2

ENTER THE OBSERVED FREQUENCY FOR CLASS   1
• 29

ENTER THE EMPIRICAL RELATIVE FREQUENCY FOR CLASS   1
AS A DECIMAL FRACTION
• .4

ENTER THE OBSERVED FREQUENCY FOR CLASS   2
• 21

ENTER THE EMPIRICAL RELATIVE FREQUENCY FOR CLASS   2
AS A DECIMAL FRACTION
• .6

CLASS	CLASS IDENTITY	OBSERVED FREQUENCY	RELATIVE FREQUENCY	EXPECTED FREQUENCY	χ^2
1	1.0	29.00	0.400	20.00	4.05
2	2.0	21.00	0.600	30.00	2.70
TOTALS		50.00	1.000	50.00	6.75

DEGREES OF FREEDOM EQUALS NUMBER OF CLASSES
MINUS ONE, OR 1.

CRITICAL CHI-SQUARE VALUE AT THE .05 LEVEL OF SIGNIFICANCE IS
3.841 FOR 1 DEGREES OF FREEDOM.

IF YOU WANT TO SOLVE ANOTHER PROBLEM, TYPE A 1. OTHERWISE, TYPE A 0.
• 0

TABLE 2A-13   SERVICE TIME UNIFORM GOFFIT TEST

GOODNESS OF FIT TESTS

DO YOU WANT INSTRUCTIONS? TYPE 1 FOR YES, 0 FOR NO.
• 0

TYPE THE NUMBER OF THE DISTRIBUTION YOU WANT TO TEST
• 4

YOU HAVE CHOSEN THE GOODNESS OF FIT TEST FOR:
4. UNIFORM DISTRIBUTION.

IF YOU WANT TO DO A CHI SQUARE TEST, TYPE A 1.
IF YOU WANT TO DO A KOLMOGOROV-SMIRNOV TEST, TYPE A 0.
• 1

   YOU HAVE CHOSEN A CHI SQUARE TEST.

ENTER THE NUMBER OF CLASSES
• 10

ENTER THE CLASS MIDPOINT FOR CLASS   1
• 15.1

ENTER THE OBSERVED FREQUENCY FOR CLASS NUMBER   1
• 13

ENTER THE CLASS MIDPOINT FOR CLASS   2
• 45.3

ENTER THE OBSERVED FREQUENCY FOR CLASS NUMBER   2
• 13

ENTER THE CLASS MIDPOINT FOR CLASS   3
• 75.5

ENTER THE OBSERVED FREQUENCY FOR CLASS NUMBER   3
• 7

ENTER THE CLASS MIDPOINT FOR CLASS   4
• 105.7

ENTER THE OBSERVED FREQUENCY FOR CLASS NUMBER   4
• 9

ENTER THE CLASS MIDPOINT FOR CLASS   5
• 135.9

TABLE 2A-13 (cont.)

ENTER THE OBSERVED FREQUENCY FOR CLASS NUMBER   5
• 2

ENTER THE CLASS MIDPOINT FOR CLASS   6
• 166.1

ENTER THE OBSERVED FREQUENCY FOR CLASS NUMBER   6
• 4

ENTER THE CLASS MIDPOINT FOR CLASS   7
• 196.3

ENTER THE OBSERVED FREQUENCY FOR CLASS NUMBER   7
• 1

ENTER THE CLASS MIDPOINT FOR CLASS   8
• 226.5

ENTER THE OBSERVED FREQUENCY FOR CLASS NUMBER   8
• 0

ENTER THE CLASS MIDPOINT FOR CLASS   9
• 256.7

ENTER THE OBSERVED FREQUENCY FOR CLASS NUMBER   9
• 3

ENTER THE CLASS MIDPOINT FOR CLASS   10
• 286.9

ENTER THE OBSERVED FREQUENCY FOR CLASS   10
• 1

CLASS	CLASS IDENTITY	OBSERVED FREQUENCY	RELATIVE FREQUENCY	EXPECTED FREQUENCY	χ^2
1	15.1	13.00	0.100	5.30	11.19
2	45.3	13.00	0.100	5.30	11.19
3	75.5	7.00	0.100	5.30	0.55
4	105.7	9.00	0.100	5.30	2.58
5	135.9	2.00	0.100	5.30	2.05
6	166.1	4.00	0.100	5.30	0.32
7	196.3	1.00	0.100	5.30	3.49
8	226.5	0.00	0.100	5.30	5.30
9	256.7	3.00	0.100	5.30	1.00
10	286.9	1.00	0.100	5.30	3.49
TOTALS		53.00	1.000	53.00	41.15

TABLE 2A-13 (cont.)

DEGREES OF FREEDOM EQUALS NUMBER OF CLASSES
MINUS ONE, OR 9.

CRITICAL CHI-SQUARE VALUE AT THE .05 LEVEL OF SIGNIFICANCE IS
16.919 FOR 9 DEGREES OF FREEDOM.

IF YOU WANT TO SOLVE ANOTHER PROBLEM, TYPE A 1. OTHERWISE, TYPE A 0.
• 1

TABLE 2A-14   SERVICE TIME NORMAL GOFFIT TEST

GOODNESS OF FIT TESTS

DO YOU WANT INSTRUCTIONS? TYPE 1 FOR YES, 0 FOR NO.
• 0

TYPE THE NUMBER OF THE DISTRIBUTION YOU WANT TO TEST
• 3

YOU HAVE CHOSEN THE GOODNESS OF FIT TEST FOR:
3. NORMAL DISTRIBUTION.

IF YOU WANT TO DO A CHI SQUARE TEST, TYPE A 1.
IF YOU WANT TO DO A KOLMOGOROV-SMIRNOV TEST, TYPE A 0.
• 1

ENTER THE MEAN
• 83.5283

ENTER THE STANDARD DEVIATION
• 72.5603

    YOU HAVE CHOSEN A CHI SQUARE TEST.

ENTER THE NUMBER OF CLASSES
• 10

ENTER THE UPPER CLASS BOUNDARY FOR CLASS   1
• −9.3489

ENTER THE OBSERVED FREQUENCY FOR CLASS   1
• 0

ENTER THE UPPER CLASS BOUNDARY FOR CLASS   2
• 22.5776

TABLE 2A-14 (cont.)

ENTER THE OBSERVED FREQUENCY FOR CLASS   2
● 11

ENTER THE UPPER CLASS BOUNDARY FOR CLASS   3
● 45.4341

ENTER THE OBSERVED FREQUENCY FOR CLASS   3
● 8

ENTER THE UPPER CLASS BOUNDARY FOR CLASS   4
● 65.1705

ENTER THE OBSERVED FREQUENCY FOR CLASS   4
● 7

ENTER THE UPPER CLASS BOUNDARY FOR CLASS   5
● 83.5283

ENTER THE OBSERVED FREQUENCY FOR CLASS   5
● 6

ENTER THE UPPER CLASS BOUNDARY FOR CLASS   6
● 101.8861

ENTER THE OBSERVED FREQUENCY FOR CLASS   6
● 6

ENTER THE UPPER CLASS BOUNDARY FOR CLASS   7
● 121.6225

ENTER THE OBSERVED FREQUENCY FOR CLASS   7
● 4

ENTER THE UPPER CLASS BOUNDARY FOR CLASS   8
● 144.4790

ENTER THE OBSERVED FREQUENCY FOR CLASS   8
● 2

ENTER THE UPPER CLASS BOUNDARY FOR CLASS   9
● 176.4055

ENTER THE OBSERVED FREQUENCY FOR CLASS   9
● 2

ENTER ● ● ● ONLY ● ● ● THE OBSERVED FREQUENCY FOR CLASS   10
● 7

TABLE 2A-14 (cont.)

CLASS	CLASS IDENTITY	OBSERVED FREQUENCY	RELATIVE FREQUENCY	EXPECTED FREQUENCY	χ^2
1	−9.3	0.00	0.100	5.31	5.31
2	22.6	11.00	0.100	5.31	6.10
3	45.4	8.00	0.099	5.26	1.42
4	65.2	7.00	0.100	5.32	0.53
5	83.5	6.00	0.100	5.29	0.09
6	101.9	6.00	0.100	5.29	0.09
7	121.6	4.00	0.100	5.32	0.33
8	144.5	2.00	0.099	5.26	2.02
9	176.4	2.00	0.100	5.31	2.06
10	999.0	7.00	0.100	5.32	0.53
TOTALS		53.00	1.000	53.00	18.50

IF YOU USED THESE OBSERVED DATA TO ESTIMATE HYPOTHESIS PARAMETERS, DEGREES OF FREEDOM EQUALS NUMBER OF CLASSES MINUS 3 OR 7.

IF YOU USED AN INDEPENDENT ESTIMATE OF THE HYPOTHESIS PARAMETERS, DEGREES OF FREEDOM EQUALS NUMBER OF CLASSES MINUS 1 OR 9.

CRITICAL CHI-SQUARE VALUES AT THE .05 LEVEL OF SIGNIFICANCE ARE
    14.067 FOR 7 DEGREES OF FREEDOM AND 16.919 FOR 9 DEGREES OF FREEDOM.

IF YOU WANT TO SOLVE ANOTHER PROBLEM, TYPE A 1. OTHERWISE, TYPE A 0.
• 1

TABLE 2A-15   SERVICE TIME NEGATIVE EXPONENTIAL GOFFIT TEST

GOODNESS OF FIT TESTS

DO YOU WANT INSTRUCTIONS? TYPE 1 FOR YES, 0 FOR NO.
• 0

TYPE THE NUMBER OF THE DISTRIBUTION YOU WANT TO TEST
• 1

YOU HAVE CHOSEN THE GOODNESS OF FIT TEST FOR:
1. NEGATIVE EXPONENTIAL DISTRIBUTION.

IF YOU WANT TO DO A CHI SQUARE TEST, TYPE A 1.
IF YOU WANT TO DO A KOLMOGOROV-SMIRNOV TEST, TYPE A 0.
• 1

TABLE 2A-15 (cont.)

ENTER THE MEAN TIME
● 83.5283

YOU HAVE CHOSEN A CHI SQUARE TEST.

ENTER THE NUMBER OF CLASSES
● 10

IF YOU WANT TO DO A GOODNESS OF FIT TEST, TYPE A 1.
IF YOU WANT TO COMPUTE BOUNDARIES, TYPE A 0.
● 0

CLASS IDENTITY	CUMULATIVE GT PROBABILITY	CLASS BOUNDARY	CLASS WIDTH
1	1.0000	0.00	8.80
2	0.9000	8.80	9.84
3	0.8000	18.64	11.15
4	0.7000	29.79	12.88
5	0.6000	42.67	15.23
6	0.5000	57.90	18.64
7	0.4000	76.54	24.03
8	0.3000	100.57	33.87
9	0.2000	134.43	57.90
10	0.1000	192.33	999.99

IF YOU WANT TO PROCEED TO A CHI SQUARE GOODNESS
OF FIT TEST OF THIS NEGATIVE EXPONENTIAL DISTRIBUTION,
TYPE A 1. OTHERWISE, TYPE A 0.
● 1

ENTER THE OBSERVED FREQUENCY FOR CLASS   1
● 4

ENTER THE OBSERVED FREQUENCY FOR CLASS   2
● 5

ENTER THE OBSERVED FREQUENCY FOR CLASS   3
● 4

ENTER THE OBSERVED FREQUENCY FOR CLASS   4
● 3

ENTER THE OBSERVED FREQUENCY FOR CLASS   5
● 8

TABLE 2A-15 (cont.)

ENTER THE OBSERVED FREQUENCY FOR CLASS   6
• 6

ENTER THE OBSERVED FREQUENCY FOR CLASS   7
• 7

ENTER THE OBSERVED FREQUENCY FOR CLASS   8
• 7

ENTER THE OBSERVED FREQUENCY FOR CLASS   9
• 5

ENTER THE OBSERVED FREQUENCY FOR CLASS   10
• 4

CLASS	CLASS IDENTITY	OBSERVED FREQUENCY	RELATIVE FREQUENCY	EXPECTED FREQUENCY	χ^2
1	8.8	4.00	0.100	5.30	0.32
2	18.6	5.00	0.100	5.30	0.02
3	29.8	4.00	0.100	5.30	0.32
4	42.7	3.00	0.100	5.30	1.00
5	57.9	8.00	0.100	5.30	1.38
6	76.5	6.00	0.100	5.30	0.09
7	100.6	7.00	0.100	5.30	0.55
8	134.4	7.00	0.100	5.30	0.55
9	192.3	5.00	0.100	5.30	0.02
10	999.0	4.00	0.100	5.30	0.32
TOTALS		53.00	1.000	53.00	4.55

IF YOU USED THESE OBSERVED DATA TO ESTIMATE HYPOTHESIS PARAMETERS,
DEGREES OF FREEDOM EQUALS NUMBER OF CLASSES MINUS 2 OR 8.

IF YOU USED AN INDEPENDENT ESTIMATE OF THE HYPOTHESIS PARAMETERS,
DEGREES OF FREEDOM EQUALS NUMBER OF CLASSES MINUS 1 OR 9.

CRITICAL CHI-SQUARE VALUES AT THE .05 LEVEL OF SIGNIFICANCE ARE
    15.507 FOR 8 DEGREES OF FREEDOM AND 16.919 FOR 9 DEGREES OF FREEDOM.

IF YOU WANT TO SOLVE ANOTHER PROBLEM, TYPE A 1. OTHERWISE, TYPE A 0.
• 1

chapter three

ANALYTICAL WAITING-LINE MODELS

ASSUMPTIONS OF ANALYTICAL MODELS

If a system can be modeled accurately without a customized simulation, the modeling effort will require much less time and money. There is a group of waiting-line models which will yield predictions of system performance given only the mean arrival rate and mean service rate, if certain assumptions about the statistical distribution of arrivals and service are true. Most of these models assume that arrival and service rates are Poisson distributed or, equivalently, that interarrival and service times are negative exponentially distributed. Often, the abbreviation $M/M/1$ is used to designate a system with negative exponential interarrival and service times and one server, while the abbreviation $M/M/c$ is used to designate a system with negative exponential interarrival and service times and c (one or more) servers. We will not attempt to derive the formulas for the performance measures, but derivations are available in books on queuing theory and operations research in general (see the bibliography).

The results of the goodness-of-fit tests discussed in Chapter 2 enable us to decide which, if any, analytical waiting-line models might be suitable for a particular system model. The simplest of the analytical models is the single-channel, single-phase model, with Poisson arrival and service rates and first-come, first-served (FIFO) queue discipline. These models make three key assumptions about the behavior of the members of the queue:

1. No balking—that is, a potential new member of the queue cannot refuse to join the queue because of its length, composition, and so on.

2. No reneging—that is, once a member has joined the queue, that member must remain in the queue and complete the service.
3. No jockeying—that is, once having joined a particular queue, the member cannot switch membership to an alternate queue which might, for example, have become shorter.

A final caveat is that the system modeled must be in steady state, that is, in its usual and customary condition and not passing through a peak-load or just-opened stage.

The analytical waiting-line formulas offer the modeler the following system performance predictions:

1. Mean (expected) number in the waiting line (queue length)
2. Mean (expected) number in the system (in the queue or being served)
3. Mean (expected) time in the waiting line
4. Mean (expected) time in the system
5. Average utilization factor (mean arrival rate/mean service rate)
6. Probability of various numbers of customers being in the system at a random point in time

AN EXAMPLE

Let us return to the example of the ticket seller given in Chapter 2. We know from calculations based on the raw observation data that the sample mean interarrival time was 101.28 seconds, that the sample mean service time was 83.528 seconds, and that we were unable to reject the null hypothesis that interarrival time and service time are negative exponentially distributed. The analytical waiting-line formulas use λ (lambda) to designate the mean arrival rate and μ (mu) to designate the mean service rate. Thus, for our ticket seller, $\lambda = 1/101.28 = 0.0098736$ arrivals per second. Similarly, $\mu = 1/83.528 = 0.011972$ customers able to be served each second. We did not observe any balking, jockeying, or reneging, and the system appeared to be in steady state, so we feel confident in attempting to evaluate average system performance measures using the analytical waiting-line formulas. Table 3-1 shows performance estimates for a process with Poisson arrival and service rates of the magnitude we observed in the sampling process.

The first set of items in Table 3-1 is the probability distribution of number of units in the system. $P(N >$ or $= c)$ is the probability that the system is busy, that is, that the number of units in the system, N, equals or exceeds the number of servers, c. This is the complement of $P(0)$ units in the system.

The implicit time unit in the ticket seller example is seconds. It would seem potentially interesting to compare the output shown in Table 2-2, which summarizes these performance measures for the sample data, with the performance predicted by an $M/M/1$ analytical model. Table 3-2 displays these measures for the sample data.

TABLE 3-1 PERFORMANCE ESTIMATES FOR THE TICKET SELLER EXAMPLE

$P(0)$	=	0.17528
$P(1)$	=	0.14455
$P(2)$	=	0.11922
$P(3)$	=	0.09832
$P(4)$	=	0.08109
$P(5)$	=	0.06688
$P(6)$	=	0.05515
$P(7)$	=	0.04549
$P(8)$	=	0.03751
$P(9)$	=	0.03094
$P(10)$	=	0.02552
$P(11)$	=	0.02104
$P(12)$	=	0.01736
$P(13)$	=	0.01431
$P(14)$	=	0.01180
$P(N \geq c)$	=	0.82472

STEADY–STATE MEAN NUMBER OF UNITS:

IN SYSTEM =	4.70529
IN QUEUE =	3.88057
IN QUEUE FOR BUSY SYSTEM =	4.70529

STEADY–STATE MEAN TIME:

IN SYSTEM =	476.55270
IN QUEUE =	393.02460
IN QUEUE FOR BUSY SYSTEM =	476.55270

TABLE 3-2 COMPARISONS BETWEEN SAMPLE PERFORMANCE AND PERFORMANCE PREDICTED BY ANALYTICAL MODELS

	SAMPLE ESTIMATE	MODEL PROJECTION
MEAN NUMBER IN QUEUE	1.608	3.88057
MEAN TIME IN QUEUE	128.804 SEC	393.0246 SEC
MEAN TRANSIT TIME	213.304 SEC	476.5527 SEC
AVERAGE SERVER IDLE TIME (%)	17.17	$(1 - 0.82472) \times 100 = 17.528$

VALIDATION OF PERFORMANCE PREDICTIONS

The sample statistics produced in Table 2-2 act as a check for reasonableness and accuracy in the use of models to describe a system. Validation is the exercise of verifying that the outputs of the model are reasonable, given the inputs and processing steps in the system. That is, validation assures the analyst and management that the model behaves just as the real system should. When the system modeled already exists, as does the ticket seller system, we may compare sample statistics with performance predictions from the model and attempt to explain any differences either by random variation or by specific ways in which the model and the real system are not identical. The differences may be great enough to induce the analyst to make significant changes to the model, or they may be small enough to discourage the analyst from incurring additional costs of custom modeling to accommodate them.

While we cannot compare all system performance estimates with sample data because we attempted to keep the data collection task to the bare minimum, we can see that certain estimates appear closer than others. We might want to test the hypothesis that the sample idle time percentage is a statistically reasonable sample from a process whose mean idle time percentage is shown, but those values are extremely close anyway, and we might assume that the hypothesis is correct. Unfortunately, it is quite difficult (and probably not worth the effort) to test this and similar hypotheses that sample waiting time is a statistically reasonable sample from a process whose mean waiting time is shown or that sample queue length or transit time are likewise reasonable draws from the steady-state process modeled. The reason for the difficulty is that each successive observation of queue length, time in queue, and transit time is dependent on its predecessors to one or another degree. That is, if the Nth customer has to wait 20 seconds or more, the chances are better than 50:50 that the $(N + 1)$th customer will also have to wait. This dependence of consecutive observations in a time series is called autocorrelation or serial correlation. If it is not detected and removed prior to a hypothesis test, it will tend to underestimate the standard deviation and variance of the process, causing the denominator of the z statistic appropriate to the hypothesis to be understated and ultimately biasing the test in the direction of not rejecting the null hypothesis when it ought to be rejected—a type II error. We shall discuss this problem and its resolution in greater detail when it becomes necessary to test sample output actually produced by simulation models.

For the moment, then, we will note the discrepancy between sample and predicted queue length, waiting time, and transit time and make some general conjectures about the cause, if other than random variation. Some possible guesses about the discrepancy might include the fact that the variance of interarrival time and the variance of service time in the sample data (see Table 2-2) are considerably smaller than the mean values. Yet, in the analytical models, perfect Poisson or negative exponential relationships in which the mean is equal to, respectively, the variance or the standard deviation, are assumed. If we use the estimate of the standard deviation of interarrival time from Table 2-3 as the best estimate of the mean, λ becomes $1/78.845 = 0.0126831$. Similarly, if we use the estimate of the standard deviation of service time from Table 2-11 as the best estimate of the mean, μ becomes $1/72.5603 = 0.0137816$. If we recalculate the system performance projections based on these new values of λ and μ, we obtain the results shown in Table 3-3.

Clearly the use of the standard deviation as an estimator of λ and μ has made the discrepancy between the sample estimates and the model estimates worse. It is a

TABLE 3-3   PERFORMANCE ESTIMATES USING
STANDARD DEVIATIONS INSTEAD OF MEANS TO
ESTIMATE λ AND μ

$P(0)$	$= 0.07971$
$P(1)$	$= 0.07335$
$P(2)$	$= 0.06751$
$P(3)$	$= 0.06213$
$P(4)$	$= 0.05717$
$P(5)$	$= 0.05262$
$P(6)$	$= 0.04842$
$P(7)$	$= 0.04456$
$P(8)$	$= 0.04101$
$P(9)$	$= 0.03774$
$P(10)$	$= 0.03473$
$P(11)$	$= 0.03197$
$P(12)$	$= 0.02942$
$P(13)$	$= 0.02707$
$P(14)$	$= 0.02492$
$P(15)$	$= 0.02293$
$P(16)$	$= 0.02110$
$P(17)$	$= 0.01942$
$P(18)$	$= 0.01787$
$P(19)$	$= 0.01645$
$P(20)$	$= 0.01514$
$P(21)$	$= 0.01393$
$P(22)$	$= 0.01282$
$P(23)$	$= 0.01180$
$P(24)$	$= 0.01086$
$P(N \geq c)$	$= 0.92029$

STEADY-STATE MEAN NUMBER OF UNITS:

IN SYSTEM =	11.54584
IN QUEUE =	10.62555
IN QUEUE FOR BUSY SYSTEM =	11.54584

STEADY-STATE MEAN TIME:

IN SYSTEM =	910.33250
IN QUEUE =	837.77220
IN QUEUE FOR BUSY SYSTEM =	910.33220

good idea to discuss these discrepancies with the management and staff of the subject system as soon as possible, so that it can be determined whether the sample data are unreliable and another sample should be taken or whether there are operating procedures and constraints not evident to the analyst which exert a strong influence on the performance of the system.

ESTIMATING THE COST OF A SYSTEM

Suppose that we want to calculate the true cost of one day's operation of the ticket seller system. Let us assume that the ticket seller is paid a wage of $4.00 per hour for an 8-hour workday. A more nebulous cost associated with the system is the opportunity cost (sometimes called shadow price) of customers' time while they are waiting and being served. The customers themselves may be someone's employees who are being paid a wage, say, of $10.00 per hour. Or they may be "spending" their own leisure time, which they may value at, say, $10.00 per hour. If service by the ticket seller is too slow, the opportunity cost of wasted time may dissuade prospective customers from purchasing tickets. A slow server causes the customers to experience both long service time and long waiting time, so the appropriate measure of the system's effect on total opportunity cost is transit time (time in the system), not just waiting time or service time.

The daily wage cost for the ticket seller is $4.00 per hour $\times$ 8 hours = $32.00. To compute the expected opportunity cost of customers in the system, we multiply λ, the customer arrival rate per hour, by the opportunity cost per hour in the system, and finally by 8 hours in a workday. On an hourly basis, λ is the previously calculated arrival rate in seconds (0.098736) multiplied by 3600 seconds in an hour, or an expected arrival rate of 35.545 customers per hour. Mean transit time from Table 3-2 is 476.5527 seconds, or 0.13238 of an hour. Thus, the daily expected opportunity cost of customers in the system is 35.545 $\times$ 8 $\times$ $10.00 $\times$ 0.13238 = $376.44.

The total cost of system operation for the day is server wage cost plus customer opportunity cost, or $32.00 + $376.44 = $408.44. The most noteworthy aspect of this calculation is that the opportunity cost component far outweighs the server wage component. A discussion with the manager may reveal that customer patronage is down from previous periods, that a number of customer complaints about slow service have been received, or that actual demand has not met forecasted levels. Still, the manager may say that the system is idle (zero customers in the system) 0.17528 of the time, as postulated in Table 3-2. For an 8-hour day, the manager is spending $4.00 $\times$ 8 $\times$ 0.17528 = $5.61 on the average in staffing the ticket sales station when there are no customers desiring to purchase tickets. The analyst should reply that any stochastic system will exhibit periods of idleness as well as periods of rather hectic activity. These are part of the cost of doing business.

CHOOSING SYSTEM ALTERNATIVES

Perhaps the manager feels that it might be preferable to cut the opportunity cost of customer transit time and balance server wage cost a bit more closely. One alternative would be to increase the efficiency of the present server or to hire a faster server. Suppose that the faster server would have to be paid an hourly wage of $4.50, or $36.00 for an 8-hour day.

TABLE 3-4 PERFORMANCE ESTIMATES FOR
THE FASTER TICKET SELLER EXAMPLE

$P(0)$	= 0.30885
$P(1)$	= 0.21346
$P(2)$	= 0.14753
$P(3)$	= 0.10197
$P(4)$	= 0.07048
$P(5)$	= 0.04871
$P(6)$	= 0.03367
$P(7)$	= 0.02327
$P(8)$	= 0.01608
$P(9)$	= 0.01112
$P(N \geqslant c)$	= 0.69115

STEADY-STATE MEAN NUMBER OF UNITS:

IN SYSTEM =	2.23784
IN QUEUE =	1.54669
IN QUEUE FOR BUSY SYSTEM =	2.23784

STEADY-STATE MEAN TIME:

IN SYSTEM =	226.64920
IN QUEUE =	156.64910
IN QUEUE FOR BUSY SYSTEM =	226.64910

The analyst can use the same set of performance formulas to determine outputs to be expected from the new system, expected customer opportunity costs, and total system costs. The only change will be in the average service rate, μ. Suppose that the new server has average service time of 70 seconds. Then, average service rate will be 0.0142857 customers per second, or $0.0142857 \times 3600 = 51.42852$ customers per hour. Performance estimates for this system are shown in Table 3-4.

Table 3-4 reveals that the new mean transit time will be 226.6492 seconds, or 0.0629581 of an hour for each customer. Daily expected opportunity cost will be $35.545 \times 8 \times \$10.00 \times 0.0629581 = \$179.03$. Total system costs for the faster-server system are $\$36.00 + \$179.03 = \$215.03$. In this system the server is idle 0.30885 of the 8-hour day, so wage costs of server idle time are $\$4.50 \times 8 \times 0.30885 = \$11.12$. The manager must decide whether the additional wage costs will be offset by better customer retention and further customer acquisition.

The analytical waiting-line formulas which yielded the preceding performance predictions are based on the values of λ and μ. The formulas are shown in Table 3-5.

The statistics for "busy" systems are collected only for those cases in which the customer had to wait. All customers who spent zero time in the queue are excluded from the calculation of number of customers in the queue and time in the queue for a busy system.

TABLE 3-5 ANALYTICAL WAITING-LINE FORMULAS FOR THE
SINGLE-CHANNEL, SINGLE-PHASE SYSTEM WITH FIFO QUEUE
DISCIPLINE IN STEADY STATE

PROBABILITY OF AN IDLE SYSTEM (ZERO UNITS IN SYSTEM) $= 1 - \lambda/\mu$

PROBABILITY OF A BUSY SYTEM $= \lambda/\mu$

PROBABILITY OF N CUSTOMERS IN THE SYSTEM $= (1 - \lambda/\mu) \times (\lambda/\mu)^N$

EXPECTED NUMBER OF CUSTOMERS IN THE SYSTEM $= \lambda/(\mu - \lambda)$

VARIANCE OF NUMBER OF CUSTOMERS IN THE SYSTEM $= \lambda\mu/(\mu - \lambda)^2$

EXPECTED NUMBER OF CUSTOMERS IN THE QUEUE $= \lambda^2/\mu(\mu - \lambda)$

EXPECTED NUMBER OF CUSTOMERS IN THE QUEUE FOR A BUSY SYSTEM $= \lambda/(\mu - \lambda)$

EXPECTED TIME IN SYSTEM (TRANSIT TIME) $= 1/(\mu - \lambda)$

VARIANCE OF TIME IN SYSTEM $= 1/(\mu - \lambda)^2$

EXPECTED TIME IN QUEUE $= \lambda/\mu(\mu - \lambda)$

EXPECTED TIME IN QUEUE FOR BUSY SYSTEM $= 1/(\mu - \lambda)$

MODELING A MULTISERVER SITUATION

The formulas shown in Table 3-5 are meant to cover only the pure single-channel, single-phase case. There are other formulas that deal with multiple-server situations in which all servers display identical properties and share a single queue, such as the bank teller arrangement, which is very common today. However, these formulas are very difficult to solve without the assistance of a computer. They are discussed in detail in the appendix to this chapter. A selection of these formulas appears in Appendix Table VI. The results given by these formulas may be approximated with the formulas in Table 3-5 by dividing the arrival rate among the various servers; the analyst should be aware, however, that the results are only approximations.

Consider again the ticket seller example. Suppose that management could choose to provide two ticket sellers of equal capability. What would be the effect on system performance? The revised interarrival time for each seller is 101.28 × 2, or 202.56 seconds. Thus, the arrival rate for each of the two ticket sellers is 0.0049368 customers per second, or 17.77248 customers per hour. The average service time for each seller remains 83.528 seconds. Table 3-6 shows the performance results for each ticket seller. The performance estimates shown in Table 3-6 are much more favorable to the customer than in either the original single-server case or in the faster-server case.

The wage cost of the two-server system is $4.00 × 8 × 2 = $64.00 per day. The overall arrival rate for the two-server system is still 35.545 customers per hour. The expected opportunity cost for the two-server system is 35.545 × 8 × $10.00 × (142.14230/3600) = $112.28 per day. The total cost of the two-server system is, therefore, $64.00 + $112.28 = $176.28 per day. This is more economical than either of the two single-server systems considered, despite the fact that the servers are idle about 59 percent of the time.

TABLE 3-6   PERFORMANCE ESTIMATES
FOR THE TICKET SELLER EXAMPLE WITH TWO
EQUIVALENT TICKET SELLERS

$P(0)$	$= 0.58764$
$P(1)$	$= 0.24232$
$P(2)$	$= 0.09992$
$P(3)$	$= 0.04120$
$P(4)$	$= 0.01699$
$P(N \geqslant c)$	$= 0.41236$

STEADY-STATE MEAN NUMBER OF UNITS:

IN SYSTEM =	0.70173
IN QUEUE =	0.28937
IN QUEUE FOR BUSY SYSTEM =	0.70173

STEADY-STATE MEAN TIME:

IN SYSTEM =	142.14230
IN QUEUE =	58.61414
IN QUEUE FOR BUSY SYSTEM =	142.14240

CONSTRAINTS ON SYSTEM OPERATION

Two common constraints on the operation of waiting-line systems are finite queue length resulting from limited waiting area and finite calling population when the server is responsible for only a specific customer constituency.

FINITE QUEUE LENGTH

Limitations on waiting area cause potential new arrivals to fail to enter the system when the limits on waiting area have been previously reached. One might guess that this limitation would cause system performance attributes to seem more favorable to the successful customer than otherwise. For example, average queue length would appear to be smaller. The analyst must remember, however, the implicit disservice done to the customer who arrives and is not allowed to enter the waiting area. In practice, such a customer may try again later or may vow never to patronize the establishment again.

   Let us see what would happen in the original ticket seller case if the queue were limited to one customer in length. Table 3-7 shows the results.

   As anticipated, this system appears to be busy only about 40 percent of the time and offers an enticing average queue length of 0.27154 rather than 3.88057 for the original system. Still, quite a few potential customers have been discouraged by the limited waiting area, and management would be well advised to estimate the probability that queue length will exceed waiting area before planning waiting facilities. Even if the waiting area is not formally limited, conditions such as a bank

TABLE 3-7 PERFORMANCE ESTIMATES FOR
THE TICKET SELLER EXAMPLE WITH FINITE
QUEUE LENGTH OF 1

WHAT IS THE MAXIMUM NUMBER ALLOWED IN SYSTEM?

$P(N \geq c)$ = 0.60078
$P(0)$ = 0.39922
$P(1)$ = 0.32925
$P(2)$ = 0.27154

STEADY-STATE MEAN NUMBER OF UNITS:
  IN SYSTEM =   0.87232
  IN QUEUE =   0.27154
  IN QUEUE FOR BUSY SYSTEM =   0.45197

STEADY-STATE MEAN TIME:
  IN SYSTEM =   121.28060
  IN QUEUE =   37.75244
  IN QUEUE FOR BUSY SYSTEM =   62.83887

drive-up window whose driveway overflows onto a busy street or a fast-foods restaurant whose waiting area overflows beyond the front door, will dissuade customers from joining the queues almost as effectively.

FINITE CALLING POPULATION

The finite calling population case is often exemplified by a person who is responsible for the repair of, say, all typewriters in a particular branch office. As more and more typewriters join the queue to be repaired, the probability that another typewriter will arrive and join the queue decreases until, when all typewriters are broken, the probability of a new arrival in the repair queue is zero.

As in the finite queue case, system performance statistics might seem deceptively favorable. Management might wish to consider the situation in which each server is designated to handle only certain classes of customers, whose approximate numbers are known. The finite calling population model permits analysis of these alternatives.

Suppose that the ticket seller is recast as a machine repairer for a large office with 50 electric typewriters. Clearly, the maximum number in the system is 50, and when that is the case, the probability that another typewriter will join the queue for service is zero. For the ticket seller data, the arrival rate and the service rate are fairly close. That is, the utilization factor for the system is high. For this to be the case for our hypothetical repairer, the calling population would have to be quite large—large enough that the population might be considered nearly infinite.* The

---

*When the finite population is quite large, say, 30 or larger, models in which population is unconstrained may be used quite accurately.

reason is that it is unlikely for any one machine, or any small group of machines, to break down so frequently that the server would be kept almost fully occupied. Because of this, we will not attempt to model a finite population case using the ticket seller data.

MODELING SELF-SERVICE SYSTEMS

Systems in which the arriving customers are also the servers are ever more common in an age when labor costs grow more expensive as equipment costs decline. Another impetus to the growth of self-service systems is the desire of customers to obtain service around the clock. Examples of self-service systems are do-it-yourself car washes, automated bank teller terminals, and automatic telephone systems.

Let us suppose that our ticket seller were to become the victim of automation and be replaced by a ticket-vending machine which could be operated by the customer. If the arrival rate and the service rate were as previously defined, the results would be as shown in Table 3-8.

TABLE 3-8 PERFORMANCE ESTIMATES FOR THE TICKET SELLER EXAMPLE WITH CUSTOMER SELF-SERVICE ASSUMED

$P(0)$	= 0.43836
$P(1)$	= 0.36152
$P(2)$	= 0.14908
$P(3)$	= 0.04098

STEADY-STATE MEAN NUMBER OF UNITS IN SYSTEM = 0.82472
STEADY-STATE MEAN TIME IN SYSTEM = 83.52823

The self-service model is actually a special case of the general $M/M/c$ model when c (the number of servers) is potentially infinite. We see from the limited statistics shown in Table 3-8 that the performance of the system is very favorable from the customer's point of view, as compared with the previous service situations. However, it is important for the analyst to realize that, while the customer needs no human intervention to perform the service in a "self-service" system, often some set of machines or devices is required. Usually, these devices are not unlimited in availability or negligible in cost (i.e., there is not an infinite number of stalls in a self-service car wash). Thus, when the number of customers exceeds some point, it is the spatial or mechanical resources which must be included in the model to project accurately system performance. Simulation is the appropriate vehicle for modeling systems in which required resources are nested. Nesting means that several resources are needed by the customer and are sought sequentially. Thus, the customer at the self-service car wash may be available to wash his or her own car, but if the stalls are all full, the availability of the human server is irrelevant to the expeditious completion of the task.

OTHER ANALYTICAL QUEUING MODELS

Some situations for which performance formulas are available are

1. Negative exponential service time with interarrival time
   a. constant
   b. Erlang*
2. Erlang service time with interarrival time
   a. negative exponential
   b. Erlang
   c. constant
3. Constant service time with interarrival time
   a. negative exponential
   b. Erlang

A convenient though less precise set of formulas belongs to the Pollaczek-Khintchine model, which assumes that either interarrival time or service time is negative exponentially distributed and that the other time is defined only by its mean and variance, not its distribution type. That is, the model is either $M/G/1$ or $G/M/1$, with one factor generalized and the other negative exponential.

There are also queuing models which incorporate priority queue disciplines and other queuing models with state-dependent service rates. A state-dependent service rate occurs when the server is able to see a long waiting line and consequently speeds up service or when the server sees a very short waiting line and slows down as a result.

MODELING MULTIPHASE SYSTEMS

In some instances, modeling of multiphase systems can be accomplished using the analytical waiting-line models. If the analyst has gathered data separately on arrival and service patterns of the two phases, the formulas appropriate to each phase may be applied and the performance estimates noted. In the special case when interarrival time and service time are both negative exponentially distributed in the first phase, then it may be assumed that output of the first phase and subsequent input to the second phase is also negative exponentially distributed with the same mean interarrival time.

CONCLUSION

It is clear that the analyst and management can derive great benefit from the performance estimates provided by analytical queuing models, when the system satisfies the assumptions of the model chosen. Before embarking on a custom simulation

*The Erlang distribution is a generalization of the negative exponential distribution.

project, the analyst should be certain that there are no available analytical models which might serve the purpose at low cost.

SUMMARY

There are many analytical waiting-line models which can provide the analyst with projections about system performance. Performance attributes which are estimated include mean number in the queue and in the system, mean waiting time and mean transit time, the probability distribution of the number of customers in the system, and the system utilization proportion. Most of the models assume that arrival rate and service rate are Poisson distributed, with means respectively designated as λ and μ.

There are models for single-channel, single-phase systems with one server ($M/M/1$) and with several servers ($M/M/c$). Certain models cover the situation when the waiting area is limited; these are finite queue models. Other models cover the situation when the server is responsible for a limited number of potential customers; these are finite calling population models. Another model is the self-service model, in which the customer is also the server.

Many other types of analytical waiting-line models are available and are discussed in texts on queuing theory and operations research. Some models use formulas which are relatively convenient for computation; others require the assistance of a computer for solution.

Since the use of general models is almost always less expensive than building a custom simulation model, the analyst is advised to seek such models whenever they might be appropriate for the subject system.

QUESTIONS

1. Define the following terms:
   (a) queue
   (b) analytical model
   (c) transit time
   (d) $M/M/c$
   (e) opportunity cost
   (f) calling population
2. Why is there likely to be a difference between the mean time in queue overall and the mean time in queue for a busy system? Which of these will probably be larger? Why?
3. Is it reasonable to assume that the average number in the queue is equal to the average number in the system minus 1 (the customer being served)? Why or why not? Explain.
4. What effect would you expect the following restrictions to have on the average number in the queue and the overall desirability of the system:
   (a) a finite queue
   (b) a finite calling population

   Explain your logic in each case.

PROBLEMS

Given the sample data and the hypothesis tests in the problem section at the end of Chapter 2, assume that interarrival time and service time are negative exponentially distributed with means of, respectively, 421.75 and 316.769 seconds.

   1. Use the analytical queuing formulas in Chapter 3 to estimate performance attributes for a single-channel, single-phase system with FIFO queue discipline. Calculate the following:
   - **(a)** the probability of the system being idle
   - **(b)** the utilization factor
   - **(c)** the expected length of the waiting line
   - **(d)** the expected number in the system
   - **(e)** the expected time in the waiting line
   - **(f)** the expected transit time

   2. Suppose that the server receives a wage rate of $8.50 per hour and that the opportunity cost of customer time is assumed to be $15.00. Calculate
   - **(a)** expected daily wage cost
   - **(b)** expected daily opportunity cost
   - **(c)** expected daily total cost
   - **(d)** expected wage cost of idle time

   Are the costs in this system balanced well? Why or why not? If not, what might you recommend?

   3. Suppose that you can hire a more efficient server whose average time is only 250 seconds and who earns $10.00 per hour. Recompute the answers to Problems 1 and 2 on this assumption.

   *4. Suppose that the system described just prior to Problem 1 is augmented so that two equivalent servers are available. Use the *M/M/1* formulas to approximate system performance estimates for such a system. Compare the results with those in Problem 1.

   5. Show that the results in Problem 1 would be the same if the average arrival rate and service rate in the formulas were given on a per hour rather than on a per second basis.

   *†6. Suppose that the system described just prior to Problem 1 was limited to a waiting area large enough for only five customers. Use the QMODEL program to recompute system performance estimates and compare them with the results in Problem 1.

   †7. Use the *M/M/c* formulas in QMODEL to calculate accurate estimates of performance of the two-server system described in Problem 5. Comment on the differences between the results in Problems 5 and 6.

   *†8. Suppose that the data given just before Problem 1 described a self-service system. Calculate performance estimates for this system and compare them with the results from Problem 1.

   *Answers to problems preceded by * can be found in the Selected Numerical Answers.

   †Problems designated by † can be solved easily with the SIMSTAT programs but are quite tedious to solve otherwise.

*9. Suppose that the data given just before Problem 1 applied to a system with a finite calling population of size 20. Use the QMODEL program to compute system performance estimates.

*10. Suppose that the system you wanted to model had an arrival pattern as specified just before Problem 1 but that the service time distribution did not correspond to any identifiable pattern. Service time has a mean of 316.769 and a variance of 67606. Use the Pollaczek-Khintchine formulas for the $M/G/1$ system to estimate performance. Compare your results with those from Problem 1.

APPENDIX

The program which is used to produce the output shown in Chapter 3 is called QMODEL. In addition to the outputs shown for unrestricted $M/M/1$ and $M/M/c$ systems and self-service systems, QMODEL offers the analyst the option of calling models for finite queue or finite calling population as well.

QMODEL also incorporates the Pollaczek-Khintchine model covering the case when interarrival time is negative exponential but the service time distribution is specified only by its mean and variance, not by its distribution type.

The formulas upon which QMODEL calculations are based are shown in Appendix Table VI.

Table 3A-1 shows the QMODEL dialogue which produced the output in Table 3-1; this is an unconstrained $M/M/1$ system. Table 3A-2 shows the comparable dialogue for Table 3-3. Table 3A-3 is analogous to Table 3-4, Table 3A-4 to Table 3-6, Table 3A-5 to Table 3-7, and Table 3A-6 to Table 3-8.

The user should be careful to note that rate data, not time data, are requested by the programs, so it may be necessary to take the reciprocals of mean times to obtain mean rates for input. Also, the finite queue variation asks for the maximum number in the system; this will be one more than the maximum queue length to include the customer potentially being served in an $M/M/1$ system or c more than the maximum queue length for an $M/M/c$ system.

TABLE 3A-1   PERFORMANCE ESTIMATES FOR
THE TICKET SELLER EXAMPLE

QUEUING MODELS

DO YOU WANT INSTRUCTIONS?
TYPE 1 FOR YES, 0 FOR NO
• 1
THIS PROGRAM WORKS INTERACTIVELY WITH THE
USER TO PERFORM QUEUING CALCULATIONS.

TABLE 3A-1 (cont.)

THE FOLLOWING MODELS ARE AVAILABLE:

1. FINITE QUEUE M/M/1 MODEL.
2. FINITE POPULATION M/M/1 MODEL.
3. STANDARD M/M/C MODEL.
4. FINITE QUEUE M/M/C MODEL.
5. FINITE POPULATION M/M/C MODEL.
6. SELF-SERVICE MODEL.
7. POLLACZEK-KHINTCHINE (P-K) FORMULA.
    M/G/1 SYSTEM.
TYPE THE APPROPRIATE NUMBER.

   WHICH MODEL DO YOU WISH TO USE?
● 3

   YOU HAVE CHOSEN
3. STANDARD M/M/C MODEL.

WHAT IS THE MEAN ARRIVAL RATE?
● .0098736

WHAT IS THE MEAN SERVICE RATE?
● .011972

WHAT IS THE NUMBER OF SERVERS?
● 1

P (0) =	0.17528
P (1) =	0.14455
P (2) =	0.11922
P (3) =	0.09832
P (4) =	0.08109
P (5) =	0.06688
P (6) =	0.05515
P (7) =	0.04549
P (8) =	0.03751
P (9) =	0.03094
P (10) =	0.02552
P (11) =	0.02104
P (12) =	0.01736
P (13) =	0.01431
P (14) =	0.01180
P (N > OR =C) =	0.82472

STEADY-STATE MEAN NUMBER OF UNITS:

IN SYSTEM =	4.70529
IN QUEUE =	3.88057

TABLE 3A-1 (cont.)

IN QUEUE FOR BUSY SYSTEM =	4.70529
STEADY-STATE MEAN TIME:	
IN SYSTEM =	476.55270
IN QUEUE =	393.02460
IN QUEUE FOR BUSY SYSTEM =	476.55270

DO YOU WANT TO TRY ANOTHER PROBLEM?
IF YES, TYPE 1. IF NO, TYPE 0.
● 0

TABLE 3A-2 PERFORMANCE ESTIMATES USING STANDARD DEVIATIONS INSTEAD OF MEANS TO ESTIMATE λ AND μ

QUEUING MODELS

DO YOU WANT INSTRUCTIONS?
TYPE 1 FOR YES, 0 FOR NO
● 0

WHICH MODEL DO YOU WISH TO USE?
● 3

YOU HAVE CHOSEN
3. STANDARD M/M/C MODEL.

WHAT IS THE MEAN ARRIVAL RATE?
● .0126831

WHAT IS THE MEAN SERVICE RATE?
● .0137816

WHAT IS THE NUMBER OF SERVERS?
● 1

P(0) =		0.07971
P(1) =		0.07335
P(2) =		0.06751
P(3) =		0.06213
P(4) =		0.05717
P(5) =		0.05262
P(6) =		0.04842
P(7) =		0.04456
P(8) =		0.04101
P(9) =		0.03774

TABLE 3A-2 (cont.)

P (10) =	0.03473	
P (11) =	0.03197	
P (12) =	0.02942	
P (13) =	0.02707	
P (14) =	0.02492	
P (15) =	0.02293	
P (16) =	0.02110	
P (17) =	0.01942	
P (18) =	0.01787	
P (19) =	0.01645	
P (20) =	0.01514	
P (21) =	0.01393	
P (22) =	0.01282	
P (23) =	0.01180	
P (24) =	0.01086	

P (N > OR =C) =   0.92029

STEADY-STATE MEAN NUMBER OF UNITS:

IN SYSTEM =   11.54584
IN QUEUE =   10.62555
IN QUEUE FOR BUSY SYSTEM =   11.54584

STEADY-STATE MEAN TIME:

IN SYSTEM =   910.33250
IN QUEUE =   837.77220
IN QUEUE FOR BUSY SYSTEM =   910.33220

DO YOU WANT TO TRY ANOTHER PROBLEM?
IF YES, TYPE 1. IF NO, TYPE 0.
● 0

TABLE 3A-3   PERFORMANCE ESTIMATES FOR
THE FASTER TICKET SELLER EXAMPLE

QUEUING MODELS

DO YOU WANT INSTRUCTIONS?
TYPE 1 FOR YES, 0 FOR NO
● 0

WHICH MODEL DO YOU WISH TO USE?
● 3

YOU HAVE CHOSEN
3.   STANDARD M/M/C MODEL.

WHAT IS THE MEAN ARRIVAL RATE?
● .0098736

TABLE 3A-3 (cont.)

WHAT IS THE MEAN SERVICE RATE?
- .0142857

WHAT IS THE NUMBER OF SERVERS?
- 1

P(0) =		0.30885
P(1) =		0.21346
P(2) =		0.14753
P(3) =		0.10197
P(4) =		0.07048
P(5) =		0.04871
P(6) =		0.03367
P(7) =		0.02327
P(8) =		0.01608
P(9) =		0.01112
P(N > OR = C) =		0.69115

STEADY-STATE MEAN NUMBER OF UNITS:

IN SYSTEM =	2.23784
IN QUEUE =	1.54669
IN QUEUE FOR BUSY SYSTEM =	2.23784

STEADY-STATE MEAN TIME:

IN SYSTEM =	226.64920
IN QUEUE =	156.64910
IN QUEUE FOR BUSY SYSTEM =	226.64910

DO YOU WANT TO TRY ANOTHER PROBLEM?
IF YES, TYPE 1. IF NO, TYPE 0.
- 0

TABLE 3A-4 PERFORMANCE ESTIMATES FOR THE TICKET SELLER EXAMPLE WITH TWO EQUIVALENT TICKET SELLERS

QUEUING MODELS

DO YOU WANT INSTRUCTIONS?
TYPE 1 FOR YES, 0 FOR NO
- 0

WHICH MODEL DO YOU WISH TO USE?
- 3

YOU HAVE CHOSEN
3. STANDARD M/M/C MODEL.

TABLE 3A-4 (cont.)

WHAT IS THE MEAN ARRIVAL RATE?
- 0.0049368

WHAT IS THE MEAN SERVICE RATE?
- 0.011972

WHAT IS THE NUMBER OF SERVERS?
- 1

P (0) =	0.58764
P (1) =	0.24232
P (2) =	0.09992
P (3) =	0.04120
P (4) =	0.01699
P (N > OR =C) =	0.41236

STEADY-STATE MEAN NUMBER OF UNITS:

IN SYSTEM =	0.70173
IN QUEUE =	0.28937
IN QUEUE FOR BUSY SYSTEM =	0.70173

STEADY-STATE MEAN TIME:

IN SYSTEM =	142.14230
IN QUEUE =	58.61414
IN QUEUE FOR BUSY SYSTEM =	142.14240

DO YOU WANT TO TRY ANOTHER PROBLEM?
IF YES, TYPE 1. IF NO, TYPE 0.
- 0

TABLE 3A-5   PERFORMANCE ESTIMATES FOR THE TICKET SELLER EXAMPLE WITH FINITE QUEUE LENGTH OF 1

QUEUING MODELS

DO YOU WANT INSTRUCTIONS?
TYPE 1 FOR YES, 0 FOR NO
- 0

WHICH MODEL DO YOU WISH TO USE?
- 1

YOU HAVE CHOSEN
1. FINITE QUEUE M/M/1 MODEL.

WHAT IS THE MEAN ARRIVAL RATE?
- .0098736

TABLE 3A-5 (cont.)

WHAT IS THE MEAN SERVICE RATE?
- .011972

WHAT IS THE MAXIMUM NUMBER ALLOWED IN SYSTEM?
- 2

P (N > OR =0) =   0.60078
P (0) =   0.39922
P (1) =   0.32925
P (2) =   0.27154
STEADY-STATE MEAN NUMBER OF UNITS:
IN SYSTEM =   0.87232
IN QUEUE =   0.27154
IN QUEUE FOR BUSY SYSTEM =   0.45197
STEADY-STATE MEAN TIME:
IN SYSTEM =   121.28060
IN QUEUE =   37.75244
IN QUEUE FOR BUSY SYSTEM =   62.83887

DO YOU WANT TO TRY ANOTHER PROBLEM?
IF YES, TYPE 1. IF NO, TYPE 0.
- 0

TABLE 3A-6   PERFORMANCE ESTIMATES
FOR THE TICKET SELLER EXAMPLE WITH
CUSTOMER SELF-SERVICE ASSUMED

QUEUING MODELS

DO YOU WANT INSTRUCTIONS?
TYPE 1 FOR YES, 0 FOR NO
- 0

WHICH MODEL DO YOU WISH TO USE?
- 6

YOU HAVE CHOSEN
6. SELF-SERVICE MODEL.

WHAT IS THE MEAN ARRIVAL RATE?
- 0.0098736

WHAT IS THE MEAN SERVICE RATE?
- 0.011972
P (0) =   0.43836
P (1) =   0.36152

TABLE 3A-6 (cont.)

P (2) =   0.14908
P (3) =   0.04098
STEADY-STATE MEAN NUMBER OF UNITS IN
SYSTEM =   0.82472
STEADY-STATE MEAN TIME IN SYSTEM =
                                  83.52823

DO YOU WANT TO TRY ANOTHER PROBLEM?
IF YES, TYPE 1. IF NO, TYPE 0.
• 0

chapter four

RANDOM NUMBER GENERATION AND ANALYSIS

THE PURPOSE OF RANDOM NUMBER GENERATION

In most real-world systems, events do not occur with complete predictability. Even the workers on a paced assembly line are allowed some flexibility to deal with contingencies or just to stretch from time to time. The term ''random'' is used colloquially to suggest that occurrences are completely unpredictable or that all possibilities are equally likely. We have already established that certain parameters of real-world systems can be defined statistically. For example, we cannot know precisely when the next customer is going to arrive at the ticket seller's station, but we may know that arrivals in general seem to follow a negative exponential distribution of interarrival times with a mean of 101.28 seconds. Having established these parameters allows the analyst to predict with absolute certainty that there will be no negative values for interarrival time and that the longer the interarrival time, the lower the probability that that interarrival time will occur. Thus, the analyst cannot assert that an interarrival time of 1000 seconds is impossible, because it is possible given the distribution which has been defined statistically, but the analyst can state the probability that the next customer will exhibit an interarrival time of 1000 or more seconds. While no one can be sure exactly what the next customer will do, given a sufficiently large sample size, the analyst can predict the relative frequency of various ranges of interarrival times with considerable accuracy after having performed goodness-of-fit tests.

When the analyst speaks of a ''random'' process, then, the process as a whole exhibits attributes that may be categorized and the relative frequencies of those attributes occurring may be tabulated. The purpose of random number generation in a simulation model is to convey to the model the nature of the statistical distribution to be modeled and to create the impression that the value of the next draw from the distribution cannot be guessed.

This impression is not strictly correct. Computer programming code is used to define statistical distributions and to select individual values from them in a simulation model. If the program code and the initial values of the random numbers are known, the analyst can duplicate exactly the so-called "random" stream of values on demand. In one sense, this ability to replicate the sequence of simulated events is advantageous to the analyst. For example, an identical pattern of interarrival times can be used as input to very different models. To the extent that model performance varies, that variance can be ascribed directly to changes in the processing of customers rather than to changes in the interarrival pattern. The ability to hold factors constant simplifies the task of the analyst to determine which systems are superior.

ATTRIBUTES OF A RANDOM NUMBER GENERATOR

In addition to replicability, there are certain other attributes of random number generators which are considered desirable:

1. Efficiency—that is, the generator produces random numbers at relatively little cost for computer time and computer workspace.
2. Uniformity—that is, approximately 10 percent of the digits generated will be zeroes, 10 percent will be ones, and so forth, up to 10 percent nines.
3. Conformity to the desired type of statistical distribution, with mean, variance and range as stipulated.
4. Independence—that is, the inability to predict the value of the $(N + 1)$th random number based on the value of the Nth random number except by examining the computer code.
5. Absence of trends—that is, generation of ascending or descending strings of values which are neither excessively long nor excessively short.
6. Long cycle length—that is, a relatively large number of numbers which can be generated before the algorithm produces a sequence identical to the previous sequence.

THE INVERSE TRANSFORMATION

Random number generation routines are often designed to produce uniformly distributed decimal values on the interval between 0 and 1. These values are called $0-1$ random numbers. This facilitates what is known as the inverse transformation from a cumulative probability distribution function to individual values of a random variable. Figure 4-1 shows the probability density function and the cumulative distribution function for a uniform distribution defined on the interval from 100 to 199, with only integer 10's values permitted.

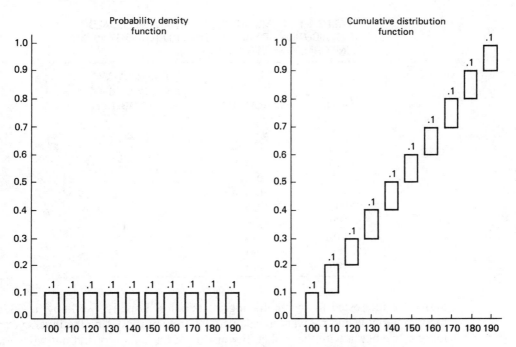

Figure 4-1 Probability density functions and cumulative distribution function for a uniform distribution on the interval between 100 and 199, integer 10's values only

Clearly, if this were a perfect uniform distribution, the probability that the next value chosen would be 100 would be 0.1, that it would be 110 would also be 0.1, and so forth. The uniform probability density function does not permit us to identify a unique value of the random variable with its probability of occurrence— indeed, in this worst case, all values of the random variable from 100 to 199 are identified with the same probability, 0.1. If we convert to the cumulative less-than-or-equals distribution function, however, we see that each unique value of the random variable is associated with a unique set of probabilities. Table 4-1 shows the set of cumulative probability values associated with each possible value of the uniformly distributed random variable.

If we could create random numbers uniformly distributed between 0 and 1, we could use the cumulative distribution function to define the value of the uniform distribution of 10's between 100 and 199 to which each 0−1 random number corresponds. For example, if we drew the 0−1 random number 0.2000, that would correspond to a value of 110. Likewise, any 0−1 random number between 0.1001 and 0.2000 would also be identified with a value of 110. There is a mutually exclusive, collectively exhaustive mapping of one distribution into another.

Random numbers can be used to select draws from any statistical distribution which can be represented in cumulative less-than-or-equals form. This method is called the inverse transformation because it does just the opposite of what we ordi-

TABLE 4-1   CUMULATIVE LESS-THAN-OR-EQUALS
DISTRIBUTION OF VALUES FROM 100 TO 199 BY 10'S,
UNIFORMLY DISTRIBUTED

VALUE	PROBABILITY	CUMULATIVE LESS-THAN-OR-EQUALS PROBABILITY
100	0.1	0.1
110	0.1	0.2
120	0.1	0.3
130	0.1	0.4
140	0.1	0.5
150	0.1	0.6
160	0.1	0.7
170	0.1	0.8
180	0.1	0.9
190	0.1	1.0

narily are interested in accomplishing in statistics. Usually, we know a value of a random variable and we want to calculate its probability. Using the inverse transformation, we begin with the probability and work backward to find out what value of the random variable corresponds to that probability.

The cumulative distribution function for either the discrete or continuous uniform distribution is easily obtained, as illustrated in Figure 4-1 and in Table 4-1. If RN is a $0-1$ random number, A is the lower limit on the interval, and B is the upper limit on the interval, the inverse transformation to obtain uniformly distributed random numbers between A and B is

$$URN = A + RN \times (B - A)$$

While not as intuitively obvious, the cumulative distribution functions for the Poisson and negative exponential distributions may be derived by means of integral calculus. The inverse transformation for the Poisson distribution is often approximated by program code which generates random variates more quickly than the inverse itself. The inverse transformation for the negative exponential distribution is

$$NERN = -\mu \times \ln(RN)$$

where μ is the mean of the negative exponential distribution and $\ln(RN)$ is the natural log of the $0-1$ random number. A table of natural logs of $0-1$ random numbers appears in Appendix Table IX. Unfortunately, the probability density function for the normal distribution cannot be integrated directly and so its inverse transformation must be accomplished by numerical methods such as Simpson's rule.*

---

*Simpson's rule is a method of approximating the area under a curve by rectangles of successively smaller width. The smaller the width of the rectangle, the better the approximation, but the greater the expenditure of computer time in making the approximation.

There is a great variety of computer-based methods for generating random draws from many different distributions. Unlike the computer code for efficient generation of $0-1$ random numbers themselves, most of the inverse transformations are machine independent and may be patched conveniently into a simulation model to be run on virtually any computer.

METHODS OF GENERATING RANDOM NUMBERS

Conceptually, the only means of generating truly random numbers is to avail ourselves of the statistician's fabled urn. We fill the urn with 10 slips of paper for our example, each slip marked with the values 100, 110, 120, 130, 140, 150, 160, 170, 180, or 190. We are careful not to peek before reaching into the urn and selecting a slip of paper. On it is our first random number. If we want to draw more than one random number, we must be careful to replace the slip from the previous draw and stir the mixture well before drawing again. This is a truly random method—there is absolutely no way to replicate the sequence of draws exactly. If the analyst is to model the ongoing activities of a complex process over time, this method is not practicable for generating the potential millions of random numbers needed. Fortunately, there is a group of computer-based techniques for generating numbers which satisfy the set of desirable properties quite well.

Every computer-based method of generating random numbers requires the initial definition of one or more constants called *seeds* which affect the magnitude of the random numbers produced. Probably the simplest method of generating random numbers is the midsquare method. The midsquare method requires the analyst to provide one seed whose size depends on the number of digits desired in the resulting random number. The seed is squared and the middle digits are extracted. These middle digits become the seed for the next random number; they are squared, the middle digits are extracted, and the process continues until enough random numbers have been produced to simulate the real events for the desired length of time. Table 4-2 shows several successive values produced by a two-digit midsquare generator.

TABLE 4-2   SELECTED OUTPUT OF A
TWO-DIGIT MIDSQUARE GENERATOR

RANDOM DIGITS	RANDOM DIGITS SQUARED
25	0625
62	3844
84	7056
05	0025
02	0004
00	0000

The random numbers in Table 4-2 increase, then decrease, and ultimately degenerate into an irretrievable condition in which the random number, 00, produces itself as its successor. The effect of this degeneration is to remove the variability or randomness from the process being modeled. Every random number generator must include a provision for reseeding when this occurs. A "good" random number generator will exhibit very long periods before the sequence of random numbers repeats itself and also will rarely degenerate into an irretrievable condition. While it is very easy to understand, the midsquare generator is very poor according to these criteria and is rarely used except as an illustration of simple random number generation.

The midproduct random number generator is somewhat more reliable. It requires two initial seeds which are multiplied together before extracting the middle digits. The extracted middle digits are multiplied by the second seed to create a third random number. Table 4-3 shows the initial seeds and several draws using a midproduct generator.

TABLE 4-3   SELECTED OUTPUT OF A
TWO-DIGIT MIDPRODUCT GENERATOR

RANDOM DIGITS	RANDOM DIGITS MULTIPLIED
25	
62	1550
55	3410
41	2255
25	1025
02	0050
05	0010
01	0005
00	0000

While it has taken a bit longer for the midproduct generator to deteriorate, eventually the same problem of reduction of variability to a constant has occurred. Increasing the number of digits in the seeds and in the resulting random numbers decreases the probability of generating a random number which is exactly zero and improves the cycle length of both the midsquare and midproduct generators. However, a greater amount of computer time and space to perform the multiplication and extraction are required.

Random number generators that offer the best combination of efficiency with the other desirable properties are multiplicative and/or additive. They often are congruential, employing mathematical congruence properties and remainder division to extract the random numbers. The generators use one constant as a multiplier of the initial seed and of subsequent random numbers, as well as another constant which is added to the product. The constants are chosen with the word size of the host computer system in mind to maximize the length of the cycle through which the generator will pass before repeating the string of random numbers previously generated.* The disadvantage of congruential methods of random number generation is that they demand detailed knowledge of the host computer system and cannot be transported for use on other computer systems with desirable results. They are also more intuitively difficult to understand. We will illustrate additive/multiplicative random number generation with a simple example. Suppose that the multiplicative constant is chosen to be 10 and that the additive constant is chosen to be 100. Table 4-4 shows a set of draws starting with a seed of 25.

TABLE 4-4   SELECTED OUTPUT OF A
TWO-DIGIT MULTIPLICATIVE/ADDITIVE
GENERATOR

RANDOM DIGITS	RANDOM CONGRUENTIAL DIGITS
25	0350
35	0450
45	0550
55	0650
65	0750
75	0850
85	0950
95	1050
05	0150

The pattern which emerges in Table 4-4 violates our notions of randomness even though it does not degenerate into a pure zero value. It is important to choose constants judiciously when building a congruential generator. A further safeguard is the use of remainder or modulus division in the random number generation algorithm, as the term "congruential" would imply.

AN EXAMPLE

Suppose that we have developed a random number generation procedure which we would like to evaluate according to the criteria previously listed. We obtain 25 single-digit random numbers from the generator. These are shown in Table 4-5.

*Random number generation routines are included with most simulation system software and scientific subroutine packages. Thus, the user typically does not have to delve into the intricacies of establishing these constants. However, it is important that the user be able to verify the accurate operation of the generator, especially when the simulation results seem not to be reasonable for no identifiable cause.

TABLE 4-5   SINGLE DIGIT
RANDOM NUMBERS

8.
6.
7.
9.
8.
8.
3.
2.
0.
1.
3.
3.
6.
5.
8.
9.
2.
4.
6.
1.
0.
3.
4.
7.
3.

To get a rough impression of the pattern of the random number data, we calculate descriptive statistics and sort the random numbers into rank order. This is displayed in Table 4-6.

TABLE 4-6   DESCRIPTIVE STATISTICS AND RANK
ORDERING OF SINGLE-DIGIT RANDOM NUMBERS

THE MEAN IS 4.6400.
THE VARIANCE IS 8.24.
THE STANDARD DEVIATION IS 2.8705.
THE SAMPLE SIZE IS 25.

SORTED DATA LIST

0.0	0.0	1.0	1.0	2.0	2.0	3.0	3.0	3.0	3.0
3.0	4.0	4.0	5.0	6.0	6.0	6.0	7.0	7.0	8.0
8.0	8.0	8.0	9.0	9.0					

GOODNESS-OF-FIT TESTS FOR RANDOM NUMBER GENERATORS

Just as we did in Chapter 2 using sample data from a real-world system, we can perform a goodness-of-fit test of the uniform distribution on these data produced by a random number generator.* The null hypothesis is that the data are uniformly distributed on the interval from 0 to 9.9, for integer values only. The alternative hypothesis is that some other distribution really underlies the output of the random number generator. The significance level will remain at 0.05.

Instead of using a χ^2 goodness-of-fit test, however, this time we select a Kolmogorov-Smirnov test. The reason is that the contention in the null hypothesis that the distribution is uniform in a particular range is not based on examination of the sample data themselves but rather on our expectations about the random number generation process. Thus, there is no issue of contamination of the null hypothesis by the sample data which necessitates correction via degrees of freedom. Since the K-S test is slightly more powerful and flexible than the χ^2 test, we select it, although the χ^2 test may be used under these circumstances with only 1 degree of freedom lost for the use of sample size, no matter what the nature of the distribution specified in the null hypothesis.†

The K-S test compares observed data with expected data values in a somewhat different way than does the χ^2 test. The data are compared on a cumulative relative frequency basis. The maximum difference between cumulative observed relative frequency and cumulative expected relative frequency is found and compared with the critical level of this statistic, which is called D. If the computed level of D is less than the critical level at the chosen level of significance, the null hypothesis cannot be rejected. If the computed level of D is greater than the critical level, the null hypothesis will be rejected.

One advantage of the K-S test is that there is no stipulation that expected class frequency should be at least 5. Classes might be defined in such a way that every item in the sample forms a separate class. This is not especially an advantage to us in testing the goodness of fit of the uniform distribution as we have defined it, since only 10 integer values (from 0 through 9) are possible and there is no purpose in defining more than 10 classes. If we used the χ^2 test, however, we could not use 10 classes because, with a sample size of 25, expected class frequency would fall to 2.5. The K-S test of our sample data with 10 classes is shown in Table 4-7. The maximum sample difference is 0.0600; the critical value of D is 0.2640 at the 0.05 level of signficance.‡

*These data happen to have been generated by a midproduct random number generator.

†The Kolmogorov-Smirnov test loses considerable power when the data are not continuous. Single-digit random numbers exhibit a decidedly discrete pattern and violate the assumptions of the test. The test is presented here because the analyst is very unlikely to use single-digit random numbers and thus may validly employ the K-S tests. Also, subsequent test of randomness—Yule's test and the gap test—are based on the K-S test.

‡Critical values of the Kolmogorov-Smirnov one-sample D statistic for samples smaller than 35 are found in Appendix Table III. This table also shows the z statistic, which gives critical values of the K-S D statistic for sample sizes of 35 or more.

TABLE 4-7 KOLMOGOROV-SMIRNOV GOODNESS-OF-FIT TEST OF THE UNIFORM DISTRIBUTION ON THE INTERVAL FROM 0 TO 9.9 FOR INTEGER VALUES ONLY

CLASS	CLASS IDENTITY	OBSERVED FREQUENCY	CUMULATIVE PROPORTION	CUMULATIVE PROPORTION	ABSOLUTE DIFFERENCE
1	0.0	2.	0.0800	0.1000	0.0200
2	1.0	2.	0.1600	0.2000	0.0400
3	2.0	2.	0.2400	0.3000	0.0600
4	3.0	5.	0.4400	0.4000	0.0400
5	4.0	2.	0.5200	0.5000	0.0200
6	5.0	1.	0.5600	0.6000	0.0400
7	6.0	3.	0.6800	0.7000	0.0200
8	7.0	2.	0.7600	0.8000	0.0400
9	8.0	4.	0.9200	0.9000	0.0200
10	9.0	2.	1.0000	1.0000	0.0000
TOTAL		25.			

There were two zero values in the 25 data items, so the observed relative frequency for the first class is 2/25, or 0.08. With 10 classes defined and a uniform distribution assumed in the null hypothesis, the expected relative frequency for this and all other classes is 0.10. The difference is (0.08 − 0.10) and the sign is not relevant for this test, so the absolute difference between observed and expected relative frequencies for class 1 is 0.02.

The second class consists of values of exactly 1. There are two 1's in the sample data, so the observed and expected relative frequencies for the second class are the same as those for the first class. However, the K-S test computes the absolute differences of the *cumulative* relative frequencies. Thus, we compare (2 × 0.08) with (2 × 0.10) and obtain (0.16 − 0.20) or 0.04 as the absolute difference for the second class. The process continues until the cumulative absolute differences for all classes have been calculated.

The computed maximum absolute difference between the observed and expected cumulative relative frequencies, 0.06, is far below the critical level of D, 0.2640, so it seems that the null hypothesis of uniform distribution on the specified interval fits very well indeed. The random number generator has passed its first test with flying colors.

If the random numbers have passed the uniform goodness-of-fit test and the inverse transformations are accurately stated, the resulting draws from other distributions should also pass their respective goodness-of-fit tests. As with χ^2, the K-S goodness-of-fit test may be used to test any statistical distribution in the null hypothesis.

At this point it is wise to be sure that the meaning of uniformity of distribution is clear. If the random number generator had produced the following string of output, it, too would have been considered to be uniformly distributed: 0, 0, 1, 1, 2, 2, 3, 3, 4, 4, 5, 5, 6, 6, 7, 7, 8, 8, 9, 9, 0, 0, 1, 1, 2. Similarly, the following pattern would have passed the uniform goodness-of-fit test: 0, 1, 2, 3, 4, 5, 6, 7, 8, 9, 0, 1, 2, 3, 4, 5, 6, 7, 8, 9, 0, 1, 2, 3, 4, 5, 6. While these strings possess the quality of uniformity, they lack much that is implied by the word ''random.'' We now consider some statistical tests which help to detect the presence of these undesirable attributes.

TESTING FOR SERIAL CORRELATION

The latter string of numbers, 0, 1, 2, . . . , 5, 6, is an example in which it is possible to predict the value of the $(N + 1)$th number with near certainty if the value of the Nth number is known. This property is called serial correlation or autocorrelation and is often, but not exclusively, found in observations of a process that have been recorded within consecutive short periods of time. The Pearson product moment correlation coefficient, commonly called **r**, is a parametric statistic which measures the degree of association between two or more variables. It may be applied to the serial correlation case by considering the two separate variables to be the data values and the data values lagged by one or more terms. The dependent or predicted variable value is the $(N + 1)$th; the independent or predictor variable is the Nth. Unfortunately, parametric statistics make certain assumptions which cannot always be met by a data set. These assumptions may include normal distribution and equal variances of the data sets. To avoid having to concern ourselves with these assumptions we will use the nonparametric equivalent of the Pearson **r**, the Spearman rank correlation coefficient, which we will also call **r**.

In the Spearman test, the null hypothesis is that there is no way to predict the value of the $(N + 1)$th item by knowing the value of the Nth item; that is, the data items are serially independent. The alternative hypothesis is that they are autocorrelated. We will allow the level of significance to remain at 0.05 as before. While the Pearson test works with the actual data values, the Spearman test first calculates the ranks of the data values and then attempts to determine the extent to which it is possible to predict the rank of the $(N + 1)$th item by knowing the rank of the Nth item. Tie values are assigned an average rank and are compensated for by means of a tie correction factor. The value of **r** can be tested for significance using a z statistic when the sample size is fairly large.* Table 4-8 shows the results of a Spearman test performed on the 25 observations from the random number generator.

Exactly what is meant by a statistically large sample is a subject of some debate. The central limit theorem suggests that samples of size 20, 25, or 30 or more may be considered large for some statistical purposes. While 30 is the commonly accepted conservative lower limit on sample size for testing with a z statistic, we will use our sample of 25 with substantial confidence in the reliability of the test. Appendix Table VIII shows a table of significant values of **r** for the Spearman rank correlation test, for various levels of significance.

*We will use the Spearman test to detect serial correlation in the output of a simulation model in a later chapter.

TABLE 4-8   SPEARMAN RANK CORRELATION TEST
OF OUTPUT OF RANDOM NUMBER GENERATOR

X VALUE	Y VALUE	X RANK	Y RANK	D	D^2
8.00	6.00	20.50	16.00	4.50	20.25
6.00	7.00	15.00	18.50	−3.50	12.25
7.00	9.00	17.50	23.50	−6.00	36.00
9.00	8.00	23.50	21.00	2.50	6.25
8.00	8.00	20.50	21.00	−0.50	0.25
8.00	3.00	20.50	9.00	11.50	132.25
3.00	2.00	8.50	5.50	3.00	9.00
2.00	0.00	5.50	1.50	4.00	16.00
0.00	1.00	1.50	3.50	−2.00	4.00
1.00	3.00	3.50	9.00	−5.50	30.25
3.00	3.00	8.50	9.00	−0.50	0.25
3.00	6.00	8.50	16.00	−7.50	56.25
6.00	5.00	15.00	14.00	1.00	1.00
5.00	8.00	13.00	21.00	−8.00	64.00
8.00	9.00	20.50	23.50	−3.00	9.00
9.00	2.00	23.50	5.50	18.00	324.00
2.00	4.00	5.50	12.50	−7.00	49.00
4.00	6.00	11.50	16.00	−4.50	20.25
6.00	1.00	15.00	3.50	11.50	132.25
1.00	0.00	3.50	1.50	2.00	4.00
0.00	3.00	1.50	9.00	−7.50	56.25
3.00	4.00	8.50	12.50	−4.00	16.00
4.00	7.00	11.50	18.50	−7.00	49.00
7.00	3.00	17.50	9.00	8.50	72.25

THE ORIGINAL SAMPLE SIZE IS 25.
THE SUM OF D^2 IS 1120.00.
SUM U = 15.00
SUM V = 17.00
R = 0.51
Z = 2.43

The Spearman test has ranked the data items in the original data set, X. The data set, Y, consists of the X items one period later. There are 24 pairs of Y and X values which result. After ranking the items in the X data set, the Spearman testing procedure contrasts the rank of the Y member with the rank of the X member of each pair. The difference, D, is squared, and the squares are summed over all pairs. The sum of D^2 is corrected for ties in the X and Y data sets, respectively, called U and V, and a value of **r** is obtained.*

*The only instance in which the Spearman test would indicate the complete absence of autocorrelation is a case in which all sample values are identical, leading to identical ranks for the Nth and $(N + 1)$th terms.

A simple formula for the Spearman rank correlation coefficient is

$$R = 1 - \frac{6 \sum D^2}{N(N^2 - 1)}$$

If there are many tie values, this formula may considerably understate the value of R. A formula which allows correction for ties is

$$R = \frac{N(N^2 - 1) - 6 \sum D^2 - 6(U + V)}{\sqrt{[N(N^2 - 1) - 12U]} \sqrt{[N(N^2 - 1) - 12V]}}$$

where

$$U = \sum I^2 - \sum I$$

and

$$V = \sum J^2 - \sum J$$

for I observations tied at a given rank in the X sample, summed over all tied ranks and, likewise, for J observations tied at a given rank in the Y sample, summed over all tied ranks.

A z value for testing the significance of **r** is found by

$$z = \mathbf{r} \times \sqrt{(N - 1)}$$

when N exceeds 30 and serves as an approximation for slightly smaller N. Special statistical tables for the evaluation of the significance of **r** from smaller samples are available.

Although the value of **r**, 0.51, is about halfway between perfect positive correlation ($\mathbf{r} = +1$) and the complete absence of correlation ($\mathbf{r} = 0$), the sample size is approximately sufficient for us to conclude that the incidence of autocorrelation in the output of the random number generator is statistically very different from zero. The value of z, 2.43, is significant well beyond the 5 percent level. Appendix Table VIII also indicates that for a sample size of 25 and a significance level of 0.05, the critical value of **r** is 0.3977, substantially lower than the computed **r** in this sample.

Apparently the random number generator has a defect—serially correlated output—which was not particularly obvious without a formal statistical test.

TESTING FOR RUNS UP AND DOWN

Another informative nonparametric test is the runs-up-and-down test. The string— 0, 0, 1, 1, 2, 2, 3, 3, 4, 4, 5, 5, 6, 6, 7, 7, 8, 8, 9, 9—is composed of a single ascending run of numbers. Tie values are not considered to be breaks in the run. If the data were truly random, an expected number of runs up and down and the standard deviation of the number of runs up and down could be calculated for statistically large sample sizes, and the actual number of runs up and down in the sample data could be compared with the expected number to ascertain if the difference between observed and expected numbers is statistically significant.

The null hypothesis in the runs-up-and-down test is that there are no more and no fewer runs in the population (in this case, output of the random number generator) than might be expected as a result of normal variation. The alternative hypothesis is that there are significantly more or fewer runs than might be attributed to chance. Again, we use the 0.05 level of significance.

If V is defined to be the total number of runs, including both runs up and runs down, the expected value of V is $(2N - 1)/3$ and the standard deviation of V is $\sqrt{(16N - 29)/90}$, where N is the sample size. A continuity correction of 0.5 may be applied when computing the value of z. The continuity correction adds 0.5 to values of V which are below the mean and subtracts 0.5 from values of V which are above the mean when calculating the z value. The effect of the continuity correction is to make the test more conservative; that is, the value of z is less likely to be statistically significant. Table 4-9 displays relevant data about the sample random numbers from a runs-up-and-down test.

TABLE 4-9   RUNS-UP-AND-DOWN TEST
OF OUTPUT OF RANDOM NUMBER
GENERATOR

THE ORIGINAL SAMPLE SIZE IS 25.
THE EFFECTIVE SAMPLE SIZE IS 23.
THE NUMBER OF RUNS IS 11.
THE EXPECTED NUMBER OF RUNS IS 15.00.
THE STANDARD DEVIATION IS 1.9408.
THE VALUE OF z IS -1.8034.

The original sample size is reduced by the number of adjacent tie values in the data set to obtain the effective sample size. It is the effective sample size which is used in the calculations. Since the value of z is -1.8034, which is greater than the critical value of -1.96 for a two-tailed test at the 0.05 level of significance, we might conclude that there seem to be somewhat fewer than the expected number of runs up and down in the random number generator output, but that the disparity is not so great that it might not be attributed to chance. However, in this instance the slightly smaller sample size than is strictly appropriate for the use of z statistics does make a difference. Tables of the runs-up-and-down distribution for small sample sizes may be found in nonparametric statistics texts. When possible, of course, it is desirable to obtain a sample size robust enough to warrant the use of the normal distribution in analyzing the significance of the number of runs up and down.

YULE'S TEST OF SUMS OF CONSECUTIVE DIGITS

Another test of the suitability of a random number generator is Yule's test of sums of consecutive digits. The test may be done for sums of two, three, or four consecutive digits. The theoretical or expected probability distribution of the values of the sums may be established from basic probability theory and the rules for permutations and combinations. For example, the sums of two digits may range from 0 (if both digits are 0) to 18 (if both digits are 9). There is only one way in which to obtain a sum of 0 or a sum of 18 when two single digits are added. The probability of a digit taking on any value at random from 0 through 9 is 0.1, since the digit values are presumed equally likely. Similarly, since the values of successive digits are presumed to be independent, the probability of a sum is the product of the probabilities of the component digits of that sum. Thus, the probability that the sum of the two digits is 0 equals $0.1 \times 0.1 = 0.01$; by the same logic, the probability of a sum of 18 is also 0.01. However, a sum of 1 could occur if the first digit were 1 and the second were a 0, or if the first digit were a 0 and the second digit were a 1. Thus, the probability of a sum with a value of 1 is $2(0.1 \times 0.1) = 0.02$.

If there is not equal likelihood of all digit values, or if the values of successive digits are not independent, the observed relative frequency of the sums will deviate from the theoretical relative frequency. Yule's test enables us to test the null hypothesis that the sums of consecutive digits are as would be expected from probability theory.

Yule's test is based on the K-S goodness-of-fit technique, comparing cumulative observed relative frequency with cumulative expected relative frequency in each class. The maximum absolute difference is noted and compared with the critical level of the D statistic. Table 4-10 displays relevant data about the sample random numbers from Yule's test. The maximum difference is 0.1900; Appendix Table III contains values of the D statistic; the critical value of D is 0.3750 at the 0.05 level of significance.

TABLE 4-10 YULE'S TEST OF SUMS OF TWO CONSECUTIVE DIGITS PRODUCED BY RANDOM NUMBER GENERATOR

SUM	OBSERVED RELATIVE FREQUENCY	EXPECTED RELATIVE FREQUENCY	CUMULATIVE OBSERVED RELATIVE FREQUENCY	CUMULATIVE EXPECTED RELATIVE FREQUENCY	ABSOLUTE DIFFERENCE
0	0.0000	0.0100	0.0000	0.0100	0.0100
1	0.0833	0.0200	0.0833	0.0300	0.0533
2	0.0000	0.0300	0.0833	0.0600	0.0233
3	0.0833	0.0400	0.1667	0.1000	0.0667
4	0.0000	0.0500	0.1667	0.1500	0.0167
5	0.0833	0.0600	0.2500	0.2100	0.0400
6	0.1667	0.0700	0.4167	0.2800	0.1367
7	0.0833	0.0800	0.5000	0.3600	0.1400
8	0.0000	0.0900	0.5000	0.4500	0.0500
9	0.0000	0.1000	0.5000	0.5500	0.0500
10	0.0000	0.0900	0.5000	0.6400	0.1400
11	0.1667	0.0800	0.6667	0.7200	0.0533

TABLE 4-10 (cont.)

12	0.0000	0.0700	0.6667	0.7900	0.1233
13	0.0000	0.0600	0.6667	0.8500	0.1833
14	0.0833	0.0500	0.7500	0.9000	0.1500
15	0.0000	0.0400	0.7500	0.9400	0.1900
16	0.1667	0.0300	0.9167	0.9700	0.0533
17	0.0833	0.0200	1.0000	0.9900	0.0100
18	0.0000	0.0100	1.0000	1.0000	0.0000

TABLE 4-11   YULE'S TEST OF SUMS OF THREE CONSECUTIVE DIGITS
PRODUCED BY RANDOM NUMBER GENERATOR

SUM	OBSERVED RELATIVE FREQUENCY	EXPECTED RELATIVE FREQUENCY	CUMULATIVE OBSERVED RELATIVE FREQUENCY	CUMULATIVE EXPECTED RELATIVE FREQUENCY	ABSOLUTE DIFFERENCE
0	0.0000	0.0010	0.0000	0.0010	0.0010
1	0.0000	0.0030	0.0000	0.0040	0.0040
2	0.0000	0.0060	0.0000	0.0100	0.0100
3	0.0000	0.0100	0.0000	0.0200	0.0200
4	0.0000	0.0150	0.0000	0.0350	0.0350
5	0.1250	0.0210	0.1250	0.0560	0.0690
6	0.0000	0.0280	0.1250	0.0840	0.0410
7	0.2500	0.0360	0.3750	0.1200	0.2550
8	0.0000	0.0450	0.3750	0.1650	0.2100
9	0.0000	0.0550	0.3750	0.2200	0.1550
10	0.0000	0.0630	0.3750	0.2830	0.0920
11	0.0000	0.0690	0.3750	0.3520	0.0230
12	0.0000	0.0730	0.3750	0.4250	0.0500
13	0.0000	0.0750	0.3750	0.5000	0.1250
14	0.1250	0.0750	0.5000	0.5750	0.0750
15	0.1250	0.0730	0.6250	0.6480	0.0230
16	0.0000	0.0690	0.6250	0.7170	0.0920
17	0.0000	0.0630	0.6250	0.7800	0.1550
18	0.0000	0.0550	0.6250	0.8350	0.2100
19	0.1250	0.0450	0.7500	0.8800	0.1300
20	0.0000	0.0360	0.7500	0.9160	0.1660
21	0.1250	0.0280	0.8750	0.9440	0.0690
22	0.0000	0.0210	0.8750	0.9650	0.0900
23	0.0000	0.0150	0.8750	0.9800	0.1050
24	0.0000	0.0100	0.8750	0.9900	0.1150
25	0.1250	0.0060	1.0000	0.9960	0.0040
26	0.0000	0.0030	1.0000	0.9990	0.0010
27	0.0000	0.0010	1.0000	1.0000	0.0000

In Table 4-11, the maximum difference is 0.2550; the critical value of D is 0.4570 at the 0.05 level of significance.

TABLE 4-12 YULE'S TEST OF SUMS OF FOUR CONSECUTIVE DIGITS PRODUCED BY RANDOM NUMBER GENERATOR

SUM	OBSERVED RELATIVE FREQUENCY	EXPECTED RELATIVE FREQUENCY	CUMULATIVE OBSERVED RELATIVE FREQUENCY	CUMULATIVE EXPECTED RELATIVE FREQUENCY	ABSOLUTE DIFFERENCE
0	0.0000	0.0001	0.0000	0.0001	0.0001
1	0.0000	0.0004	0.0000	0.0005	0.0005
2	0.0000	0.0010	0.0000	0.0015	0.0015
3	0.0000	0.0020	0.0000	0.0035	0.0035
4	0.0000	0.0035	0.0000	0.0070	0.0070
5	0.0000	0.0056	0.0000	0.0126	0.0126
6	0.0000	0.0084	0.0000	0.0210	0.0210
7	0.1667	0.0120	0.1667	0.0330	0.1337
8	0.0000	0.0165	0.1667	0.0495	0.1172
9	0.0000	0.0220	0.1667	0.0715	0.0952
10	0.0000	0.0282	0.1667	0.0997	0.0670
11	0.0000	0.0348	0.1667	0.1345	0.0322
12	0.0000	0.0415	0.1667	0.1760	0.0093
13	0.1667	0.0480	0.3333	0.2240	0.1093
14	0.1667	0.0540	0.5000	0.2780	0.2220
15	0.0000	0.0592	0.5000	0.3372	0.1628
16	0.0000	0.0633	0.5000	0.4005	0.0995
17	0.0000	0.0660	0.5000	0.4665	0.0335
18	0.0000	0.0670	0.5000	0.5335	0.0335
19	0.0000	0.0660	0.5000	0.5995	0.0995
20	0.0000	0.0633	0.5000	0.6628	0.1628
21	0.1667	0.0592	0.6667	0.7220	0.0553
22	0.0000	0.0540	0.6667	0.7760	0.1093
23	0.0000	0.0480	0.6667	0.8240	0.1573
24	0.0000	0.0415	0.6667	0.8655	0.1988
25	0.0000	0.0348	0.6667	0.9003	0.2336
26	0.0000	0.0282	0.6667	0.9285	0.2618
27	0.0000	0.0220	0.6667	0.9505	0.2838
28	0.1667	0.0165	0.8333	0.9670	0.1337
29	0.0000	0.0120	0.8333	0.9790	0.1457
30	0.1667	0.0084	1.0000	0.9874	0.0126
31	0.0000	0.0056	1.0000	0.9930	0.0070
32	0.0000	0.0035	1.0000	0.9965	0.0035
33	0.0000	0.0020	1.0000	0.9985	0.0015
34	0.0000	0.0010	1.0000	0.9995	0.0005
35	0.0000	0.0004	1.0000	0.9999	0.0001
36	0.0000	0.0001	1.0000	1.0000	0.0000

In Table 4-12, the maximum difference is 0.2838; the critical value of D is 0.5210 at the 0.05 level of significance.

Tables 4-10, 4-11 and 4-12 show, respectively, the results of Yule's Test for sums of two, three, and four consecutive digits from the output of the random number generator. Since the critical level of D always exceeds the maximum observed level, we cannot reject the null hypothesis that the generator is random in this particular respect.

THE GAP TEST

One final test which measures randomness in the random number generator output is the gap test. The gap test measures the number of digits between successive occurrences of the same digit. The distribution underlying the test is the negative binomial, an illustration of which is the number of times one must strike a match before it finally ignites. For any one digit, say a 0, the probability that it will be fol-

TABLE 4-13 GAP TEST OF SPREAD BETWEEN SUCCESSIVE LIKE DIGITS PRODUCED BY RANDOM NUMBER GENERATOR

GAP	OBSERVED FREQUENCY	CUMULATIVE OBSERVED RELATIVE FREQUENCY	NEGATIVE BINOMIAL PROBABILITY	EXPECTED FREQUENCY	CUMULATIVE EXPECTED RELATIVE FREQUENCY	ABSOLUTE DIFFERENCE
1	3	0.1200	0.1000	2.50	0.1000	0.0200
2	1	0.1600	0.0900	2.25	0.1900	0.0300
3	2	0.2400	0.0810	2.02	0.2710	0.0310
4	3	0.3600	0.0729	1.82	0.3439	0.0161
5	1	0.4000	0.0656	1.64	0.4095	0.0095
6	1	0.4400	0.0590	1.48	0.4686	0.0286
7	1	0.4800	0.0531	1.33	0.5217	0.0417
8	1	0.5200	0.0478	1.20	0.5695	0.0495
9	3	0.6400	0.0430	1.08	0.6126	0.0274
10	3	0.7600	0.0387	0.97	0.6513	0.1087
11	1	0.8000	0.0349	0.87	0.6862	0.1138
12	2	0.8800	0.0314	0.78	0.7176	0.1624
13	0	0.8800	0.0282	0.71	0.7458	0.1342
14	1	0.9200	0.0254	0.64	0.7712	0.1488
15	0	0.9200	0.0229	0.57	0.7941	0.1259
16	0	0.9200	0.0206	0.51	0.8147	0.1053
17	0	0.9200	0.0185	0.46	0.8332	0.0868
18	1	0.9600	0.0167	0.42	0.8499	0.1101
19	0	0.9600	0.0150	0.38	0.8649	0.0951
20	0	0.9600	0.0135	0.34	0.8784	0.0816
21	1	1.0000	0.0122	0.30	0.8906	0.1094

lowed by a 0 is 0.1. The probability that the first digit will be followed by some digit other than 0, but that digit will then be followed by a 0, is $(0.9 \times 0.1) = 0.09$. As with Yule's test, these theoretical probabilities are predicted on independence of successive draws from the random number generator. If the observed probability distribution of gaps between successive occurrences of a particular digit is substantially different from the theoretical or expected probability distribution, we will reject the null hypothesis that the generator is random in this sense. Again, the K-S goodness-of-fit procedure will be used (see Table 4-13).

The gap between the beginning of random number generation and the first occurrence of each digit value plus the gap between successive occurrences of each digit are included in the gap count. Since it is impossible to know how much longer the sampling would have had to continue until the next occurrence of each digit, the gap between the last sample occurrence of each digit and the end of the sampling is not counted. The last gap category in the table shows either cumulative expected relative frequency or cumulative observed relative frequency or both equal to 1.0000. If one cumulative frequency is not equal to 1.0000, there is no purpose in pursuing the details of the probabilities of subsequent possible gaps anyway because the K-S test depends on the maximum difference between the relative frequencies, and at that point, the difference cannot increase. The maximum difference is 0.1624. The critical value of D is 0.2640 at the 0.10 level of significance.

The gap test results suggest that we are unable to reject the null hypothesis of randomness as measured by the gaps between successive occurrences of a digit.

CONCLUSION

Clearly, there are many different dimensions to the property of randomness. Some of these dimensions are probably more important to the reliability of a simulation model than are others. For example, if the random number generator cannot be depended upon to produce numbers that conform to the distribution established by goodness-of-fit tests of input data, then the model will not behave comparably with the real system in terms of its performance attributes. Likewise, it is fairly reasonable to assume that the presence of serial correlation may adversely affect the predictive power of the model by creating bunching of customers where this might not be true of the real system. It is not as clear from a pragmatic point of view that a poor showing on the runs-up-and-down test, Yule's test, or the gap test will adversely affect the predictive power of the model. If all other things are equal, since the effect of these factors is not clear, we prefer but do not insist upon a random number generator which passes these and other tests.

For a large-scale simulation model, the determining factor is often cost and efficiency. If the generator passes goodness-of-fit and serial correlation tests and is relatively inexpensive to use, it will probably be accepted.

SUMMARY

The purpose of random number generation is to approximate the occurrence of events in a real-world system when using a simulation model. Uniformly distributed random numbers on the interval from 0 to 1, known as 0−1 random numbers, are used as the source materials for random number generation. These 0 to 1 random numbers themselves should exhibit desirable properties such as efficiency, independence, long cycle length, absence of trends, and provision for restart in case of degeneration to a constant value or small range of values.

The Kolmogorov-Smirnov or χ^2 goodness-of-fit tests may be used to verify that the random numbers are uniformly distributed on the stated interval. The Spearman rank correlation test is available to check for independence, the absence of serial correlation. The runs-up-and-down test checks for an unusually large or small number of sequences of ascending or descending values in the data. Yule's test of sums of consecutive digits and the gap test of spread between successive occurrences of the same digit are used to establish the presence or absence of other kinds of randomness in the data.

Inverse transformations map the 0−1 random numbers into values of random variables with other statistical distributions. The inverse transformations are based on the cumulative probability distribution functions of the distributions from which the random variates are to be drawn. The uniform and negative exponential distributions have cumulative distribution functions which are mathematically tractable, suggesting that random variates may be obtained from them by solving equations for the value of the variable given the cumulative probability. The normal distribution must be approximated by numerical computer methods, and the Poisson distribution is most conveniently reproduced in this way also.

QUESTIONS

1. Are random numbers really random? Why or why not? Discuss.
2. What is the purpose of random number generation?
3. Why is it desirable to be able to replicate strings of random numbers produced by a random number generator?
4. Define each of the following properties of random number generators, indicate whether the property is desirable or undesirable and why, and give a brief example, when possible:
   (a) efficiency
   (b) uniformity
   (c) conformity to distribution type
   (d) serial correlation
   (e) absence of trends
   (f) short cycle length
   (g) degeneration

5. For each of the following statistical tests, give the null hypothesis, alternative hypothesis, a brief discussion of the methodology of the test, and the criterion by which the result is evaluated:
   (a) Kolmogorov-Smirnov goodness-of-fit test
   (b) Spearman rank correlation test
   (c) Number of runs-up-and-down test
   (d) Yule's test
   (e) the gap test

6. Contrast the desirable and undesirable properties of the χ^2 and Kolmogorov-Smirnov goodness-of-fit tests from the point of view of random number testing.

7. What is a nonparametric statistic? Why are nonparametric statistics suggested in this chapter?

8. What is the negative binomial distribution? To which hypothesis test is it important, and why?

9. What is a $0-1$ random number? Why is it important?

10. What is an inverse transformation? Give an example of the proper usage of an inverse transformation.

PROBLEMS

The following "random" numbers should be used to solve Problems 1 through 5. When exact tables are not available or approximations are appropriate, please use the approximations and comment on their probable reliability.

$$6, 9, 5, 8, 9, 5, 5, 5, 2, 0$$

1. Use a Kolmogorov-Smirnov goodness-of-fit test to determine whether or not the digits are uniformly distributed between 0 and 9.9, for integer values only.

*2. Use a Spearman rank correlation test to determine whether or not the data exhibit serial correlation.

3. Use the runs-up-and-down test to establish whether or not the data are random in terms of the number of runs.

*4. Use Yule's test to check whether the sums of pairs of adjacent digits seem to be random.

5. Use the gap test to find out whether the spread between successive digits is random.

6. Construct a table comparable to Table 4-1 to show the results of the inverse transformation of $0-1$ random numbers into a uniform distribution which is continuous on the interval from 200 to 400. It is not necessary to show fine details.

7. Using the formula for the cumulative negative exponential distribution function and Appendix Table V, calculate the values from a distribution with a mean of 2 minutes which would be generated by an inverse transformation of the 0 to 1 random numbers 0.0, 0.1, 0.5, and 0.9. Is the random variate corresponding to 0.5 equal to 2? Explain.

*Answers to problems preceded by * can be found in the Selected Numerical Answers.

APPENDIX

The programs in the SIMSTAT package which have been used in this chapter are DESCST and GOFFIT, previously illustrated in Chapter 2, as well as SPEAR (Spearman rank correlation), RUNUPD (runs-up-and-down), YULES (sums of consecutive digits), and GAP (spread between consecutive occurrences of a digit). While SPEAR and RUNUPD read free format data from a disk file, if desired, the positioning of numbers on a line is crucial to the proper operation of YULES and GAP. All random numbers must be left justified and of equal numbers of digits in order to use YULES and GAP.

A program called RANGEN, which includes a midproduct generator and inverse transformations for the uniform, Poisson, negative exponential, and normal distributions, has been used to provide data for many of the examples and problems in this book. While the RANGEN program is not described in the chapter, the reader may wish to experiment with it.

Table 4A-1 shows the RANGEN conversation which produced the random digits for the examples in this chapter. The resulting random digits were truncated to single-digit integers before further processing.

Table 4A-2 shows the random digits after rounding, as they appeared on the disk file. Note that there are leading blanks which must be removed by text editing prior to the use of YULES and GAP.

Table 4A-3 shows the DESCST summary statistics and sorted list of random digits. Table 4A-4 displays the Kolmogorov-Smirnov option in GOFFIT. Table 4A-5 gives the SPEAR results. Table 4A-6 shows the use of RUNUPD. Table 4A-7 covers Yule's test; note the specification of the number of digits per random number. The same requirement is present in Tables 4A-8 and 4A-9, in which Yule's test is performed with sums of three and four digits, respectively. When the number of digits in each random number exceeds 1, the digits are separated row-wise within each number first, before proceeding to reading of the next number. It is vital that the numbers do not contain signs or decimal points for this test and for the gap test.

Finally, Table 4A-10 gives the interactions with the GAP test program.

TABLE 4A-1   RANGEN CONVERSATION PRODUCING RANDOM DIGITS

RANDOM NUMBER GENERATION

DO YOU WANT INSTRUCTIONS? TYPE 1 FOR YES, OR 0 FOR NO.
• 1

THIS PROGRAM GENERATES RANDOM VARIATES
FROM YOUR CHOICE OF THE FOLLOWING DISTRIBUTIONS:
1. UNIFORM
2. POISSON

TABLE 4A-1 (cont.)

3. NEGATIVE EXPONENTIAL
4. NORMAL

IF YOU WANT TO ENTER A RANDOM NUMBER SEED, TYPE IT AS
A FOUR-DIGIT INTEGER. OTHERWISE, TYPE −1.
● −1

HOW MANY RANDOM NUMBERS DO YOU WANT?
● 25

WHICH DISTRIBUTION DO YOU WANT TO USE?
● 1

IF YOU WANT YOUR OUTPUT PRINTED IN LINES OF
10 NUMBERS EACH, TYPE A 1. IF YOU WANT YOUR OUTPUT
PRINTED ONE NUMBER PER LINE, TYPE A 0.
● 1

IF YOU WANT YOUR RANDOM NUMBERS TO PRINT IN SORTED
ORDER, TYPE A 1. OTHERWISE, TYPE A 0.
● 0

YOU HAVE CHOSEN THE UNIFORM DISTRIBUTION.
PLEASE ENTER THE LOWER LIMIT.
● 0
PLEASE ENTER THE UPPER LIMIT.
● 10

IF YOU WANT YOUR UNIFORM RANDOM NUMBERS ROUNDED
TO THE NEAREST INTEGER, TYPE A 1. OTHERWISE, TYPE A 0.
● 0

DO YOU WANT OUTPUT WRITTEN ON DISK?
IF YES, TYPE 1. IF NO, TYPE 0.
● 1

```
8.13  6.49  7.54  9.13  8.31  8.70  3.14  2.89  0.75  1.85
3.96  3.20  6.73  5.33  8.62  9.32  2.84  4.44  6.12  1.97
0.69  3.53  4.19  7.96  3.80
```

TABLE 4A-2 RANDOM DIGITS
PRODUCED BY RANGEN

1.0000	8.
2.0000	6.
3.0000	7.
4.0000	9.

TABLE 4A-2 (cont.)

5.0000	8.
6.0000	8.
7.0000	3.
8.0000	2.
9.0000	0.
10.0000	1.
11.0000	3.
12.0000	3.
13.0000	6.
14.0000	5.
15.0000	8.
16.0000	9.
17.0000	2.
18.0000	4.
19.0000	6.
20.0000	1.
21.0000	0.
22.0000	3.
23.0000	4.
24.0000	7.
25.0000	3.

TABLE 4A-3 DESCST SUMMARY STATISTICS AND SORTED LIST OF RANDOM DIGITS

DESCST DESCRIPTIVE STATISTICS

DO YOU WANT INSTRUCTIONS? TYPE 1 FOR YES, 0 FOR NO.
• 0

WHICH OPTION? TYPE 1 OR 2?
• 1

IF INPUT IS TO BE PROVIDED FROM A DISK
FILE, TYPE A 1. OTHERWISE TYPE A 0.
• 1

DO YOU WANT TO SEE THE RAW DATA?
IF YES, TYPE A 1, IF NO, TYPE A 0.
• 1

8.0	6.0	7.0	9.0	8.0	8.0	3.0	2.0	0.0	1.0
3.0	3.0	6.0	5.0	8.0	9.0	2.0	4.0	6.0	1.0
0.0	3.0	4.0	7.0	3.0					

TABLE 4A-3 (cont.)

THE MEAN IS   4.6400
THE VARIANCE IS   8.24
THE STANDARD DEVIATION IS   2.8705
THE SAMPLE SIZE IS   25

WOULD YOU LIKE TO SEE A SORTED DATA LIST?
IF YES, TYPE 1. IF NO, TYPE 0.
• 1

```
.0.0    0.0    1.0    1.0    2.0    2.0    3.0    3.0    3.0    3.0
 3.0    4.0    4.0    5.0    6.0    6.0    6.0    7.0    7.0    8.0
 8.0    8.0    8.0    9.0    9.0
```

TABLE 4A-4   KOLMOGOROV-SMIRNOV OPTION
IN GOODNESS-OF-FIT TESTS

DO YOU WANT INSTRUCTIONS? TYPE 1 FOR YES, 0 FOR NO.
• 0

TYPE THE NUMBER OF THE DISTRIBUTION YOU WANT TO TEST
• 4

YOU HAVE CHOSEN THE GOODNESS OF FIT TEST FOR:
4. UNIFORM DISTRIBUTION.

IF YOU WANT TO DO A CHI SQUARE TEST, TYPE A 1.
IF YOU WANT TO DO A KOLMOGOROV–SMIRNOV TEST, TYPE A 0.
• 0

YOU HAVE CHOSEN A KOLMOGOROV–SMIRNOV TEST.
ENTER THE NUMBER OF CLASSES
• 10

ENTER THE CLASS MIDPOINT FOR CLASS   1
• 0

ENTER THE OBSERVED FREQUENCY FOR CLASS NUMBER   1
• 2

ENTER THE CLASS MIDPOINT FOR CLASS   2
• 1

ENTER THE OBSERVED FREQUENCY FOR CLASS NUMBER   2
• 2

TABLE 4A-4 (cont.)

ENTER THE CLASS MIDPOINT FOR CLASS   3
• 2

ENTER THE OBSERVED FREQUENCY FOR CLASS NUMBER   3
• 2

ENTER THE CLASS MIDPOINT FOR CLASS   4
• 3

ENTER THE OBSERVED FREQUENCY FOR CLASS NUMBER   4
• 5

ENTER THE CLASS MIDPOINT FOR CLASS   5
• 4

ENTER THE OBSERVED FREQUENCY FOR CLASS NUMBER   5
• 2

ENTER THE CLASS MIDPOINT FOR CLASS   6
• 5

ENTER THE OBSERVED FREQUENCY FOR CLASS NUMBER   6
• 1

ENTER THE CLASS MIDPOINT FOR CLASS   7
• 6

ENTER THE OBSERVED FREQUENCY FOR CLASS NUMBER   7
• 3

ENTER THE CLASS MIDPOINT FOR CLASS   8
• 7

ENTER THE OBSERVED FREQUENCY FOR CLASS NUMBER   8
• 2

ENTER THE CLASS MIDPOINT FOR CLASS   9
• 8

ENTER THE OBSERVED FREQUENCY FOR CLASS NUMBER   9
• 4

ENTER THE CLASS MIDPOINT FOR CLASS   10
• 9

ENTER THE OBSERVED FREQUENCY FOR CLASS NUMBER   10
• 2

TABLE 4A-4 (cont.)

CLASS	CLASS IDENT	OBSERVED FREQUENCY	OBSERVED CUMULATIVE PROPORTION	THEORETICAL CUMULATIVE PROPORTION	ABSOLUTE DIFFERENCE
1	0.0	2.	0.0800	0.1000	0.0200
2	1.0	2.	0.1600	0.2000	0.0400
3	2.0	2.	0.2400	0.3000	0.0600
4	3.0	5.	0.4400	0.4000	0.0400
5	4.0	2.	0.5200	0.5000	0.0200
6	5.0	1.	0.5600	0.6000	0.0400
7	6.0	3.	0.6800	0.7000	0.0200
8	7.0	2.	0.7600	0.8000	0.0400
9	8.0	4.	0.9200	0.9000	0.0200
10	9.0	2.	1.0000	1.0000	0.0000
TOTAL		25.			

THE MAXIMUM SAMPLE DIFFERENCE IS   0.0600

THE CRITICAL VALUE OF D IS   0.2640 AT THE .05 LEVEL OF SIGNIFICANCE.

IF YOU WANT TO SOLVE ANOTHER PROBLEM, TYPE A 1. OTHERWISE, TYPE A 0.
● 0

TABLE 4A-5   SPEAR TEST AND RESULTS

SPEARMAN RANK CORRELATION

DO YOU WANT INSTRUCTIONS? TYPE 1 FOR YES, 0 FOR NO
● 1

THIS PROGRAM TESTS A SET OF DATA FOR AUTOCORRELATION USING THE
SPEARMAN RANK CORRELATION COEFFICIENT. IT ALSO GIVES A Z
STATISTIC FOR THE SIGNIFICANCE OF R.
THE SAMPLE SIZE SHOULD EXCEED 30.

IF INPUT IS FROM A DISK FILE, TYPE A 1.
IF DATA ARE ENTERED DIRECTLY, TYPE A 0.
● 1

IF YOU WANT TO SEE THE INDIVIDUAL DATA ITEMS,
TYPE A 1. OTHERWISE, TYPE A 0.
● 1

TABLE 4A-5 (cont.)

Y VALUE	X VALUE	Y RANK	X RANK	D	D2
8.00	6.00	20.50	16.00	4.50	20.25
6.00	7.00	15.00	18.50	−3.50	12.25
7.00	9.00	17.50	23.50	−6.00	36.00
9.00	8.00	23.50	21.00	2.50	6.25
8.00	8.00	20.50	21.00	−0.50	0.25
8.00	3.00	20.50	9.00	11.50	132.25
3.00	2.00	8.50	5.50	3.00	9.00
2.00	0.00	5.50	1.50	4.00	16.00
0.00	1.00	1.50	3.50	−2.00	4.00
1.00	3.00	3.50	9.00	−5.50	30.25
3.00	3.00	8.50	9.00	−0.50	0.25
3.00	6.00	8.50	16.00	−7.50	56.25
6.00	5.00	15.00	14.00	1.00	1.00
5.00	8.00	13.00	21.00	−8.00	64.00
8.00	9.00	20.50	23.50	−3.00	9.00
9.00	2.00	23.50	5.50	18.00	324.00
2.00	4.00	5.50	12.50	−7.00	49.00
4.00	6.00	11.50	16.00	−4.50	20.25
6.00	1.00	15.00	3.50	11.50	132.25
1.00	0.00	3.50	1.50	2.00	4.00
0.00	3.00	1.50	9.00	−7.50	56.25
3.00	4.00	8.50	12.50	−4.00	16.00
4.00	7.00	11.50	18.50	−7.00	49.00
7.00	3.00	17.50	9.00	8.50	72.25

THE ORIGINAL SAMPLE SIZE IS   25

THE SUM OF D SQUARED IS   1120.00

SUMU=   15.00
SUMV=   17.00

R=   0.51

Z=   2.43

DO YOU WANT TO SOLVE ANOTHER PROBLEM?
IF YES, TYPE 1. IF NO, TYPE 0.
• 1

TABLE 4A-6   THE RUNS-UP-AND-DOWN TEST
AND RESULTS

IF INPUT IS TO COME FROM A DISK FILE, TYPE A 1.
OTHERWISE, TYPE A 0.
• 1

THE ORIGINAL SAMPLE SIZE IS   25

THE EFFECTIVE SAMPLE SIZE IS   23

THE NUMBER OF RUNS IS   11
THE EXPECTED NUMBER OF RUNS IS   15.00
THE STANDARD DEVIATION IS   1.9408
THE VALUE OF Z IS   −1.8034

TABLE 4A-7   YULE'S TEST OF SUMS OF CONSECUTIVE DIGITS AND RESULTS*

HOW MANY RANDOM NUMBERS ARE THERE?
• 25

HOW MANY DIGITS ARE IN EACH NUMBER? YOU MAY CHOOSE 1, 2, 3 OR 4.
• 1

IF INPUT IS FROM DISK, TYPE A 1. IF INPUT IS FROM A TERMINAL, TYPE A 0.
• 1

HOW MANY DIGITS ARE USED FOR EACH SUM? YOU MAY CHOOSE 2, 3, OR 4.
• 2

SUM	OBSERVED RELATIVE FREQUENCY	EXPECTED RELATIVE FREQUENCY	CUMULATIVE OBSERVED RELATIVE FREQUENCY	CUMULATIVE EXPECTED RELATIVE FREQUENCY	DIFFERENCE
0	0.0000	0.0100	0.0000	0.0100	0.0100
1	0.0833	0.0200	0.0833	0.0300	0.0533
2	0.0000	0.0300	0.0833	0.0600	0.0233
3	0.0833	0.0400	0.1667	0.1000	0.0667
4	0.0000	0.0500	0.1667	0.1500	0.0167
5	0.0833	0.0600	0.2500	0.2100	0.0400
6	0.1667	0.0700	0.4167	0.2800	0.1367

TABLE 4A-7 (cont.)

7	0.0833	0.0800	0.5000	0.3600	0.1400
8	0.0000	0.0900	0.5000	0.4500	0.0500
9	0.0000	0.1000	0.5000	0.5500	0.0500
10	0.0000	0.0900	0.5000	0.6400	0.1400
11	0.1667	0.0800	0.6667	0.7200	0.0533
12	0.0000	0.0700	0.6667	0.7900	0.1233
13	0.0000	0.0600	0.6667	0.8500	0.1833
14	0.0833	0.0500	0.7500	0.9000	0.1500
15	0.0000	0.0400	0.7500	0.9400	0.1900
16	0.1667	0.0300	0.9167	0.9700	0.0533
17	0.0833	0.0200	1.0000	0.9900	0.0100
18	0.0000	0.0100	1.0000	1.0000	0.0000

THE MAXIMUM DIFFERENCE IS   0.1900

THE CRITICAL VALUE OF D IS   0.3750 AT THE .05 LEVEL OF SIGNIFICANCE

*A Kolmogorov-Smirnov Goodness-of-Fit-Test is used.

TABLE 4A-8   YULE'S TEST OF SUMS OF CONSECUTIVE DIGITS AND RESULTS*

HOW MANY RANDOM NUMBERS ARE THERE?
- 25

HOW MANY DIGITS ARE IN EACH NUMBER? YOU MAY CHOOSE 1, 2, 3 OR 4.
- 1

IF INPUT IS FROM DISK, TYPE A 1. IF INPUT IS FROM A TERMINAL, TYPE A 0.
- 1

HOW MANY DIGITS ARE USED FOR EACH SUM? YOU MAY CHOOSE 2, 3 OR 4.
- 3

SUM	OBSERVED RELATIVE FREQUENCY	EXPECTED RELATIVE FREQUENCY	CUMULATIVE OBSERVED RELATIVE FREQUENCY	CUMULATIVE EXPECTED RELATIVE FREQUENCY	DIFFERENCE
0	0.0000	0.0010	0.0000	0.0010	0.0010
1	0.0000	0.0030	0.0000	0.0040	0.0040
2	0.0000	0.0060	0.0000	0.0100	0.0100
3	0.0000	0.0100	0.0000	0.0200	0.0200
4	0.0000	0.0150	0.0000	0.0350	0.0350
5	0.1250	0.0210	0.1250	0.0560	0.0690
6	0.0000	0.0280	0.1250	0.0840	0.0410
7	0.2500	0.0360	0.3750	0.1200	0.2550
8	0.0000	0.0450	0.3750	0.1650	0.2100

TABLE 4A-8 (cont.)

9	0.0000	0.0550	0.3750	0.2200	0.1550
10	0.0000	0.0630	0.3750	0.2830	0.0920
11	0.0000	0.0690	0.3750	0.3520	0.0230
12	0.0000	0.0730	0.3750	0.4250	0.0500
13	0.0000	0.0750	0.3750	0.5000	0.1250
14	0.1250	0.0750	0.5000	0.5750	0.0750
15	0.1250	0.0730	0.6250	0.6480	0.0230
16	0.0000	0.0690	0.6250	0.7170	0.0920
17	0.0000	0.0630	0.6250	0.7800	0.1550
18	0.0000	0.0550	0.6250	0.8350	0.2100
19	0.1250	0.0450	0.7500	0.8800	0.1300
20	0.0000	0.0360	0.7500	0.9160	0.1660
21	0.1250	0.0280	0.8750	0.9440	0.0690
22	0.0000	0.0210	0.8750	0.9650	0.0900
23	0.0000	0.0150	0.8750	0.9800	0.1050
24	0.0000	0.0100	0.8750	0.9900	0.1150
25	0.1250	0.0060	1.0000	0.9960	0.0040
26	0.0000	0.0030	1.0000	0.9990	0.0010
27	0.0000	0.0010	1.0000	1.0000	0.0000

THE MAXIMUM DIFFERENCE IS   0.2550

THE CRITICAL VALUE OF D IS   0.4570 AT THE .05 LEVEL OF SIGNIFICANCE

*A Kolmogorov-Smirnov Goodness-of-Fit-Test is used.

TABLE 4A-9   YULE'S TEST OF SUMS OF CONSECUTIVE DIGITS AND RESULTS*

HOW MANY RANDOM NUMBERS ARE THERE?
- 25

HOW MANY DIGITS ARE IN EACH NUMBER? YOU MAY CHOOSE 1, 2, 3 OR 4.
- 1

IF INPUT IS FROM DISK, TYPE A 1. IF INPUT IS FROM A TERMINAL, TYPE A 0.
- 1

HOW MANY DIGITS ARE USED FOR EACH SUM? YOU MAY CHOOSE 2, 3 OR 4.
- 4

SUM	OBSERVED RELATIVE FREQUENCY	EXPECTED RELATIVE FREQUENCY	CUMULATIVE OBSERVED RELATIVE FREQUENCY	CUMULATIVE EXPECTED RELATIVE FREQUENCY	DIFFERENCE
0	0.0000	0.0001	0.0000	0.0001	0.0001
1	0.0000	0.0004	0.0000	0.0005	0.0005
2	0.0000	0.0010	0.0000	0.0015	0.0015

TABLE 4A-9 (cont.)

3	0.0000	0.0020	0.0000	0.0035	0.0035
4	0.0000	0.0035	0.0000	0.0070	0.0070
5	0.0000	0.0056	0.0000	0.0126	0.0126
6	0.0000	0.0084	0.0000	0.0210	0.0210
7	0.1667	0.0120	0.1667	0.0330	0.1337
8	0.0000	0.0165	0.1667	0.0495	0.1172
9	0.0000	0.0220	0.1667	0.0715	0.0952
10	0.0000	0.0282	0.1667	0.0997	0.0670
11	0.0000	0.0348	0.1667	0.1345	0.0322
12	0.0000	0.0415	0.1667	0.1760	0.0093
13	0.1667	0.0480	0.3333	0.2240	0.1093
14	0.1667	0.0540	0.5000	0.2780	0.2220
15	0.0000	0.0592	0.5000	0.3372	0.1628
16	0.0000	0.0633	0.5000	0.4005	0.0995
17	0.0000	0.0660	0.5000	0.4665	0.0335
18	0.0000	0.0670	0.5000	0.5335	0.0335
19	0.0000	0.0660	0.5000	0.5995	0.0995
20	0.0000	0.0633	0.5000	0.6628	0.1628
21	0.1667	0.0592	0.6667	0.7220	0.0553
22	0.0000	0.0540	0.6667	0.7760	0.1093
23	0.0000	0.0480	0.6667	0.8240	0.1573
24	0.0000	0.0415	0.6667	0.8655	0.1988
25	0.0000	0.0348	0.6667	0.9003	0.2336
26	0.0000	0.0282	0.6667	0.9285	0.2618
27	0.0000	0.0220	0.6667	0.9505	0.2838
28	0.1667	0.0165	0.8333	0.9670	0.1337
29	0.0000	0.0120	0.8333	0.9790	0.1457
30	0.1667	0.0084	1.0000	0.9874	0.0126
31	0.0000	0.0056	1.0000	0.9930	0.0070
32	0.0000	0.0035	1.0000	0.9965	0.0035
33	0.0000	0.0020	1.0000	0.9985	0.0015
34	0.0000	0.0010	1.0000	0.9995	0.0005
35	0.0000	0.0004	1.0000	0.9999	0.0001
36	0.0000	0.0001	1.0000	1.0000	0.0000

THE MAXIMUM DIFFERENCE IS   0.2838

THE CRITICAL VALUE OF D IS   0.5210 AT THE .05 LEVEL OF SIGNIFICANCE

*A Kolmogorov-Smirnov Goodness-of-Fit-Test is used.

TABLE 4A-10   THE GAP TEST OF SPREAD BETWEEN CONSECUTIVE DIGITS AND RESULTS*

HOW MANY RANDOM NUMBERS ARE THERE?
● 25

HOW MANY DIGITS ARE IN EACH NUMBER? YOU MAY CHOOSE 1, 2, 3, 4 OR 5.
● 1

IF INPUT IS FROM DISK, TYPE A 1. IF INPUT IS FROM A TERMINAL, TYPE A 0.
● 1

IF YOU WANT TO SEE THE DETAILS OF THE TEST, TYPE A 1. OTHERWISE, TYPE A 0.
● 1

GAP	OBSERVED FREQUENCY	CUM. OBS. REL. FREQ.	NEGATIVE BIN. PROB.	EXPECTED FREQUENCY	CUM. EXP. REL. FREQ.	ABSOLUTE DIFFERENCE
1	3	0.1200	0.1000	2.50	0.1000	0.0200
2	1	0.1600	0.0900	2.25	0.1900	0.0300
3	2	0.2400	0.0810	2.02	0.2710	0.0310
4	3	0.3600	0.0729	1.82	0.3439	0.0161
5	1	0.4000	0.0656	1.64	0.4095	0.0095
6	1	0.4400	0.0590	1.48	0.4686	0.0286
7	1	0.4800	0.0531	1.33	0.5217	0.0417
8	1	0.5200	0.0478	1.20	0.5695	0.0495
9	3	0.6400	0.0430	1.08	0.6126	0.0274
10	3	0.7600	0.0387	0.97	0.6513	0.1087
11	1	0.8000	0.0349	0.87	0.6862	0.1138
12	2	0.8800	0.0314	0.78	0.7176	0.1624
13	0	0.8800	0.0282	0.71	0.7458	0.1342
14	1	0.9200	0.0254	0.64	0.7712	0.1488
15	0	0.9200	0.0229	0.57	0.7941	0.1259
16	0	0.9200	0.0206	0.51	0.8147	0.1053
17	0	0.9200	0.0185	0.46	0.8332	0.0868
18	1	0.9600	0.0167	0.42	0.8499	0.1101
19	0	0.9600	0.0150	0.38	0.8649	0.0951
20	0	0.9600	0.0135	0.34	0.8784	0.0816
21	1	1.0000	0.0122	0.30	0.8906	0.1094

THE LAST GAP CATEGORY CONTAINS GREATER-THAN-OR-EQUALS INFORMATION.
FREQUENCIES MAY NOT ACCUMULATE TO 1.0000 FOR THAT CLASS, HOWEVER.

THE MAXIMUM DIFFERENCE IS   0.1624

THE CRITICAL VALUE OF D IS   0.2640 AT THE .10 LEVEL OF SIGNIFICANCE

*A Kolmogorov-Smirnov Goodness-of-Fit-Test is used.

chapter five

INTRODUCTION TO GPSS

COMPARISON WITH OTHER LANGUAGES

Although the GPSS language is expressly designed for simulation, it shares many features of high-level, general-purpose programming languages. There are mechanisms for initializing constants and producing values for functions and variables. There are conditional and unconditional branching statements. There are methods of printing the results of the simulation run and indicating when it is to terminate. GPSS programs may be divided into definitional or "housekeeping" sections, sections which specify processing to occur only at the beginning or at the end of the simulation run, as well as routine, "loop"-type processing. An individual who is conversant with any high-level programming language will find it relatively easy to learn the grammatical aspects of GPSS, since everyone is familiar with the particular context of waiting-line systems which the program will eventually describe. As with any high-level language, GPSS statements or "blocks" must be compiled or translated into numerous assembly or machine language instructions before the program can be executed or run.

Certain elements which are common to all simulation languages are found in GPSS. There is an internal simulator clock which monitors the length of the run in units which the modeler has defined, and which schedules the occurrence of events such as completion of service. There is a mechanism for generating temporary entities or transactions and also a mechanism for removing them from the system when their processing has been completed. There are ways of ascribing attributes or parameters to transactions which identify them as similar to or different from other transactions. Permanent system entities which service the temporary entities can also be defined easily. There are routines which print the summarized results of the simulation run, with values of generally useful performance parameters such as transit time and queue length; there are also provisions for printing nonstandard items and placing output on magnetic tape for certain purposes.

GPSS is a very powerful and comprehensive language. In this chapter and the two chapters that follow, we will attempt to cover those features of GPSS which are most often provided in compilers and are most often used. GPSS action statements are called blocks. There are numerous blocks and special features of blocks which are beyond the scope of this book. The reader is referred to the user's manuals prepared by computer system vendors or to the several excellent books devoted exclusively to GPSS (see the bibliography). The reader is also cautioned that GPSS is not an ANSI (American National Standards Institute) standard language. We will use a version of GPSS which is format-free; many versions are restricted in formatting. Certain blocks are available in some versions of the language and not in others. Some block names differ from version to version, and some features of certain blocks are not available in all versions.

THE TICKET SELLER IN GPSS

A GPSS program which will simulate the ticket seller system until 100 customers have completed service is shown in Table 5-1. To simplify the program temporarily, interarrival and service times have been considered to be uniformly distributed rather than negative exponentially distributed. The mean and the modifier, which is half the width of the uniform distribution centered on the mean, have been made equal to each other to approximate the negative exponential distribution as closely as possible.

TABLE 5-1 GPSS MODEL OF TICKET SELLER SYSTEM WITH UNIFORMLY DISTRIBUTED INTERARRIVAL AND SERVICE TIMES

```
JOB
GENERATE 101, 101
QUEUE TKTQ
SEIZE TKTF
DEPART TKTQ
ADVANCE 84, 84
RELEASE TKTF
TERMINATE 1
START 100
END
```

There are three "housekeeping" statements in the GPSS program. The JOB statement defines the beginning of a new GPSS program, and the END statement indicates the end of GPSS compilation. The START block is always the last entry in a model and in this case tells us that the model is to run until 100 customers have completed processing. The rest of the program, from the GENERATE block through the TERMINATE block, constitutes a repetitive sequence through which all customers (transactions) will pass.

The GENERATE block creates transactions in the model. The first value shown is the mean interarrival time. Since time in GPSS is integral, when the mean time has a decimal component, the mean must be rounded to the appropriate integer value. When there is no second value, interarrival time is assumed to be constant at this first value. When the second value is an integer, as in this case, the distribution of interarrival time is assumed to be uniform, centered on the mean value, with a modifier or spread on either side of the mean equal to the second value. In GPSS we refer to the possible fields after the block name with the letters ''A,'' ''B,'' ''C,'' and so forth. Our GENERATE block has an A-field value of 101 (the mean) and a B-field value of 101 also (the modifier). Thus, the range of the uniform distribution is from 0 to 202 seconds interarrival time.

The QUEUE block marks a point at which it is *possible* but not inevitable that the transaction will have to wait for a server to become available before it can be processed. The QUEUE block and its companion, the DEPART block, often precede the blocks which define service facilities. The reason for the pairing of blocks in GPSS is to permit activities to take place between the commencement and termination of some phase. Thus, we may model other activities of the transaction which are carried out while the transaction is waiting for service; these other activities would be described in steps intervening between the QUEUE and DEPART blocks. Each QUEUE block and its associated DEPART block are identified; depending on the version of GPSS, this identifier, which appears in the A-field, may be numeric, alphanumeric, or a choice of either numeric or alphanumeric. The QUEUE and DEPART blocks define waiting lines which have first-come, first-served queue discipline, with no balking, reneging, or jockeying permitted. As we shall see in Chapter 7, there are more complex modeling tools available for use when these restrictions are not valid.

Names in GPSS are restricted to three to six characters in length; the first three characters must be alphabetic. It is wise to make sure that any names chosen do not bear a close resemblance to GPSS block names and other standard GPSS features to avoid compiler misinterpretations.

The SEIZE block, which appears between the QUEUE block and the DEPART block in the example, is the point at which the transaction is able to gain control of the server. The transaction cannot exit the QUEUE block until the next sequential block, the SEIZE block, is ready to accept it. It is important to realize that a queue would build just prior to the SEIZE block whether or not the QUEUE and DEPART blocks were in place. The QUEUE and DEPART blocks simply offer a vehicle for capturing statistics about this queue. The other portion of the pair of blocks delineating usage of the server is the RELEASE block. When the transaction passes through the RELEASE block, the service facility is made available to the next transaction. Before this can occur, however, the current transaction must spend an amount of time defined by a random draw from the distribution specified in the ADVANCE block. The service time distribution is stipulated in much the same fashion as was interarrival time in the GENERATE block. A uniform distribution of service time is specified, with a mean of 84 seconds and a range of 84 + or − 84, or from 0 to 168 seconds.

After service is completed, the transaction passes through the RELEASE block and enters the TERMINATE block. This discontinues the passage of the transaction through the system. The "1" in the A-field of the TERMINATE block indicates that the number of transactions left to be completed before the run is to end is decremented by 1 each time a transaction passes this block. Since the A-field in the START statement gives the run length as 100 transactions, there will be 99 terminations to go after the first transaction reaches the TERMINATE block and before the run ends.

To summarize the passage of a typical transaction through the model, we might develop the following narrative. The transaction follows its predecessor (or the initiation of the run) by an interarrival time drawn at random from a uniform distribution with a mean of 101 seconds and a range from 0 to 202 seconds. If service facility 1 is not busy, the transaction will pass through the staging area for queue TKTQ, perhaps spending some time waiting, and seize control of facility TKTF. If facility TKTF is busy at the time the transaction is generated, the transaction will not leave the QUEUE block, seize the facility, and depart its place at the top of the queue until the facility is released by the previous transaction. The transaction then spends an amount of time being served, which is defined by a random draw from a uniform distribution with a mean of 84 and a range from 0 to 168 seconds. When this amount of time has elapsed, the transaction releases the facility and is terminated from the model, decrementing the START count by 1. The procedure continues until the START count has been decremented to 0 (i.e., 100 customers have completed processing).

OUTPUT OF THE TICKET SELLER PROGRAM

Now let us examine the printed output which is routinely produced by the GPSS compiler without any specific instructions from the user. Table 5-2 displays the output of the ticket seller model after a run of 100 transactions.

TABLE 5-2 OUTPUT OF THE GPSS PROGRAM FOR THE TICKET SELLER SYSTEM WITH UNIFORMLY DISTRIBUTED INTERARRIVAL AND SERVICE TIMES

BLOCK NUMBER	*LOC	NAME	A, B, C, D, E, F, G	COMMENTS	CARD
		JOB			00001
00001		GENERATE 101, 101			00002
00002		QUEUE TKTQ			00003
00003		SEIZE TKTF			00004

TABLE 5-2 (cont.)

00004	DEPART TKTQ	00005
00005	ADVANCE 84, 84	00006
00006	RELEASE TKTF	00007
00007	TERMINATE 1	00008
	START 100	00009
	END	00010

RELATIVE CLOCK TIME: 10737
ABSOLUTE CLOCK TIME: 10737

BLOCK COUNTS

BLOCK	CURR.	TOTAL	BLOCK	CURR.	TOTAL	BLOCK	CURR.	TOTAL
1	0	104	2	4	100	3	0	100
4	0	100	5	0	100	6	0	100
7	0	100						

FACILITY ID	AVERAGE UTILIZATION	NUMBER ENTRIES	AVERAGE TIME/TRANS	SEIZING TRNSACTION	PREEMPTING TRNSACTION
TKTF	0.852	100	91.470	0	0

QUEUE ID	MAXIMUM CONTENTS	AVERAGE CONTENTS	TOTAL ENTRIES	ZERO ENTRIES	% −ZERO ENTRIES	AVERAGE TIME/TRANS	NZ–AV TM/TN	CURR. CONT.
TKTQ	4	0.698	104	23	22.11	72.09	92.56	4

THERE ARE NO INTERNAL TABLES

The GPSS output numbers the statements in the GPSS program in two different ways. On the far left, a block number is assigned *only* to processing blocks, not to housekeeping or descriptive blocks; thus, there are no block numbers for the JOB, START, or END statements. However, on the far right in the column labeled ''CARD'' (since this compiler has existed since the days of exclusively punched card processing), a consecutive sequence number is assigned to every line in the GPSS program. A listing of the source program itself is provided. Since there were no compilation or execution errors,* the run concludes normally in 10737 simulated seconds. We will temporarily defer the distinction between relative and absolute clock time, which are identical for the program being analyzed.

    The next portion of output gives the number of transactions which were located at each processing block at the time of completion of the run, as well as the total number of transactions which passed through the block previously during the

---

*Some GPSS compilers will not proceed from compilation to execution unless a SIMULATE statement is included in the program. The reader should check the user's manual for the version of GPSS available to him or her before attempting to run a program.

course of the run. All blocks are empty at the conclusion of the run except for the QUEUE block, which shows 4 customers waiting for service. Since it appears that block 5, the ADVANCE block, has become free, if the run had continued, the first of those 4 customers would immediately have been accepted for service. Note that it was necessary to generate 104 customers to obtain 100 terminations because of the delays in processing.

Directly below the block counts are summary statistics on utilization of the service facility. Facility 1, as we have elected to call it, was busy 85.2 percent of the time. A total of 100 customers made use of the service facility during the course of the run, with an average service time of 91.470 seconds. At the time the run ended, no transactions were using the facility, although the next customer's arrival was imminent.

The last set of summary statistics for our GPSS program refers to the operation of the waiting line consisting of customers desiring to use facility TKTF. We have called this queue TKTQ. There were at most 4 customers in the queue, and in fact, there were 4 customers waiting at the time the run concluded. The average number of customers in the waiting line was 0.698. A total of 104 customers passed through the waiting area, and of those, 23 spent no time at all waiting. Those 23 represent 22.11 percent of all those who passed through the waiting area. The average customer spent 72.087 seconds waiting in queue TKTQ; for those who spent a nonzero amount of time waiting, the average wait was 92.56 seconds.

CHECKING OUTPUT FOR ERRORS

It is possible to detect certain types of errors in the simulation by examining the GPSS output carefully. One possible error is suggested by the discrepancy between the mean service time requested (84 seconds) and the "observed" mean service time for the sample of 100 simulated customers (91.47 seconds). While this might indicate an error, more likely the sample size was too small for the true mean and the sample mean to coincide closely.

A common error is to omit one of the several pairs of blocks which are required for queuing and service. If a QUEUE block is used without a DEPART block, transactions will be unable to leave the queue, and the run will end either because there were too many transactions in process or because the user-specified task time expired. If a DEPART block is used without a QUEUE block, an error message will appear to indicate that a transaction is trying to leave a queue which it never joined. If a SEIZE block is used without a RELEASE block, the facility will be permanently occupied and the run will end for the same reasons as when the DEPART block is missing. If a RELEASE block is used without a SEIZE block, a message comparable to that for the solo DEPART block will appear.

Block counts offer clues to each of these problems as well. When transaction processing is impeded by the absence of a DEPART or RELEASE block, there will be many more transactions currently residing at the QUEUE block than there are at other blocks in the model. Of course, large queues may occur by chance also.

If there is no TERMINATE block or if there is no nonzero value in the A-field of the TERMINATE block, the decrementing of the START count cannot be accomplished, and either the run will terminate in an assembly error or the run will continue until the task time allotted by the operating system has elapsed and the job is canceled. In this case there will be little or no output from the run.

USING SUCCESSIVE FACILITIES

Suppose that every ticket customer subsequently buys a program from a server other than the ticket seller. The time required to sell a program is uniformly distributed with a mean of 50 seconds and a modifier of 30 seconds; that is, possible program purchase times range from 20 to 80 seconds, with all values in this range equally likely to occur. Table 5-3 shows the GPSS program for this modified ticket seller system.

TABLE 5-3 GPSS MODEL OF TICKET SELLER SYSTEM WITH UNIFORMLY DISTRIBUTED INTERARRIVAL TIMES AND SERVICE TIMES FOR TICKET SALES AND PROGRAM SALES

```
JOB
GENERATE 101,101
QUEUE TKTQ
SEIZE TKTF
DEPART TKTQ
ADVANCE 84,84
RELEASE TKTF
QUEUE PGMQ
SEIZE PGMF
DEPART PGMQ
ADVANCE 50,30
RELEASE PGMF
TERMINATE 1
START 100
END
```

Output for this program is comparable to that in Table 5-2, except that there are more blocks, more block counts, and queue and facility statistics for two independent servers instead of one. The output is shown in Table 5-4.

TABLE 5-4   OUTPUT OF THE GPSS PROGRAM FOR THE TICKET SELLER SYSTEM WITH UNIFORMLY DISTRIBUTED INTERARRIVAL AND SERVICE TIMES FOR TICKET SALES AND PROGRAM SALES

BLOCK NUMBER	*LOC	NAME	A, B, C, D, E, F, G	COMMENTS	CARD
		JOB			00001
00001		GENERATE 101, 101			00002
00002		QUEUE TKTQ			00003
00003		SEIZE TKTF			00004
00004		DEPART TKTQ			00005
00005		ADVANCE 84, 84			00006
00006		RELEASE TKTF			00007
00007		QUEUE PGMQ			00008
00008		SEIZE PGMF			00009
00009		DEPART PGMQ			00010
00010		ADVANCE 50, 30			00011
00011		RELEASE PGMF			00012
00012		TERMINATE 1			00013
		START 100			00014
		END			00015

RELATIVE CLOCK TIME: 10992
ABSOLUTE CLOCK TIME: 10992

BLOCK COUNTS

BLOCK	CURR.	TOTAL	BLOCK	CURR.	TOTAL	BLOCK	CURR.	TOTAL
1	0	108	2	7	101	3	0	101
4	0	101	5	1	100	6	0	100
7	0	100	8	0	100	9	0	100
10	0	100	11	0	100	12	0	100

FACILITY ID	AVERAGE UTILIZATION	NUMBER ENTRIES	AVERAGE TIME/TRANS	SEIZING TRNSACTION	PREEMPTING TRNSACTION
TKTF	0.847	101	92.198	101	0
PGMF	0.415	100	45.590	0	0

QUEUE ID	MAXIMUM CONTENTS	AVERAGE CONTENTS	TOTAL ENTRIES	ZERO ENTRIES	% – ZERO ENTRIES	AVERAGE TIME/TRANS	NZ–AV TM/TN	CURR. CONT.
TKTQ	8	2.202	108	24	22.22	224.13	288.17	7
PGMQ	1	0.046	100	81	81.00	5.03	26.47	0

THERE ARE NO INTERNAL TABLES

USING MULTISERVER FACILITIES

Suppose that the ticket booth is staffed by two equally efficient servers who occupy a common ticket counter and deal with a common queue. The GPSS program to display this alternative is shown in Table 5-5.

TABLE 5-5 GPSS MODEL OF TICKET SELLER SYSTEM WITH UNIFORMLY DISTRIBUTED INTERARRIVAL AND SERVICE TIMES AND TWO EQUIVALENT SERVERS SHARING A COMMON QUEUE

```
        JOB
        GENERATE 101, 101
        QUEUE TKTQ
        ENTER TKTS
        DEPART TKTQ
        ADVANCE 84, 84
        LEAVE TKTS
        TERMINATE 1
TKTS STORAGE 2
        START 100
        END
```

A multiserver facility in which the servers draw customers from a common queue and in which the servers are virtually identical in performance capability is called a "STORAGE." The paired statements for gaining and relinquishing control of a storage are ENTER and LEAVE rather than SEIZE and RELEASE. In addition, a statement must be included in the housekeeping section below the TERMINATE block which indicates the storage name or number and the number of servers in the storage.

Note the character string TKTS beginning in column 1 of the STORAGE statement. It is a mnemonic label identifying the storage. The "2" to the right of the STORAGE statement indicates that the storage is to have two servers. The name or number in the A-Field of the ENTER and LEAVE blocks must correspond to each other and also to the label in the STORAGE statement. The output is shown in Table 5-6.

Note that the categories of summary statistics for a storage are almost identical with those for a facility. There is a field for the maximum contents of the storage during the run—a number which could range from zero to the storage capacity. Average utilization is equal to average contents of the storage divided by the storage capacity.

TABLE 5-6 OUTPUT OF THE GPSS PROGRAM FOR THE TICKET SELLER SYSTEM WITH UNIFORMLY DISTRIBUTED INTERARRIVAL AND SERVICE TIMES AND TWO EQUIVALENT SERVERS

BLOCK NUMBER	*LOC	NAME	A, B, C, D, E, F, G	COMMENTS	CARD
		JOB			00001
00001		GENERATE 101, 101			00002
00002		QUEUE TKTQ			00003
00003		ENTER TKTS			00004
00004		DEPART TKTQ			00005
00005		ADVANCE 84, 84			00006
00006		LEAVE TKTS			00007
00007		TERMINATE 1			00008
		TKTS STORAGE 2			00009
		START 100			00010
		END			00011

RELATIVE CLOCK TIME: 10418
ABSOLUTE CLOCK TIME: 10418

BLOCK COUNTS

BLOCK	CURR.	TOTAL	BLOCK	CURR.	TOTAL	BLOCK	CURR.	TOTAL
1	0	103	2	2	101	3	0	101
4	0	101	5	1	100	6	0	100
7	0	100						

STORAGE ID	CAPACITY	AVERAGE CONTENTS	AVERAGE UTILIZATION	NUMBER ENTRIES	AVERAGE TIME/TRANS	CURR. CONTS.	MAX. CONTS.
TKTS	2	0.834	0.417	101	86.000	1	2

QUEUE ID	MAXIMUM CONTENTS	AVERAGE CONTENTS	TOTAL ENTRIES	ZERO ENTRIES	% −ZERO ENTRIES	AVERAGE TIME/TRANS	NZ-AV TM/TN	CURR. CONT.
TKTQ	2	0.006	103	95	92.23	0.60	7.75	2

THERE ARE NO INTERNAL TABLES

BRANCHING STATEMENTS

All the models you have seen so far have been straight-line programs with no branching. In a sense, these programs include a "loop," since the series of instructions is repeated for the number of transactions specified in the START statement. However, no deviation from the routine processing path is permitted for any transaction.

We now consider an entity common to every programming language, the unconditional branching statement and a simple conditional branching statement. The branching statement we will use is called "TRANSFER."

TABLE 5-7   GPSS MODEL OF TICKET SELLER SYSTEM
WITH UNIFORMLY DISTRIBUTED INTERARRIVAL AND SERVICE TIMES
AND PRIORITY QUEUE DISCIPLINE FOR ELDERLY CUSTOMERS

```
          JOB
          GENERATE 101, 101
          TRANSFER . 900, ELD, REG
     ELD PRIORITY 1
          QUEUE TKTQ
          SEIZE TKTF
          DEPART TKTQ
          ADVANCE 84, 84
          RELEASE TKTF
          TRANSFER , TERM1
     REG QUEUE TKTQ
          SEIZE TKTF
          DEPART TKTQ
          ADVANCE 84, 84
          RELEASE TKTF
     TERM1 TERMINATE 1
          START 100
          END
```

The program shown in Table 5-7 models a case in which the ticket seller is especially fond of elderly people. When one or more elderly people are in the queue, they advance to the head of the queue in FIFO order within their own set of elderly. They do not, however, interrupt the service of anyone who may be currently purchasing a ticket, whether or not the purchaser is elderly. The PRIORITY block is the mechanism for ensuring that certain types of transactions will move ahead of other types of transactions in a waiting line. All transactions are "born" with priority of 0.* If transactions do not pass through a PRIORITY block containing a value from 1 to 127, their priority remains at 0. The larger the priority number, the higher the priority of the transaction.

There is a single GENERATE block whose interarrival time specifications are for *all* customers, not just those who are or are not elderly. This eliminates the necessity of recording separate sets of interarrival time data for different classes of customers. We have presumably indicated on the data collection sheet which customers appeared to be elderly and received priority service; this system could be formalized by posting a notice that those with a Medicare card should come to the head of the waiting line. It was found that 90 percent of the customers do not fall in the

---

*There are several fields in the GENERATE block which we have not discussed. One of these fields may be used to set the initial priority of transactions at a value other than 0.

elderly category. The TRANSFER statement will distribute 90 percent of all customers to the program segment beginning with the label REG and the remaining 10 percent of the customers to the program segment with the label ELD (for elderly). The grammar of the TRANSFER statement requires that the proportion of customers to be diverted to the block label found in field C must be specified as a three-digit decimal number in field A.

The only difference between the two program segments for regular and elderly customers is the presence of the PRIORITY block which is labeled ELD and precedes entry into the queue. This will presumably cause the transit time of elderly customers to be shorter, on the average, than that of regular customers, even though they face equivalent service time distributions.

While the conditional TRANSFER block splits the original transaction stream into two parts, the unconditional TRANSFER block causes them to be reunited, terminating together at the block labeled TERM1. Table 5-8 shows the GPSS compilation and output for the GPSS model given in Table 5-7.

TABLE 5-8 OUTPUT OF THE GPSS PROGRAM FOR THE TICKET SELLER SYSTEM WITH UNIFORMLY DISTRIBUTED INTERARRIVAL AND SERVICE TIMES AND PRIORITY QUEUE DISCIPLINE FOR ELDERLY CUSTOMERS

BLOCK NUMBER	*LOC	NAME	A, B, C, D, E, F, G	COMMENTS	CARD
		JOB			00001
00001		GENERATE 101, 101			00002
00002		TRANSFER . 900, ELD, REG			00003
00003	ELD	PRIORITY 1			00004
00004		QUEUE TKTQ			00005
00005		SEIZE TKTF			00006
00006		DEPART TKTQ			00007
00007		ADVANCE 84, 84			00008
00008		RELEASE TKTF			00009
00009		TRANSFER , TERM1			00010
00010	REG	QUEUE TKTQ			00011
00011		SEIZE TKTF			00012
00012		DEPART TKTQ			00013
00013		ADVANCE 84, 84			00014
00014		RELEASE TKTF			00015
00015	TERM1	TERMINATE 1			00016
		START 100			00017
		END			00018

RELATIVE CLOCK TIME: 9998
ABSOLUTE CLOCK TIME: 9998

TABLE 5-8 (cont.)

BLOCK ENTITY TAGS

SYMBOL	VALUE	CARDS REFERENCED BY THIS SYMBOL
ELD	3	3
REG	10	3
TERM1	15	10

BLOCK COUNTS

BLOCK	CURR.	TOTAL	BLOCK	CURR.	TOTAL	BLOCK	CURR.	TOTAL
1	0	103	2	0	103	ELD	0	9
4	1	8	5	0	8	6	0	8
7	0	8	8	0	8	9	0	8
REG	2	92	11	0	92	12	0	92
13	0	92	14	0	92	TERM1	0	100

FACILITY ID	AVERAGE UTILIZATION	NUMBER ENTRIES	AVERAGE TIME/TRANS	SEIZING TRNSACTION	PREEMPTING TRNSACTION
TKTF	0.932	100	93.140	0	0

QUEUE ID	MAXIMUM CONTENTS	AVERAGE CONTENTS	TOTAL ENTRIES	ZERO ENTRIES	%−ZERO ENTRIES	AVERAGE TIME/TRANS	NZ-AV TM/TN	CURR. CONT.
TKTQ	5	1.012	103	20	19.41	98.25	121.93	3

THERE ARE NO INTERNAL TABLES

INTERRUPTING FACILITY USAGE

An even more expediting form of priority is offered by the PREEMPT and RETURN blocks. PREEMPT and RETURN take the place of SEIZE and RELEASE for transactions using facilities. A transaction which encounters a PREEMPT block will immediately acquire use of the facility if the facility is either idle or in use by a transaction which gained control of the facility through a SEIZE block. If the current transaction also gained control through a PREEMPT block, the new transaction must continue to wait at the top of the queue until processing is complete. Ordinarily, a preempted transaction will regain control of the facility and take up where it left off as soon as the preempting transaction has completed service. However, with some forms of the PREEMPT block, it may be preempted again by a higher-rated transaction. It is possible to cause preempted transactions to terminate permanently rather than to return to complete processing. It is also possible to cause the preempted transaction to repeat some portion of its previous processing time—the ''Now where were we?''—situation.

Table 5-9 shows the ticket seller model output with elderly and regular customers using PREEMPT and RETURN to expedite the processing of elderly customers even more than the PRIORITY block was able to do.

TABLE 5-9   OUTPUT OF THE GPSS PROGRAM WITH PRIORITY AND PREEMPTION
POSSIBILITIES FOR ELDERLY CUSTOMERS

BLOCK NUMBER	*LOC	NAME	A, B, C, D, E, F, G	COMMENTS	CARD
		JOB			00001
00001		GENERATE 101, 101			00002
00002		TRANSFER . 900, ELD, REG			00003
00003		ELD PRIORITY 1			00004
00004		QUEUE TKTF			00005
00005		PREEMPT TKTF			00006
00006		DEPART TKTQ			00007
00007		ADVANCE 84, 84			00008
00008		RETURN TKTF			00009
00009		TRANSFER , TERM1			00010
00010		REG QUEUE TKTQ			00011
00011		SEIZE TKTF			00012
00012		DEPART TKTQ			00013
00013		ADVANCE 84, 84			00014
00014		RELEASE TKTF			00015
00015		TERM1 TERMINATE 1			00016
		START 100			00017
		END			00018

RELATIVE CLOCK TIME: 10067
ABSOLUTE CLOCK TIME: 10067

BLOCK ENTITY TAGS

SYMBOL	VALUE	CARDS REFERENCED BY THIS SYMBOL
ELD	3	3
REG	10	3
TERM1	15	10

BLOCK COUNTS

BLOCK	CURR.	TOTAL	BLOCK	CURR.	TOTAL	BLOCK	CURR.	TOTAL
1	0	100	2	0	100	ELD	0	6
4	0	6	5	0	6	6	0	6
7	0	6	8	0	6	9	0	6
REG	0	94	11	0	94	12	0	94
13	0	94	14	0	94	TERM1	0	100

FACILITY ID	AVERAGE UTILIZATION	NUMBER ENTRIES	AVERAGE TIME/TRANS	SEIZING TRNSACTION	PREEMPTING TRNSACTION
TKTF	0.909	100	91.490	0	0

TABLE 5-9 (cont.)

QUEUE ID	MAXIMUM CONTENTS	AVERAGE CONTENTS	TOTAL ENTRIES	ZERO ENTRIES	% − ZERO ENTRIES	AVERAGE TIME/TRANS	NZ−AV TM/TN	CURR. CONT.
TKTQ	3	0.717	100	30	30.00	72.21	103.16	0

THERE ARE NO INTERNAL TABLES

STANDARD NUMERICAL ATTRIBUTES

Although the programs in Tables 5-7, 5-8, and 5-9 are technically correct in expediting the progress of elderly customers, the statistical output does not permit us to confirm that effect. We would like to obtain statistics on the transit time of elderly versus regular customers, but these statistics are not part of the prepackaged GPSS output. It is possible to obtain these transit time distributions as well as a wide variety of other types of output because GPSS internally records many interesting aspects of the simulation run and makes them available for recall. These internally recorded aspects of the run are called standard numerical attributes or SNAs.

SNAs are retrieved by referencing mnemonics which must be used for that purpose only and not for naming other entities or labels. Some SNAs measure systemwide attributes such as the current time on the simulator clock (C1). Others measure attributes of permanent entities in the system such as the length of queue 1 (Q1).* Still others measure attributes of temporary entities in the system such as the transit time of transactions (M1).

Table 5-10 displays a modified GPSS program for the ticket seller system with priority and preemption privileges given to elderly customers. The unconditional TRANSFER block has been removed and replaced with a TERMINATE block for the exclusive use of elderly customers. Since there is no "1" in the A-field of this TABULATE block, elderly customers will not decrement the START count; that is, the model will run until 100 *regular* customers have completed service.

Each customer passes through a TABULATE block which will record that customer's transit time just before termination. Every TABULATE block must be paired with a TABLE block in the housekeeping section below the processing loops. If the A-field of the TABULATE block contains a number, the label field of the TABLE block must contain the same number; if the A-FIELD of the TABULATE block contains a name, the label field of the TABLE block must contain the same name. Names and numbers should not be mixed.

The A-field of the TABLE block indicates which SNA is to be tabulated; in this case it is M1, customer transit time. The B-field of the TABLE card gives the upper limit of the first class of a frequency distribution of values of M1; since transit

---

*For those versions of GPSS which permit the assignment of either names or numbers for facilities, storages, and queues, the modeler should take care to use either all names or all numbers for each entity type to avoid compiler confusion.

TABLE 5-10 TICKET SELLER PROGRAM
WITH TRANSIT TIMES TABULATED

```
JOB
GENERATE 101, 101
TRANSFER . 900, ELD, REG
ELD PRIORITY 1
QUEUE TKTQ
PREEMPT TKTF
DEPART TKTQ
ADVANCE 84, 84
RETURN TKTF
TABULATE TRELD
TERMINATE
REG QUEUE TKTQ
SEIZE TKTF
DEPART TKTQ
ADVANCE 84, 84
RELEASE TKTF
TABULATE TRREG
TERMINATE 1
TRELD TABLE M1, 20, 20, 20
TRREG TABLE M1, 20, 20, 20
START 100
END
```

time can never be negative, 0 is a safe choice. The B-field of the TABLE block states the width of each class in the frequency distribution of transit times. The C-field specifies how many classes, at most, are to be printed; if there are no values of M1 to be tallied in the higher classes, these classes will be omitted by many versions of GPSS.

The output of the program in Table 5-10 will be the same as the output of the program in Table 5-9 except for the printing of the tables of transit time frequencies. Those tables are shown in Table 5-11.

TABLE 5-11 TRANSIT TIME TABLES FOR TICKET SELLER GPSS MODEL

TABLE NUMBER	1					
ENTRIES IN TABLE		MEAN ARGUMENT		STANDARD DEVIATION		SUM OF ARGUMENTS
8		79.625		44.883		637.000

UPPER LIMIT	OBSERVED FREQUENCY	PER CENT OF TOTAL	CUMULATIVE PERCENTAGE	CUMULATIVE REMAINDER	MULTIPLE OF MEAN	DEVIATION FROM MEAN
20	1	12.50	12.50	87.50	0.251	1.328−
40	1	12.50	25.00	75.00	0.502	0.882−
60	1	12.50	37.50	62.50	0.754	0.437−

TABLE 5-11 (cont.)

80	1	12.50	50.00	50.00	1.005	0.008
100	1	12.50	62.50	37.50	1.256	0.453
120	1	12.50	75.00	25.00	1.507	0.899
140	1	12.50	87.50	12.50	1.758	1.345
160	1	12.50	100.00	00.00	2.009	1.790
180	0	00.00	100.00	00.00	2.261	2.236
200	0	00.00	100.00	00.00	2.512	2.681
220	0	00.00	100.00	00.00	2.763	3.127
240	0	00.00	100.00	00.00	3.014	3.573
260	0	00.00	100.00	00.00	3.265	4.018
280	0	00.00	100.00	00.00	3.516	4.464
300	0	00.00	100.00	00.00	3.768	4.909
320	0	00.00	100.00	00.00	4.019	5.355
340	0	00.00	100.00	00.00	4.270	5.801
360	0	00.00	100.00	00.00	4.521	6.246
380	0	00.00	100.00	00.00	4.772	6.692
400	0	00.00	100.00	00.00	5.024	7.138

TABLE NUMBER   2

ENTRIES IN TABLE	MEAN ARGUMENT	STANDARD DEVIATION	SUM OF ARGUMENTS
100	169.560	90.807	16956.000

UPPER LIMIT	OBSERVED FREQUENCY	PER CENT OF TOTAL	CUMULATIVE PERCENTAGE	CUMULATIVE REMAINDER	MULTIPLE OF MEAN	DEVIATION FROM MEAN
20	3	3.00	3.00	97.00	0.118	1.657−
40	4	4.00	7.00	93.00	0.236	1.426−
60	6	6.00	13.00	87.00	0.354	1.206−
80	0	0.00	13.00	87.00	0.472	0.986−
100	4	4.00	17.00	83.00	0.590	0.766−
120	14	14.00	31.00	69.00	0.708	0.545−
140	6	6.00	37.00	63.00	0.826	0.325−
160	14	14.00	51.00	49.00	0.944	0.105−
180	12	12.00	63.00	37.00	1.062	0.114
200	7	7.00	70.00	30.00	1.180	0.335
220	9	9.00	79.00	21.00	1.297	0.555
240	5	5.00	84.00	16.00	1.415	0.775
260	4	4.00	88.00	12.00	1.533	0.995
280	1	1.00	89.00	11.00	1.651	1.216
300	1	1.00	90.00	10.00	1.769	1.436
320	0	0.00	90.00	10.00	1.887	1.656
340	2	2.00	92.00	8.00	2.005	1.876
360	2	2.00	94.00	6.00	2.123	2.097
380	4	4.00	98.00	2.00	2.241	2.317
400	0	0.00	98.00	2.00	2.359	2.537

OVERFLOW   2

AVERAGE VALUE OF OVERFLOW: 431.000

It is now possible to establish that the elderly customers whose transit times were collected and summarized in GPSS Table 1 in Table 5-11 completed processing much sooner than did the other customers whose transit times are collected and summarized in GPSS Table 2. Eight elderly people had a mean transit time of 79.625 time units with standard deviation 44.883, while 100 regular customers had a mean transit time of 169.560 time units with standard deviation 90.807. The frequency distributions show that the elderly customers cluster in the top or low-transit-time classes, whereas the regular customers are distributed more evenly over the classes, clustering toward the center classes. It is not necessary for all tables in a GPSS model to use the same class structure. If warranted, one table might have 20 classes which are 20 time units wide, and another table might have 10 classes 5 time units wide. Often, it is impossible for the modeler to ascertain in advance of the GPSS run just how the data will fall within the classes, so if more than just mean transit time is desired and a well-structured table is helpful, a second run with the TABLE statement respecified may be worthwhile. Generally, however, the mean, standard deviation, and observed frequencies are the most valuable information provided by a table. Note that when some data items exceed the range of the highest class specified, they are included in an overflow category; the number of items and their average value are shown.

SPECIFYING FUNCTIONS

We are now able to introduce the negative exponential distribution into our ticket seller model and observe the "correct" simulation results. The reason for deferring discussion of this topic was that the negative exponential function must be specified point by point, unlike the uniform distribution, which is fully defined by its range. Also, it is necessary to use the SNA for functions to reference the function in the program.

Two types of statements are needed to define functions. Both are placed in the top housekeeping section of the GPSS program, after the JOB statement and before the first GENERATE block. Every function is labeled with a name or number and is further defined by the random number generator from which it draws (GPSS may have eight independent random number generators, each called by the SNA named RN), whether the distribution is continuous or discrete, and the number of points of the function which are to be specified. The second type of statement gives those points, with pairs of X and Y values separated from other pairs by slashes.

The negative exponential distribution is especially convenient to use because it has the property that percentage points of any negative exponential distribution can be created by multiplying the mean of that distribution by the percentage points of the negative exponential distribution with a mean of 1. Table 5-12 shows a GPSS program which simulates the simple ticket seller system with negative exponential interarrival and service times as suggested by the sample data evaluated in Chapter 2.

TABLE 5-12   GPSS MODEL OF TICKET SELLER SYSTEM WITH
NEGATIVE EXPONENTIAL INTERARRIVAL AND SERVICE TIMES

```
JOB
EXPN1 FUNCTION RN1, C24
0, 0/. 1, . 104/. 2, . 222/. 3, . 355/. 4, . 509/. 5, . 69
. 6, . 915/. 7, 1. 2/. 75, 1. 38/. 8, 1. 6/. 84, 1. 83/. 88, 2. 12
. 9, 2. 3/. 92, 2. 52/. 94, 2. 81/. 95, 2. 99/. 96, 3. 2/. 97, 3. 5
. 98, 3. 9/. 99, 4. 6/. 995, 5. 3/. 998, 6. 2/. 999, 7. 0/. 9997, 8. 0
EXPN2 FUNCTION RN2, C24
0, 0/. 1, . 104/. 2, . 222/. 3, . 355/. 4, . 509/. 5, . 69
. 6, . 915/. 7, 1. 2/. 75, 1. 38/. 8, 1. 6/. 84, 1. 83/. 88, 2. 12
. 9, 2. 3/. 92, 2. 52/. 94, 2. 81/. 95, 2. 99/. 96, 3. 2/. 97, 3. 5
. 98, 3. 9/. 99, 4. 6/. 995, 5. 3/. 998, 6. 2/. 999, 7. 0/. 9997, 8. 0
GENERATE 101, FN$EXPN1
QUEUE TKTQ
SEIZE TKTF
DEPART TKTQ
ADVANCE 84, FN$EXPN2
RELEASE TKTF
TERMINATE 1
START 100
END
```

Two separate but identical negative exponential distribution functions are specified by 24 pairs of points. The first function, named EXPN1, is a continuous function which draws values from random number generator 1. The second function, named EXPN2, is a continuous function which draws values from random number generator 2. Function EXPN1 will supply interarrival time values; function EXPN2 will supply service time values. Although both interarrival times and service times could be supplied by the same function, it is desirable to separate them for statistical purposes, such as optimizing run length, which are beyond the scope of this book. We will separate them in all examples.

The output of the GPSS program in Table 5-12 is shown in Table 5-13.

TABLE 5-13   OUTPUT OF THE GPSS PROGRAM FOR THE TICKET SELLER SYSTEM
WITH NEGATIVE EXPONENTIAL INTERARRIVAL AND SERVICE TIMES

BLOCK NUMBER	*LOC	NAME	A, B, C, D, E, F, G	COMMENTS	CARD
	JOB				00001
	EXPN1 FUNCTION RN1, C24				00002
	0, 0/. 1, . 104/. 2, . 222/. 3, . 355/. 4, . 509/. 5, . 69				00003
	. 6, . 915/. 7, 1. 2/. 75, 1. 38/. 8, 1. 6/. 84, 1. 83/. 88, 2. 12				00004
	. 9, 2. 3/. 92, 2. 52/. 94, 2. 81/. 95, 2. 99/. 96, 3. 2/. 97, 3. 5				00005
	. 98, 3. 9/. 99, 4. 6/. 995, 5. 3/. 998, 6. 2/. 999, 7. 0/. 9997, 8. 0				00006

TABLE 5-13 (cont.)

	EXPN2 FUNCTION RN2, C24	00007
	0, 0/. 1, . 104/. 2, . 222/. 3, . 355/. 4, . 509/. 5, . 69	00008
	. 6, . 915/. 7, 1. 2/. 75, 1. 38/. 8, 1. 6/. 84, 1. 83/. 88, 2. 12	00009
	. 9, 2. 3/. 92, 2. 52/. 94, 2. 81/. 95, 2. 99/. 96, 3. 2/. 97, 3. 5	00010
	. 98, 3. 9/. 99, 4. 6/. 995, 5. 3/. 998, 6. 2/. 999, 7. 0/. 9997, 8. 0	00011
00001	GENERATE 101, FN$EXPN1	00012
00002	QUEUE TKTQ	00013
00003	SEIZE TKTF	00014
00004	DEPART TKTQ	00015
00005	ADVANCE 84, FN$EXPN2	00016
00006	RELEASE TKTF	00017
00007	TERMINATE 1	00018
	START 100	00019
	END	00020

RELATIVE CLOCK TIME: 10056
ABSOLUTE CLOCK TIME: 10056

FUNCTION ENTITY TAGS

SYMBOL	VALUE	CARDS	REFERENCED BY THIS SYMBOL
EXPN1	1	2,	12
EXPN2	2	7,	16
TERM1	15	10	

BLOCK COUNTS

BLOCK	CURR.	TOTAL	BLOCK	CURR.	TOTAL	BLOCK	CURR.	TOTAL
1	0	115	2	15	100	3	0	100
4	0	100	5	0	100	6	0	100
7	0	100						

FACILITY ID	AVERAGE UTILIZATION	NUMBER ENTRIES	AVERAGE TIME/TRANS	SEIZING TRNSACTION	PREEMPTING TRNSACTION
TKTF	0. 976	100	98. 170	0	0

QUEUE ID	MAXIMUM CONTENTS	AVERAGE CONTENTS	TOTAL ENTRIES	ZERO ENTRIES	%−ZERO ENTRIES	AVERAGE TIME/TRANS	NZ−AV TM/TN	CURR. CONT.
TKTQ	15	5. 275	115	4	3. 47	461. 243	477. 865	15

THERE ARE NO INTERNAL TABLES

JOINING THE SHORTEST QUEUE

Let us return to the earlier situation in which it was assumed that two servers shared a common queue and service area. This arrangement has become commonplace in banks and post offices. However, the case in which each server has an independent queue and work area remains much more common, even though the single-queue setup has been proven to yield shorter transit times.

When customers enter an area in which there are several queues, they are forced to substitute a realistic but undesirable goal for an unrealistic but desirable goal. The customer would like to join the queue in which the wait for service would be shortest, but there is no way politely to discern what sort of work those ahead of the customer have for the server to do. Thus, the customer substitutes the goal, "join the shortest queue in length," for the goal, "join the shortest queue in time." As we all have found to our chagrin, these two goals are often not congruent.

It is possible to model the "join the shortest queue" policy using the SNA for queue length and another form of branching statement, the TEST block. The TEST block compares the length of each queue with the lengths of the others. If the condition specified in the test (i.e., equal, not equal, less than, greater than, less than or equal, or greater than or equal) is met, the transaction proceeds ahead to the next sequential block. If the test condition is not met, the transaction either waits until the condition is met or branches to the block whose label is found in the C-field of the TEST block, if there is such a label provided. Table 5-14 shows a program in which customers will join the shortest of two queues.

TABLE 5-14   TICKET SELLER PROGRAM WITH GOAL TO JOIN THE SHORTEST OF TWO INDEPENDENT QUEUES

```
    JOB
EXPN1 FUNCTION RN1, C24
0, 0/.1, .104/.2, .222/.3, .355/.4, .509/.5, .69
.6, .915/.7, 1.2/.75, 1.38/.8, 1.6/.84, 1.83/.88, 2.12
.9, 2.3/.92, 2.52/.94, 2.81/.95, 2.99/.96, 3.2/.97, 3.5
.98, 3.9/.99, 4.6/.995, 5.3/.998, 6.2/.999, 7.0/.9997, 8.0
EXPN2 FUNCTION RN2, C24
0, 0/.1, .104/.2, .222/.3, .355/.4, .509/.5, .69
.6, .915/.7, 1.2/.75, 1.38/.8, 1.6/.84, 1.83/.88, 2.12
.9, 2.3/.92, 2.52/.94, 2.81/.95, 2.99/.96, 3.2/.97, 3.5
.98, 3.9/.99, 4.6/.995, 5.3/.998, 6.2/.999, 7.0/.9997, 8.0
EXPN3 FUNCTION RN3, C24
0, 0/.1, .104/.2, .222/.3, .355/.4, .509/.5, .69
.6, .915/.7, 1.2/.75, 1.38/.8, 1.6/.84, 1.83/.88, 2.12
.9, 2.3/.92, 2.52/.94, 2.81/.95, 2.99/.96, 3.2/.97, 3.5
.98, 3.9/.99, 4.6/.995, 5.3/.998, 6.2/.999, 7.0/.9997, 8.0
    GENERATE 101, FN$EXPN1
    TEST L Q$TKTQ1, Q$TKTQ2, TST2
```

TABLE 5-14 (cont.)

```
TST1 QUEUE TKTQ1
  SEIZE TKTF1
  DEPART TKTQ1
  ADVANCE 84, FN$EXPN2
  RELEASE TKTF1
  TERMINATE 1
TST2 TEST L Q$TKTQ2, Q$TKTQ1, TST3
TST4 QUEUE TKTQ2
  SEIZE TKTF2
  DEPART TKTQ2
  ADVANCE 84, FN$EXPN3
  RELEASE TKTF2
  TERMINATE 1
TST3 TRANSFER .500, TST1, TST4
  START 100
  END
```

Table 5-15 reveals that the two queues have been balanced almost perfectly. Queue TKTQ1 had 50 entries whereas queue TKTQ2 had 51.

TABLE 5-15 OUTPUT OF THE GPSS PROGRAM FOR TICKET SELLER SYSTEM WITH NEGATIVE EXPONENTIAL INTERARRIVAL AND SERVICE TIMES AND TWO INDEPENDENT QUEUES

BLOCK NUMBER	*LOC	NAME	A, B, C, D, E, F, G	COMMENTS	CARD
	JOB				00001
	EXPN1	FUNCTION RN1, C24			00002
	0,0/.1,.104/.2,.222/.3,.355/.4,.509/.5,.69				00003
	.6,.915/.7,1.2/.75,1.38/.8,1.6/.84,1.83/.88,2.12				00004
	.9,2.3/.92,2.52/.94,2.81/.95,2.99/.96,3.2/.97,3.5				00005
	.98,3.9/.99,4.6/.995,5.3/.998,6.2/.999,7.0/.9997,8.0				00006
	EXPN2	FUNCTION RN2, C24			00007
	0,0/.1,.104/.2,.222/.3,.355/.4,.509/.5,.69				00008
	.6,.915/.7,1.2/.75,1.38/.8,1.6/.84,1.83/.88,2.12				00009
	.9,2.3/.92,2.52/.94,2.81/.95,2.99/.96,3.2/.97,3.5				00010
	.98,3.9/.99,4.6/.995,5.3/.998,6.2/.999,7.0/.9997,8.0				00011
	.				
	.				
	.				
00001		GENERATE 101, FN$EXPN1			00012
00002		TEST L Q1, Q2, TST2			00013
00003	TST1	QUEUE TKTQ1			00014
00004		SEIZE TKTF TKTF1			00015
00005		DEPART TKTQ1			00016

TABLE 5-15 (cont.)

00006	ADVANCE 84 , FN$EXPN2	00017
00007	RELEASE TKTF1	00018
00008	TERMINATE 1	00019
00009	TST2 TEST L Q2, Q1, TST3	00020
00010	TST4 QUEUE TKTQ2	00021
00011	SEIZE TKTF2	00022
00012	TKTQ2 DEPART 2	00023
00013	ADVANCE 84, FN$EXPN3	00024
00014	RELEASE TKTF2	00025
00015	TERMINATE 1	00026
00016	TST3 TRANSFER . 500, TST1, TST4	00027
	START 100	00028
	END	00029

RELATIVE CLOCK TIME: 8794
ABSOLUTE CLOCK TIME: 8794

BLOCK ENTITY TAGS

SYMBOL	VALUE	CARDS REFERENCED BY THIS SYMBOL
TST1	3	27
TST2	9	13
TST3	16	20
TST4	10	27

FUNCTION ENTITY TAGS

SYMBOL	VALUE	CARDS	REFERENCED BY THIS SYMBOL	
EXPN1	1	2,	12	
EXPN2	2	7,	17,	24

BLOCK COUNTS

BLOCK	CURR.	TOTAL	BLOCK	CURR.	TOTAL	BLOCK	CURR.	TOTAL
1	0	101	2	0	101	TST1	0	50
4	0	50	5	0	50	6	1	49
7	0	49	8	0	49	TST2	0	88
TST4	0	51	11	0	51	12	0	51
13	0	51	14	0	51	15	0	51
TST3	0	73						

FACILITY ID	AVERAGE UTILIZATION	NUMBER ENTRIES	AVERAGE TIME/TRANS	SEIZING TRNSACTION	PREEMPTING TRNSACTION
TKTF1	0. 465	50	81. 800	100	0
TKTF2	0. 496	51	85. 510	0	0

TABLE 5-15 (cont.)

QUEUE ID	MAXIMUM CONTENTS	AVERAGE CONTENTS	TOTAL ENTRIES	ZERO ENTRIES	%−ZERO ENTRIES	AVERAGE TIME/TRANS	NZ-AV TM/TN	CURR. CONT.
TKTQ1	2	0.188	50	31	62.00	33.140	87.211	0
TKTQ2	2	0.204	51	29	56.86	35.176	81.545	0

THERE ARE NO INTERNAL TABLES

SUMMARY

The GPSS language enables the analyst to model a waiting-line system without concern for details such as the construction and updating of a clock, accumulation and reduction of performance statistics, or low-level programming.

An elementary GPSS model will include one or more GENERATE blocks to create transactions; one or more TERMINATE blocks to remove transactions from the model; a JOB, START, and END statement to define the length of the run and the limits of the model; and one or more processing blocks. Processing blocks include QUEUE and DEPART to attach and detach transactions from FIFO waiting lines, SEIZE and RELEASE to gain and relinquish control of a single-server facility, ENTER and LEAVE to gain and relinquish control of a single- or multiple-server facility (storage), or PREEMPT and RETURN to gain and relinquish control of a single-server facility which may or may not be already occupied with processing a less important transaction. The PRIORITY block defines the hierarchy within which transactions will be ranked at QUEUE blocks. The ADVANCE block indicates the amount of time a transaction is scheduled to spend at each processing step.

As do most programming languages, GPSS offers both unconditional and conditional branching statements. The TRANSFER block can cause transactions to branch unconditionally to other blocks which are appropriately labeled, but can also direct transactions to alternative destinations on a proportionate basis.

GPSS stores standard numerical attribute (SNA) values which describe properties of the system such as clock time, properties of transactions such as transit time, and properties of queues, facilities, and storages such as current contents. SNAs may be printed during or after the run, used to create tabular output, or employed to control transaction processing during the run. An example of the latter usage is reference to the SNA for queue length in the GPSS program to cause a transaction to join the shortest queue. The TEST block is a branching statement which incorporates SNAs to direct transactions to appropriate destinations such as the shortest queue.

GPSS functions allow the modeler to define interarrival and service time distributions either based on theoretical models such as the uniform or negative exponential or based on empirically defined data sets.

GPSS output is formatted so that most statistics of interest are provided either automatically or with little work on the part of the programmer. Some output statistics which are printed routinely are the average utilization of facilities and storages, the average and maximum length of queues as well as the average time in queues, and the number of transactions which passed through each block in the model. Examination of these statistics permits the analyst to draw tentative conclusions about system performance as well as to isolate certain common types of errors.

QUESTIONS

1. What are the common attributes of simulation languages?
2. Describe the three sections which characterize most GPSS programs.
3. Define the purpose and give the basic grammatical structure of each of the following GPSS blocks or statements:
   (a) JOB
   (b) END
   (c) START
   (d) GENERATE
   (e) TERMINATE
   (f) QUEUE
   (g) DEPART
   (h) SEIZE
   (i) RELEASE
   (j) ENTER
   (k) LEAVE
   (l) PREEMPT
   (m) RETURN
   (n) PRIORITY
   (o) FUNCTION
   (p) TABULATE
   (q) TABLE
   (r) ADVANCE
   (s) TRANSFER
   (t) TEST
4. What is a GPSS standard numerical attribute? Define the concept in general and give some examples.
5. What outputs are provided routinely by GPSS? What other outputs might the analyst desire to request?
6. What is the difference between a facility and a storage?
7. What is the difference between the definition of priority and its effects and the definition of preemption and its effects?
8. What is the difference between conditional and unconditional transfer? What other block accomplishes branching?
9. Enumerate some common GPSS errors and how they might be detected.

PROBLEMS

Assume that the data collected in the problem section of Chapter 2 and applied to the problem section of the waiting-line models in Chapter 3 are applicable to the following cases. In each problem, run the model until 100 transactions have passed through the system. (Remember that GPSS accepts only integer values for interarrival and service time.) The mean interarrival time was 421.75 sec and mean service time was 316.769 sec.

1. Prepare a GPSS model of the $M/M/1$ system just described. Compare your results with those predicted by the $M/M/1$ analytical model in Chapter 3.

2. Suppose that the interarrival time was uniform on the interval from 0 to 844 seconds and that service time was uniform on the interval from 0 to 634 seconds. Prepare a GPSS model of this system and compare your results with those for Problem 1.

3. Suppose that the service described in Problem 1 is preceded by a brief talk with a receptionist whose service time is negative exponential with a mean of 40 seconds. Prepare a GPSS model of this system.

4. Suppose that the situation in Problem 3 prevails but that only 30 percent of the customers need to stop and talk with the receptionist. The others go directly to the service queue. Model this variation.

5. Suppose that there are two identical servers for the situation described in Problem 1. Model this if there are (is)
   (a) Two separate queues and the customer joins the shortest queue
   (b) One common queue

6. Suppose that there are two classes of customers, one of whom is given priority over the other (but not preemption rights). Model the situation in Problem 1 with this change, if the priority customers comprise 25 percent of the population.

7. Suppose that there are two classes of customers and one class has both priority and preemption rights. Model the situation in Problem 1 with this change, if the priority customers comprise 25 percent of the population.

8. Modify the program for Problem 7 to include a pair of tables of transit time for the two classes of customers.

chapter six

INTERMEDIATE GPSS

In Chapter 5 some basic GPSS concepts were introduced which will now be expanded. These are the specification of statistical distribution functions, branching statements, and the classification of different kinds of transactions. We shall also consider some circumstances in which it is useful to refer to the simulator clock time, C1.

USER-DEFINED FUNCTIONS

Some versions of GPSS include prewritten functions which are negative exponential, normal, or uniform. These versions require the user to provide only the mean, the standard deviation, and/or the range, as appropriate. Other versions of GPSS consider these functional forms as unique, special cases whose points must be painstakingly provided by the user. The more points provided by the user, the more accurate the functional representation will be. The functions may be continuous, as in the negative exponential function shown in Chapter 5, or they may be discrete. For continuous functions, GPSS will make a linear interpolation between successive points of the function based on the value of the random number drawn. For discrete functions, all values of the random number above the previous point and up to and including the next point will return the Y-value of the next point. Except for the negative exponential function which is used as a modifier in the GENERATE and ADVANCE blocks, references to functions usually will appear in the mean or A-field of the GENERATE and ADVANCE blocks.

Suppose that the ticket seller exhibits a very simple distribution of service times that do not conform to any common theoretical model such as the negative exponential. The service time is found to be 50 seconds for 20 percent of the time, 70 seconds for 55 percent of the time, and 100 seconds for the remaining 25 percent of the time. This is a discrete function with three pairs of points specified. We call a distribution which follows no theoretical pattern but instead is derived entirely from observation an "empirically defined distribution." Table 6-1 gives the probability distribution of service times as well as the cumulative distribution function points.

TABLE 6-1 EMPIRICALLY DEFINED PROBABILITY DISTRIBUTION OF SERVICE TIMES

SERVICE TIME, X	PROBABILITY OF SERVICE TIME, X	CUMULATIVE PROBABILITY OF SERVICE TIME, X
50	0.200	0.200
70	0.550	0.750
100	0.250	1.000

To specify this function in GPSS, we let the Y-coordinate of each pair of points be the service time, and we let the X-coordinate be the cumulative less-than-or-equals probability of that service time. The function is discrete; that is, there are assumed to be no service time values between 50 and 70 seconds or between 70 and 100 seconds. The function definition statement would be, for example,

EMPIR FUNCTION RN2,D3

where the name of the function is EMPIR, the X-coordinates are bounds for draws from random number generator 2, and there are three pairs of discrete points. Following the function definition statement, the three pairs of coordinates would appear as

.200,50/.750,70/1.000,100

If the service times were continuous and could be interpolated between 0 and 100 seconds, the function definition statement would be:

EMPIR FUNCTION RN2,C4

with a fourth point serving as the origin for interpolation purposes. The pairs of coordinates, then, would be:

0,0/.200,50/.750,70/1.000,100

Based on draws from random number generator 2, then, there would be a 20 percent chance of drawing a service time between 0 and 50, a 55 percent chance of drawing a service time above 50 but less than or equal to 70, and a 25 percent chance of drawing a service time above 70 but less than or equal to 100 seconds. For a continuous distribution, the probability that service time would precisely equal any set value such as 70 is effectively zero; probabilities are attached to ranges of values, not to individual values of service time.

The ADVANCE block which calls this service time function would appear as follows:

```
ADVANCE FN$EMPIR
```

Note that the function reference is in the A-field, not in the B-field, as was the case for the general negative exponential function with a mean of 1.

PARAMETERS

In Chapter 5, we observed how different types of transactions could be separated from the generated stream of customers by using a TRANSFER block. There are two potential disadvantages to this procedure. First, it may be necessary to reproduce virtually all of a series of blocks through which the various types of transactions will pass. Second, if the separated transactions are remerged to avoid the duplication of blocks, there is no way in which to distinguish the transactions at a later time.

One way of avoiding these problems is the use of parameters or descriptive tags which may be attached to transactions and scanned as the transactions proceed through the model to effect differential processing. The value of parameters may be used to divert transactions from the general stream of processing or to determine the length of time a transaction will require to be processed. While different versions of GPSS permit different numbers of parameters to be defined for each transaction, it may be assumed that at least 100 parameter values are possible for each transaction.

Let us reconsider the case in Chapter 5 where it was desirable to separate the elderly from the regular customers of the ticket seller. In that case, the objective was to offer priority processing for the elderly. Table 6-2 restates the material from Table 5-8, the GPSS program to process the elderly customers with a higher priority. Notice that the blocks from QUEUE through RELEASE are duplicated from the regular and elderly branches of transaction processing.

TABLE 6-2   GPSS MODEL OF TICKET SELLER SYSTEM
WITH UNIFORMLY DISTRIBUTED INTERARRIVAL
AND SERVICE TIMES AND PRIORITY QUEUE
DISCIPLINE FOR ELDERLY CUSTOMERS

```
       JOB
       GENERATE 101,101
       TRANSFER .900,ELD,REG
   ELD PRIORITY 1
       QUEUE TKTQ
       SEIZE TKTF
```

TABLE 6-2 (cont.)

```
DEPART TKTQ
ADVANCE 84, 84
RELEASE TKTF
TRANSFER , TERM1
REG QUEUE TKTQ
SEIZE TKTF
DEPART TKTQ
ADVANCE 84, 84
RELEASE TKTF
TERM1 TERMINATE 1
START 100
END
```

Suppose that the ticket seller offered a discount to elderly customers who displayed a Medicare card. This procedure might require slightly longer service times, perhaps a uniform distribution from 0 to 200 instead of 0 to 168 seconds. If we remove the prioritizing of queue discipline for the elderly customers and introduce the differential service times, the resulting program can be condensed as shown in Table 6-3.

TABLE 6-3 GPSS MODEL OF TICKET SELLER SYSTEM WITH DIFFERENT SERVICE TIMES FOR ELDERLY AND REGULAR CUSTOMERS

```
JOB
GENERATE 101, 101
TRANSFER . 900, ELD, REG
ELD ASSIGN 1, 100
TRANSFER , NEXT
REG ASSIGN 1, 84
NEXT QUEUE TKTQ
SEIZE TKTF
DEPART TKTQ
ADVANCE P1, P1
RELEASE TKTF
TERMINATE 1
START 100
END
```

Parameters must be called by number rather than by name. In this example, the mean and spread of service time has been stored in parameter number 1. This is accomplished by separating the two transaction streams temporarily and using the ASSIGN block to place different values in parameter number 1 for elderly and regular customers. The ADVANCE block references this mean and spread using the standard numerical attribute for parameters, P, followed by the specific parameter number desired, parameter 1.

TESTING STATUS OF SERVERS

In Chapter 5, we observed the use of the TRANSFER block in diverting some transactions from a common path based on a condition such as the length of a queue.
Another type of conditional branching statement is available which examines the
idle or busy status of facilities and storages and directs transactions accordingly.
This block is called the GATE block. The grammatical structure of the GATE block
is similar to that of the TEST block. While the TEST block uses comparison operators such as E, NE, LE, LT, GT, and GE, the GATE block employs operators such
as U, NU, SNF, SF, SE, and SNE. These operators stand for, respectively, facility
in use, facility not in use, storage not full, storage full, storage empty, and storage
not empty. The name or number of the facility or storage whose content is to be
examined is given in the A-field of the GATE block. If the transaction is to be
detained at the GATE block until the condition becomes true, there is no entry in the
B-field. If the transaction is to be redirected to another block when the condition is
false, the label of that alternate destination is given in the B-field.

The GATE block is particularly appropriate when a transaction is unwilling to
join a queue upon finding a facility or storage busy. Suppose, for example, that our
ticket seller's customers must first park in a parking lot with five spaces. If there are
no spaces left in the lot, the customer does not wait but drives away. Table 6-4
shows the GPSS program which models this situation.

TABLE 6-4   TICKET SELLER MODEL WITH
FIVE-SPACE PARKING LOT AND CUSTOMERS
WHO REFUSE TO WAIT FOR PARKING SPACES

```
          JOB
          GENERATE 101, 101
          GATE SNF LOT, BYE
          ENTER LOT
          QUEUE TKTQ
          SEIZE TKTF
          DEPART TKTQ
          ADVANCE 84, 84
          RELEASE TKTF
          LEAVE LOT
          TERMINATE 1
      BYE TERMINATE
      LOT STORAGE 5
          START 100
          END
```

The B-field operand is the destination for transactions which are denied entry
to STORAGE LOT. Since there is a separate TERMINATE block labeled BYE, the
block count for BYE will tell us how many transactions were denied entry to LOT.

Since the TERMINATE block labeled BYE does not have a 1 in the A-field, transactions which are denied entry to LOT will not decrement the count in the START statement.

ENDING THE RUN BASED ON TIME

It is possible to end the simulation run after a stipulated number of simulated seconds have elapsed rather than after a stipulated number of transactions have completed processing. To accomplish this, we remove any entries from the A-field of the TERMINATE blocks in our model and introduce a new GENERATE and TERMINATE block as well. The new GENERATE and TERMINATE blocks serve as a timing mechanism. The GENERATE block produces a dummy transaction precisely when the number of clock units we desire has elapsed. The dummy transaction proceeds directly to the TERMINATE block which has a ''1'' in the A-field. The START statement has only a ''1'' in the A-field, so this single dummy transaction will terminate the run after the desired amount of clock time. Table 6-5 shows the ticket seller model with a run length of 10,000 seconds.

TABLE 6-5 TICKET SELLER MODEL
WITH RUN LENGTH OF 10,000 SEC

```
JOB
GENERATE 101, 101
QUEUE TKTQ
SEIZE TKTF
DEPART TKTQ
ADVANCE 84, 84
RELEASE TKTF
TERMINATE
GENERATE 10000
TERMINATE 1
START 1
END
```

SEGMENTING A SIMULATION RUN

Sometimes it is desirable to segment a simulation run into an initial portion and a later portion on the supposition that some time will elapse before the model is in typical or steady-state condition. The beginning portion of the run may exhibit transient conditions such as an abnormally low average time in queue because the run begins with no one in the system.* To initiate the execution of successive segments of a

*It is possible to load the system initially with a nonzero number of transactions which conforms more closely to steady-state conditions. However, these will form a queue and will require some time to disperse throughout the system.

run, it is necessary to use more than one START statement. If successive START statements are immediately adjacent to each other, the latest set of run statistics will include the latest run segment results as well as the preceding run segment results on a cumulative basis. If successive START statements are separated by a CLEAR statement, the latest run segment will begin in the empty state, just as the prior run segments did, but with a different random number stream. If successive START statements are separated by a RESET statement, the latest run segment will begin with the conditions existing at the end of the prior segment, such as three people waiting, but summary statistics for the latest run segment will not include any data from prior run segments. The RESET and CLEAR statements may be used either with run length based on a completed number of transactions or an amount of elapsed simulated time.

Table 6-6 shows the ticket seller system output for two successive periods of 10,000 seconds.* Table 6-7 shows the ticket seller system output for two successive periods of 10,000 seconds, each beginning with the empty state. Table 6-8 shows the ticket seller system output for two successive periods of 10,000 seconds, but with data collection and reduction separate for each period.

TABLE 6-6   TICKET SELLER MODEL AND OUTPUT FOR TWO SUCCESSIVE TIME PERIODS WITH THE SECOND OUTPUTS INCLUSIVE OF THE FIRST OUTPUTS

BLOCK NUMBER	*LOC	NAME	A, B, C, D, E, F, G	COMMENTS	CARD
		JOB			0001
00001		GENERATE 101, 101			0002
00002		QUEUE TKTQ			0003
00003		SEIZE TKTF			0004
00004		DEPART TKTQ			0005
00005		ADVANCE 84, 84			0006
00006		RELEASE TKTF			0007
00007		TERMINATE			0008
00008		GENERATE 10000			0009
00009		TERMINATE 1			0010
		START 1			0011
		START 1			0012
		END			0013

RELATIVE CLOCK TIME: 10000
ABSOLUTE CLOCK TIME: 10000

---

*The unit of measure is known to be seconds only because the data defining the model were originally measured in that scale. Since GPSS usually requires numeric values to be integers, it may be necessary to specify all time values in the smallest unit of measure used anywhere in the model.

TABLE 6-6 (cont.)

BLOCK COUNTS

BLOCK	CURR.	TOTAL	BLOCK	CURR.	TOTAL	BLOCK	CURR.	TOTAL
1	0	94	2	2	92	3	0	92
4	0	92	5	1	91	6	0	91
7	0	91	8	0	1	9	0	1

FACILITY ID	AVERAGE UTILIZATION	NUMBER ENTRIES	AVERAGE TIME/TRANS	SEIZING TRNSACTION	PREEMPTING TRNSACTION
TKTF	0.839	92	91.217	93	0

QUEUE ID	MAXIMUM CONTENTS	AVERAGE CONTENTS	TOTAL ENTRIES	ZERO ENTRIES	PCT-ZERO ENTRIES	AVERAGE TIME/TRANS	NZ-AV TM/TN	CURR. CONT.
TKTQ	3	0.620	94	23	24.46	65.915	87.27	2

THERE ARE NO INTERNAL TABLES

RELATIVE CLOCK TIME: 20000
ABSOLUTE CLOCK TIME: 20000

BLOCK COUNTS

BLOCK	CURR.	TOTAL	BLOCK	CURR.	TOTAL	BLOCK	CURR.	TOTAL
1	0	199	2	0	199	3	0	199
4	0	199	5	1	198	6	0	198
7	0	198	8	0	2	9	0	2

FACILITY ID	AVERAGE UTILIZATION	NUMBER ENTRIES	AVERAGE TIME/TRANS	SEIZING TRNSACTION	PREEMPTING TRNSACTION
TKTF	0.901	199	90.543	201	0

QUEUE ID	MAXIMUM CONTENTS	AVERAGE CONTENTS	TOTAL ENTRIES	ZERO ENTRIES	PCT-ZERO ENTRIES	AVERAGE TIME/TRANS	NZ-AV TM/TN	CURR. CONT.
TKTQ	8	1.534	199	31	15.57	154.156	182.60	0

THERE ARE NO INTERNAL TABLES

TABLE 6-7   TICKET SELLER MODEL AND OUTPUT FOR TWO SUCCESSIVE TIME PERIODS WITH THE TWO PERIODS INDEPENDENT AND SPANNING AN INITIAL PHASE

BLOCK NUMBER	*LOC	NAME	A, B, C, D, E, F, G	COMMENTS	CARD
		JOB			0001
00001		GENERATE 101, 101			0002
00002		QUEUE TKTQ			0003

TABLE 6-7 (cont.)

00003	SEIZE TKTF	0004
00004	DEPART TKTQ	0005
00005	ADVANCE 84, 84	0006
00006	RELEASE TKTF	0007
00007	TERMINATE	0008
00008	GENERATE 10000	0009
00009	TERMINATE 1	0010
	START 1	0011
	CLEAR	0012
	START 1	0013
	END	0014

RELATIVE CLOCK TIME: 10000
ABSOLUTE CLOCK TIME: 10000

BLOCK COUNTS

BLOCK	CURR.	TOTAL	BLOCK	CURR.	TOTAL	BLOCK	CURR.	TOTAL
1	0	94	2	2	92	3	0	92
4	0	92	5	1	91	6	0	91
7	0	91	8	0	1	9	0	1

FACILITY ID	AVERAGE UTILIZATION	NUMBER ENTRIES	AVERAGE TIME/TRANS	SEIZING TRNSACTION	PREEMPTING TRNSACTION
TKTF	0.839	92	91.217	93	0

QUEUE ID	MAXIMUM CONTENTS	AVERAGE CONTENTS	TOTAL ENTRIES	ZERO ENTRIES	PCT-ZERO ENTRIES	AVERAGE TIME/TRANS	NZ-AV TM/TN	CURR. CONT.
TKTQ	3	0.620	94	23	24.46	65.915	87.27	2

THERE ARE NO INTERNAL TABLES

RELATIVE CLOCK TIME: 10000
ABSOLUTE CLOCK TIME: 20000

BLOCK COUNTS

BLOCK	CURR.	TOTAL	BLOCK	CURR.	TOTAL	BLOCK	CURR.	TOTAL
1	0	104	2	3	101	3	0	101
4	0	101	5	1	100	6	0	100
7	0	100	8	0	1	9	0	1

FACILITY ID	AVERAGE UTILIZATION	NUMBER ENTRIES	AVERAGE TIME/TRANS	SEIZING TRNSACTION	PREEMPTING TRNSACTION
TKTF	0.833	101	82.455	102	0

TABLE 6-7 (cont.)

QUEUE ID	MAXIMUM CONTENTS	AVERAGE CONTENTS	TOTAL ENTRIES	ZERO ENTRIES	PCT-ZERO ENTRIES	AVERAGE TIME/TRANS	NZ-AV TM/TN	CURR. CONT.
TKTQ	5	1.245	104	31	29.80	119.731	170.58	3

THERE ARE NO INTERNAL TABLES

TABLE 6-8 TICKET SELLER MODEL AND OUTPUT FOR TWO SUCCESSIVE TIME PERIODS WITH THE SECOND FOLLOWING THE FIRST BUT WITH SEPARATE DATA COLLECTION

BLOCK NUMBER	*LOC	NAME	A, B, C, D, E, F, G	COMMENTS	CARD
		JOB			0001
00001		GENERATE 101, 101			0002
00002		QUEUE TKTQ			0003
00003		SEIZE TKTF			0004
00004		DEPART TKTQ			0005
00005		ADVANCE 84, 84			0006
00006		RELEASE TKTF			0007
00007		TERMINATE			0008
00008		GENERATE 10000			0009
00009		TERMINATE 1			0010
		START 1			0011
		RESET			0012
		START 1			0013
		END			0014

RELATIVE CLOCK TIME: 10000
ABSOLUTE CLOCK TIME: 10000

BLOCK COUNTS

BLOCK	CURR.	TOTAL	BLOCK	CURR.	TOTAL	BLOCK	CURR.	TOTAL
1	0	94	2	2	92	3	0	92
4	0	92	5	1	91	6	0	91
7	0	91	8	0	1	9	0	1

FACILITY ID	AVERAGE UTILIZATION	NUMBER ENTRIES	AVERAGE TIME/TRANS	SEIZING TRNSACTION	PREEMPTING TRNSACTION
TKTF	0.839	92	91.217	93	0

QUEUE ID	MAXIMUM CONTENTS	AVERAGE CONTENTS	TOTAL ENTRIES	ZERO ENTRIES	PCT-ZERO ENTRIES	AVERAGE TIME/TRANS	NZ-AV TM/TN	CURR. CONT.
TKTQ	3	0.620	94	23	24.46	65.915	87.27	2

TABLE 6-8 (cont.)

THERE ARE NO INTERNAL TABLES

RELATIVE CLOCK TIME: 10000
ABSOLUTE CLOCK TIME: 20000

BLOCK COUNTS

BLOCK	CURR.	TOTAL	BLOCK	CURR.	TOTAL	BLOCK	CURR.	TOTAL
1	0	105	2	0	107	3	0	107
4	0	107	5	1	107	6	0	107
7	0	107	8	0	1	9	0	1

FACILITY ID	AVERAGE UTILIZATION	NUMBER ENTRIES	AVERAGE TIME/TRANS	SEIZING TRNSACTION	PREEMPTING TRNSACTION
TKTF	0.961	108	88.963	201	0

QUEUE ID	MAXIMUM CONTENTS	AVERAGE CONTENTS	TOTAL ENTRIES	ZERO ENTRIES	PCT-ZERO ENTRIES	AVERAGE TIME/TRANS	NZ-AV TM/TN	CURR. CONT.
TKTQ	8	2.444	107	8	7.47	228.458	246.92	0

THERE ARE NO INTERNAL TABLES

While all three programs run for 20,000 clock units, the relative and absolute clock times for the second run segment differ when a RESET or CLEAR statement is used, while the relative and absolute clock times are the same when a START statement immediately follows another START statement. The block counts for the second run segment also reflect the difference between using a RESET or CLEAR statement and simply using two consecutive START statements.

In Table 6-6, the second run segment includes all information from the first run segment, so we cannot obtain any impression of what the second period alone was like. We hypothesize that the first segment contained transients and the second segment was closer to steady state by noting the considerable difference between the average and maximum queue contents in the two segments. However, this might be attributable to the rather high service times during the two periods.

To check the assumption that the results in Table 6-6 were not representative of typical system performance due to inordinately high service times, we may observe the result of using the CLEAR statement in Table 6-7 to get a new representation of the situation using a different random number stream. The second segment of Table 6-7 shows the *first* 10,000 clock units for the system, since the CLEAR statement erases all statistics from the first segment. Here we observe an average service time which is below the average we requested, but not as far below it as the prior examples were above it. There does not seem to be a definite discrepancy between the two samples of the initial conditions of the ticket seller system.

Table 6-8 repeats the same initial period but displays statistics for the second period which are totally independent of the first period except for retaining the transactions in process in the model at the time the first segment ended; this includes both transactions being served and those waiting for service. This time we note that average service time for the second segment is again above the stipulated mean but by a lesser amount. There is a marked difference in average waiting times and percentage of transactions which did not have to wait, since the second segment's statistics are not clouded by inclusion with those of the first period.

VARIABLES AND THE CLOCK SNA

We have seen that the GPSS compiler provides access to the clock time throughout a simulation run so that it can be determined when a transaction will appear in the model and when its service begins and ends. These conditions are established automatically by GPSS. However, there are other types of occurrences which are related to specific values of the simulation clock which may be programmed by the analyst. One common situation is to prevent the passage of a transaction at a particular point in the model until the clock time has achieved some specific value. Usually this clock value is a relative one which will occur at periodic intervals throughout the simulation run.

An example might be that the ticket seller does not begin to sell tickets until 9 A.M. each day. Customers begin to arrive at the ticket window at 8:30 A.M. and wait until the window opens to purchase their tickets. Suppose that we model this system from 8:30 A.M. until 6:00 P.M., assuming that customers are not permitted to join the ticket queue after 5 P.M.* The number of seconds between 8:30 A.M. and 6:00 P.M. is $9\frac{1}{2}$ hours $\times$ 3600 sec per hour = 34,200. Table 6-9 shows a GPSS program with these characteristics which is to run for one working day.

TABLE 6-9 TICKET SELLER MODEL AND OUTPUT FOR ONE WORKING DAY IN WHICH CUSTOMERS MAY QUEUE BEFORE AND AFTER THE OFFICIAL OPENING AND CLOSING OF THE SYSTEM

BLOCK NUMBER	*LOC	NAME	A, B, C, D, E, F, G	COMMENTS	CARD
		JOB			0001
		VARTM VARIABLE C1@34200			0002
00001		GENERATE 101, 101			0003
00002		QUEUE ERLYQ			0004
00003		TEST GE V$VARTM, K1800			0005
00004		DEPART ERLYQ			0006

*We will assume temporarily that any customers in the waiting line at 5 P.M. may be serviced before 6 P.M. A more complex model could verify this assumption. It is desirable to model only the portions of the day in which events of interest to us can occur. Otherwise, the simulation becomes more costly without offering more information about the system.

TABLE 6-9 (cont.)

00005	TEST LE V$VARTM, K30600, BYE	0007
00006	QUEUE TKTQ	0008
00007	SEIZE TKTF	0009
00008	DEPART TKTQ	0010
00009	ADVANCE 84, 84	0011
00010	RELEASE TKTF	0012
00011	TERMINATE	0013
00012	GENERATE 34200	0014
00013	TERMINATE 1	0015
00014	BYE TERMINATE	0016
	START 1	0017
	END	0018

RELATIVE CLOCK TIME: 34200
ABSOLUTE CLOCK TIME: 34200

BLOCK COUNTS

BLOCK	CURR.	TOTAL	BLOCK	CURR.	TOTAL	BLOCK	CURR.	TOTAL
1	0	355	2	0	355	3	0	355
4	0	355	5	0	355	6	0	319
7	0	319	8	0	319	9	0	319
10	0	319	11	0	319	12	0	1
13	0	1	BYE	0	36			

FACILITY ID	AVERAGE UTILIZATION	NUMBER ENTRIES	AVERAGE TIME/TRANS	SEIZING TRNSACTION	PREEMPTING TRNSACTION
TKTF	0.798	319	85.564	0	0

QUEUE ID	MAXIMUM CONTENTS	AVERAGE CONTENTS	TOTAL ENTRIES	ZERO ENTRIES	PCT-ZERO ENTRIES	AVERAGE TIME/TRANS	NZ-AV TM/TN	CURR. CONT.
ERLYQ	17	0.513	355	338	95.21	49.380	1031.18	0
TKTQ	18	3.489	319	23	7.21	374.075	403.14	0

THERE ARE NO INTERNAL TABLES

This program illustrates the pairing of the TEST block with the VARIABLE statement. A variable in GPSS is the result of one or more calculations in an expression which may include the symbols for addition (+), subtraction (−), multiplication (∗), and division (/). Although expressions are generally evaluated in left-to-right order, terms in parentheses receive priority treatment and multiplication and division are completed before addition and subtraction. The @ symbol is the operator for modulus division, in which the remainder rather than the quotient is retained as the result of the operation.*

In Table 6-9 the current clock time (C1) is divided by 34,200 seconds in a simulated day of $9^{1}/2$ hours. Whenever a transaction is produced by the GENERATE block, the first TEST block compares the remainder resulting from the division, stored in VARIABLE VARTM, with the hour at which the ticket window opens. If the window is not yet open, the transactions must wait in QUEUE ERLYQ until it is 9 A.M. This unconditional use of the TEST block does not offer a C-field exit when the test condition is false. The SNA for the value of the variable is V followed by a variable number of V$ followed by a variable name.

At 9:00 A.M., any transactions which were detained at the TEST block are permitted to join the regular queue, TKTQ. At 5:00 P.M., no further customers are permitted to join TKTQ; since the window will be closed overnight, any latecomers do not wait but are diverted to the TERMINATE block labeled BYE. The evening TEST block is a conditional TEST block with a C-field exit.

Although this program will run for only one day (34,200 seconds), the logic and grammar permit it to be rerun for any number of additional days because of the remainder division. At 9:00 A.M. on the first day, the value of C1 will be 1800 seconds, and the result of dividing 1800 by 34,200 will be 0 with a remainder of 1800. At 9:00 A.M. on a second day, C1 would be 36,000, and the result of dividing 36,000 by 34,200 would be 1 with a remainder of 1800. Since the remainders are identical, the treatement of transactions at the TEST block would be the same in both cases.

Similar logic applies to the TEST block which closes the queue for the day. At 5:00 P.M. on the first day, C1 is 30,600, the divisor is 34,200, the quotient is 0, and the remainder is 30,600. At 5:00 P.M. on a second day, C1 would be 64,800, the divisor would be 34,200, the quotient would be 1, and the remainder would again be 30,600.

How many customers had to wait in the morning before the ticket window opened? The statistics for queue ERLYQ tell us that $(355 - 338 = 17)$ customers arrived too early. They spent an average of 1031 seconds or 17 minutes waiting. This information might convince management to offer an earlier opening of the ticket window to promote goodwill.

How many customers were turned away after 5:00 P.M.? The block count at the block labeled BYE tells us that 36 customers arrived between 5:00 P.M. and 6:00 P.M. and were denied service. This, too, might prompt management to extend business hours.

---

*While many GPSS compilers permit floating-point arithmetic, integer arithmetic is preferred because it is more economical of space. The analyst should be cautious when performing arithmetic in variable expressions because truncation can distort the results.

EFFICIENT TERMINATION OF A RUN

In the illustration in Table 6-9, we attempted to economize on computer time to run the simulation by modeling only the portion of the day from 8:30 A.M. to 6:00 P.M. It would be even more efficient to terminate the run when the last customer who arrived by 5:00 P.M. completes service and the ticket seller goes home. We must devise a way to signal when this event occurs.

Let us modify the program in Table 6-9 so that the run ends as soon as all customers have completed service. If the clock time remainder is at least 30,600 and if no customers remain in the queue or in service, the ticket window may close for the evening. Table 6-10 shows a GPSS program containing TEST and GATE blocks which implement the desired stopping rule.

The clock time at the end of the run is 30,731 seconds. This is 131 seconds or about 2 minutes past the 5:00 P.M. equivalent of 30,600 seconds, but it is considerably less than the 6:00 P.M. equivalent of 34,200 seconds. The difference is $(34,200 - 30,731) = 3469$ seconds or about one-tenth of the run which is saved.

TABLE 6-10   GPSS PROGRAM AND OUTPUT TO CLOSE TICKET WINDOW AS SOON AS POSSIBLE AFTER 5:00 P.M.

BLOCK NUMBER	*LOC	NAME	A, B, C, D, E, F, G	COMMENTS	CARD
		JOB			0001
		VARTM VARIABLE C1@34200			0002
00001		GENERATE 101, 101			0003
00002		QUEUE ERLYQ			0004
00003		TEST GE V$VARTM, K1800			0005
00004		DEPART ERLYQ			0006
00005		TEST LE V$VARTM, K30600, BYE			0007
00006		QUEUE TKTQ			0008
00007		SEIZE TKTF			0009
00008		DEPART TKTQ			0010
00009		ADVANCE 84, 84			0011
00010		RELEASE TKTF			0012
00011		TERMINATE			0013
00012		GENERATE 30600			0014
00013		GATE NU TKTF			0015
00014		TERMINATE 1			0016
00015		BYE TERMINATE			0017
		START 1			0018
		END			0019

RELATIVE CLOCK TIME:	30731
ABSOLUTE CLOCK TIME:	30731

TABLE 6-10 (cont.)

BLOCK COUNTS

BLOCK	CURR.	TOTAL	BLOCK	CURR.	TOTAL	BLOCK	CURR.	TOTAL
1	0	314	2	0	314	3	0	314
4	0	314	5	0	314	6	0	312
7	0	312	8	0	312	9	0	312
10	0	312	11	0	312	12	0	1
13	0	1	14	0	1	BYE	0	2

FACILITY ID	AVERAGE UTILIZATION	NUMBER ENTRIES	AVERAGE TIME/TRANS	SEIZING TRNSACTION	PREEMPTING TRNSACTION
TKTF	0.920	312	90.628	0	0

QUEUE ID	MAXIMUM CONTENTS	AVERAGE CONTENTS	TOTAL ENTRIES	ZERO ENTRIES	PCT-ZERO ENTRIES	AVERAGE TIME/TRANS	NZ-AV TM/TN	CURR. CONT.
ERLYQ	19	0.583	314	295	93.94	57.032	942.53	0
TKTQ	25	11.366	312	12	3.84	1119.538	1164.32	0

THERE ARE NO INTERNAL TABLES

DEBUGGING AIDS

When writing programs in most programming languages, it is helpful to print intermediate results of calculations to locate the source of subsequent errors. GPSS offers a block called PRINT which permits the analyst to display any of several categories of output during the course of a simulation run. This may replace or supplement standard GPSS summary statistics. For example, the analyst may want to know the length of one or all queues or the contents of one or more storages or facilities at selected times during the simulated day or whenever selected conditions occur. This avoids the printing of complete GPSS summary statistics when only specific portions of the output are needed.

Suppose that we want to display the length of all queues every 1000 time units for the ticket seller model. Table 6-11 shows the GPSS code which will produce this result.

TABLE 6-11 GPSS PROGRAM AND OUTPUT FOR TICKET SELLER SYSTEM WITH QUEUE STATISTICS PRINTED EVERY 1000 TIME UNITS

BLOCK NUMBER	*LOC	NAME	A, B, C, D, E, F, G	COMMENTS	CARD
		JOB			0001
00001		GENERATE 101, 101			0002
00002		QUEUE 1			0003

TABLE 6-11 (cont.)

00003	SEIZE 1	0004
00004	DEPART 1	0005
00005	ADVANCE 84, 84	0006
00006	RELEASE 1	0007
00007	TERMINATE 1	0008
00008	GENERATE 1000	0009
00009	PRINT , , Q	0010
00010	TERMINATE	0011
	START 100	0012
	END	0013

QUEUE ID	MAXIMUM CONTENTS	AVERAGE CONTENTS	TOTAL ENTRIES	ZERO ENTRIES	PCT-ZERO ENTRIES	AVERAGE TIME/TRANS	NZ-AV TM/TN	CURR. CONT.
TKTQ	2	0.724	10	2	20.00	72.400	90.50	1

QUEUE ID	MAXIMUM CONTENTS	AVERAGE CONTENTS	TOTAL ENTRIES	ZERO ENTRIES	PCT-ZERO ENTRIES	AVERAGE TIME/TRANS	NZ-AV TM/TN	CURR. CONT.
TKTQ	2	0.640	18	3	16.67	71.167	85.40	1

QUEUE ID	MAXIMUM CONTENTS	AVERAGE CONTENTS	TOTAL ENTRIES	ZERO ENTRIES	PCT-ZERO ENTRIES	AVERAGE TIME/TRANS	NZ-AV TM/TN	CURR. CONT.
TKTQ	3	0.639	28	6	21.42	68.429	87.091	3

QUEUE ID	MAXIMUM CONTENTS	AVERAGE CONTENTS	TOTAL ENTRIES	ZERO ENTRIES	PCT-ZERO ENTRIES	AVERAGE TIME/TRANS	NZ-AV TM/TN	CURR. CONT.
TKTQ	3	0.650	36	8	22.22	72.222	92.86	0

QUEUE ID	MAXIMUM CONTENTS	AVERAGE CONTENTS	TOTAL ENTRIES	ZERO ENTRIES	PCT-ZERO ENTRIES	AVERAGE TIME/TRANS	NZ-AV TM/TN	CURR. CONT.
TKTQ	3	0.550	46	13	28.26	59.826	83.39	1

QUEUE ID	MAXIMUM CONTENTS	AVERAGE CONTENTS	TOTAL ENTRIES	ZERO ENTRIES	PCT-ZERO ENTRIES	AVERAGE TIME/TRANS	NZ-AV TM/TN	CURR. CONT.
TKTQ	3	0.518	54	15	27.77	57.556	79.69	1

QUEUE ID	MAXIMUM CONTENTS	AVERAGE CONTENTS	TOTAL ENTRIES	ZERO ENTRIES	PCT-ZERO ENTRIES	AVERAGE TIME/TRANS	NZ-AV TM/TN	CURR. CONT.
TKTQ	7	1.024	68	15	22.05	105.368	135.189	3

QUEUE ID	MAXIMUM CONTENTS	AVERAGE CONTENTS	TOTAL ENTRIES	ZERO ENTRIES	PCT-ZERO ENTRIES	AVERAGE TIME/TRANS	NZ-AV TM/TN	CURR. CONT.
TKTQ	7	1.094	76	16	21.05	115.211	145.933	0

QUEUE ID	MAXIMUM CONTENTS	AVERAGE CONTENTS	TOTAL ENTRIES	ZERO ENTRIES	PCT-ZERO ENTRIES	AVERAGE TIME/TRANS	NZ-AV TM/TN	CURR. CONT.
TKTQ	7	1.031	89	19	21.34	104.225	132.514	1

TABLE 6-11 (cont.)

QUEUE ID	MAXIMUM CONTENTS	AVERAGE CONTENTS	TOTAL ENTRIES	ZERO ENTRIES	PCT-ZERO ENTRIES	AVERAGE TIME/TRANS	NZ-AV TM/TN	CURR. CONT.
TKTQ	7	0.973	96	22	22.91	101.323	131.45	0

In addition to the output shown in Table 6-11, the regular GPSS summary statistical output would appear at the conclusion of the run. All the statistics given by the PRINT block are cumulative and therefore not particularly useful except for the "CURRENT CONTENTS" field.

If only a subset of a group of queues is of interest, the A- and B-fields, deleted in the PRINT block shown in Table 6-11, may be used to specify the lower and upper limit of the queues for which output is to be printed.

In addition to queue information, the PRINT block can produce other statistics periodically. These include statistics about facilities (PRINT ,,F), storages (PRINT ,,S, block counts (PRINT ,,B), and savevalues or constants (PRINT ,,X).

Another helpful diagnostic tool is the TRACE and UNTRACE combination of blocks. When a TRACE block is inserted in a GPSS program, a full line of statistical information is printed every time a transaction proceeds from any block within the TRACE-UNTRACE limits to any other block within the TRACE-UNTRACE limits. If there is no UNTRACE block, transactions will be traced until they enter a TERMINATE block. Table 6-12 shows a GPSS program with a TRACE and UNTRACE block included, as well as a portion of the trace mechanism.

TABLE 6-12 GPSS PROGRAM AND OUTPUT FOR TICKET SELLER SYSTEM INCLUDING TRACE AND UNTRACE

BLOCK NUMBER	*LOC	NAME	A, B, C, D, E, F, G	COMMENTS	CARD
		JOB			0001
00001		GENERATE 101, 101			0002
00002		TRACE			0003
00003		QUEUE TKTQ			0004
00004		SEIZE TKTF			0005
00005		DEPART TKTQ			0006
00006		ADVANCE 84, 84			0007
00007		RELEASE TKTF			0008
00008		UNTRACE			0009
00009		TERMINATE 1			0010
		START 100			0011
		END			0012

TRANS. 1 MOVED FROM BLOCK 2 TO 3; TIME UNITS 147; 100 TERMINATION (S) TO GO.
TRANS. 1 MOVED FROM BLOCK 3 TO 4; TIME UNITS 147; 100 TERMINATION (S) TO GO.
TRANS. 1 MOVED FROM BLOCK 4 TO 5; TIME UNITS 147; 100 TERMINATION (S) TO GO.

TABLE 6-12 (cont.)

TRANS.	1 MOVED FROM BLOCK 5 TO 6; TIME UNITS 147; 100 TERMINATION (S) TO GO.
TRANS.	1 MOVED FROM BLOCK 6 TO 7; TIME UNITS 238; 100 TERMINATION (S) TO GO.
TRANS.	1 MOVED FROM BLOCK 7 TO 8; TIME UNITS 238; 100 TERMINATION (S) TO GO.
TRANS.	2 MOVED FROM BLOCK 2 TO 3 ; TIME UNITS 297; 99 TERMINATION (S) TO GO.
TRANS.	2 MOVED FROM BLOCK 3 TO 4 ; TIME UNITS 297; 99 TERMINATION (S) TO GO.
TRANS.	2 MOVED FROM BLOCK 4 TO 5 ; TIME UNITS 297; 99 TERMINATION (S) TO GO.
TRANS.	2 MOVED FROM BLOCK 5 TO 6 ; TIME UNITS 297; 99 TERMINATION (S) TO GO.

$$\vdots$$

TRANS.	99 MOVED FROM BLOCK 6 TO 7 ; TIME UNITS 10580; 2 TERMINATION (S) TO GO.
TRANS.	99 MOVED FROM BLOCK 7 TO 8 ; TIME UNITS 10580; 2 TERMINATION (S) TO GO.
TRANS.	100 MOVED FROM BLOCK 3 TO 4 ; TIME UNITS 10580; 1 TERMINATION (S) TO GO.
TRANS.	100 MOVED FROM BLOCK 4 TO 5 ; TIME UNITS 10580; 1 TERMINATION (S) TO GO.
TRANS.	100 MOVED FROM BLOCK 5 TO 6 ; TIME UNITS 10580; 1 TERMINATION (S) TO GO.
TRANS.	102 MOVED FROM BLOCK 2 TO 3 ; TIME UNITS 10590; 1 TERMINATION (S) TO GO.
TRANS.	103 MOVED FROM BLOCK 2 TO 3 ; TIME UNITS 10643; 1 TERMINATION (S) TO GO.
TRANS.	104 MOVED FROM BLOCK 2 TO 3 ; TIME UNITS 10645; 1 TERMINATION (S) TO GO.
TRANS.	100 MOVED FROM BLOCK 6 TO 7 ; TIME UNITS 10737; 1 TERMINATION (S) TO GO.
TRANS.	100 MOVED FROM BLOCK 7 TO 8 ; TIME UNITS 10737; 1 TERMINATION (S) TO GO.

Table 6-12 displays the GPSS program which caused the trace as well as the first and last 10 lines of trace output. Each line gives the transaction serial number, the block from which it came, the block to which it went, the simulator clock unit at which the move occurred, and the START statement count of terminations left to go before the run ends. Each transaction is moved as far as it can go in any one clock unit. With eight blocks included in the trace and 100 transactions to be terminated, there will be a minimum of 800 lines of trace output, so the analyst should choose carefully just which blocks should be included in the trace. In the example in Table 6-12, 104 transactions appear in the trace because of delays in the processing of transaction 100, which ends the run when it enters the TERMINATE block.

The TRACE-UNTRACE mechanism is very handy in a complex model with many branches where transactions may seem to disappear until they ultimately influence block counts at the end of the run. It is possible to verify that a transaction did or did not take the intended path.

CONSTANTS AND COUNTERS IN GPSS

Constants or very simple variables which require only addition or subtraction can be specified in GPSS with the INITIAL and SAVEVALUE blocks. Initialization is accomplished with the INITIAL statement and subsequent updating and use of the contents is done with the SAVEVALUE statement.

Suppose that we want to trigger printing of queue statistics based on the number of transactions which have appeared at some point in the model rather than on clock time. Table 6-13 shows a GPSS program which accomplishes this.

TABLE 6-13 GPSS PROGRAM
WITH PRINT BLOCK ACTIVE EVERY
TENTH TRANSACTION

```
JOB
INITIAL X1, 0
GENERATE 101, 101
QUEUE TKTQ
SEIZE TKTF
DEPART TKTQ
ADVANCE 84, 84
RELEASE TKTF
SAVEVALUE 1+, K1
TEST E X1, K10, TERM1
PRINT , , Q
SAVEVALUE 1, K0
TERM1 TERMINATE 1
START 100
END
```

The program produces output comparable to that in Table 6-11, except that the
PRINT block output is provided for only 1 out of 10 transactions. SAVEVALUE 1
is initialized at zero. Then, in the repetitive portion of the program, SAVEVALUE 1
is incremented by one each time a transaction completes service. If it is the tenth
transaction since the beginning of the run or since the last printing, a line of queue
statistics will appear for each queue in the model. After this line is printed,
SAVEVALUE 1 is reset to zero and the transaction terminates from the model.
Transactions other than the tenth terminate immediately upon releasing control of
the facility.

It is possible to print values of any SNA by first reading the SNA into a
SAVEVALUE and then printing the SAVEVALUE. Suppose that we wanted to print
the transit time of every tenth transaction. Table 6-14 shows the program and output
to accomplish this.

TABLE 6-14 GPSS PROGRAM AND OUTPUT WITH TRANSIT TIME PRINTED FOR
EVERY TENTH TRANSACTION

BLOCK NUMBER	*LOC	NAME	A, B, C, D, E, F, G	COMMENTS	CARD
		JOB			0001
		INITIAL X1, 0			0002
		INITIAL X2, 0			0003
00001		GENERATE 101, 101			0004
00002		QUEUE TKTQ			0005
00003		SEIZE TKTF			0006
00004		DEPART TKTQ			0007

TABLE 6-14 (cont.)

00005	ADVANCE 84, 84	0008
00006	RELEASE TKTF	0009
00007	SAVEVALUE 2+, K1	0010
00008	TEST E X2, K10, TERM1	0011
00009	SAVEVALUE 1, M1	0012
00010	PRINT , , X	0013
00011	SAVEVALUE 2, K0	0014
00012	TERM1 TERMINATE 1	0015
	START 100	0016
	END	0017

FULLWORD
SAVEVALUES

NUMBER	CONTENT	NUMBER	CONTENT
1	+232	2	+10

FULLWORD
SAVEVALUES

NUMBER	CONTENT	NUMBER	CONTENT
1	+121	2	+10

FULLWORD
SAVEVALUES

NUMBER	CONTENT	NUMBER	CONTENT
1	+150	2	+10

FULLWORD
SAVEVALUES

NUMBER	CONTENT	NUMBER	CONTENT
1	+145	2	+10

FULLWORD
SAVEVALUES

NUMBER	CONTENT	NUMBER	CONTENT
1	+392	2	+10

FULLWORD
SAVEVALUES

NUMBER	CONTENT	NUMBER	CONTENT
1	+214	2	+10

FULLWORD
SAVEVALUES

NUMBER	CONTENT	NUMBER	CONTENT
1	+77	2	+10

FULLWORD
SAVEVALUES

NUMBER	CONTENT	NUMBER	CONTENT
1	+231	2	+10

TABLE 6-14 (cont.)

FULLWORD SAVEVALUES			
NUMBER	CONTENT	NUMBER	CONTENT
1	+55	2	+10
FULLWORD SAVEVALUES			
NUMBER	CONTENT	NUMBER	CONTENT
1	+118	2	+10

In addition to the intermediate output shown, standard GPSS summary statistics will be printed. These will include a repetition of the last savevalue line since the number of terminations is evenly divisible by 10. The example in Table 6-14 makes use of two savevalues, the first to enable the printing of transit time for every tenth transaction and the second to act as a counter for finding every tenth transaction. Since savevalues always include a sign except with a value of zero, it is possible to write a program to extract the savevalue containing the quantity of interest by searching for a sign followed by a value, if the values are always nonzero. The purpose of doing this will emerge in Chapter 8.*

SUMMARY

When the simulation analyst has collected data and performed goodness-of-fit tests, only to find that none of the theoretical models fit the data, it is still possible to specify the distributions in GPSS with user-defined functions. User-defined functions may be either continuous or discrete and may be precisely stated with many pairs of points or merely approximations.

Every GPSS transaction may carry descriptors called parameters, numeric values which in some sense convey information about the transaction. A common use of parameters is to define service time for each of several types of customers and retain the ability to associate the service time with the correct customer type even though the types of transactions have been intermingled in the model.

Simulation runs may be ended either after some specified number of transactions have been processed or after some specified time period has elapsed. To end the run based on clock time, it is necessary to include dummy GENERATE and TERMINATE blocks in the model and to alter the START statement so that the beginning count reflects the number of time periods rather than the number of transactions for which the model is to be run.

It is frequently desirable to segment a simulation run into several components to observe the change in system attributes over time. Successive START statements

*See sample SIGN program on p. 443.

will yield separate sets of system statistics, but the later run segments will include the statistics for the earlier segments as well. The use of a CLEAR statement between START statements will return the model to its initial state but with a new random number stream. The use of a RESET statement between START statements will begin the later run segment with all transactions which were in process at the time the earlier segment ended but will not collect statistics which include data from the prior period. These techniques help the analyst to determine whether the model accurately portrays typical performance of the system.

The GPSS SNA for simulator clock time is C1. It may be referenced anywhere in the model, but is particularly convenient for checking whether a particular time of day has arrived. The clock time is used as the dividend in a VARIABLE statement with the number of clock units in a simulated day as the divisor. The quotient is ignored and the remainder resulting from the division is retained as the value of the variable.

Occasionally, it is desirable to report intermediate results during the course of a simulation run. This may be done to help locate a logical error in the model or to create a stream of output data items for statistical analysis. The PRINT block serves these purposes well. It gives statistics for queues, facilities, storages, blocks, and savevalues whenever appropriate in the same format as at the end of the simulation run.

Another helpful debugging tool is the TRACE-UNTRACE pair of blocks. After a transaction passes a TRACE block, it will trigger the printing of one line of output every time it passes another block, until it reaches either an UNTRACE or a TERMINATE block. This procedure lets the analyst observe in detail the course taken by each transaction. While this technique assists in finding logical errors, it can create voluminous output which may be counterproductive.

The INITIAL and SAVEVALUE blocks are used to initialize constants or simple variables and to change their values during the simulation run. Savevalues allow the analyst to print values of an SNA during the simulation run and facilitate extraction of these values from the image of GPSS output on disk. It is possible to change, increase or decrease, the contents of an existing savevalue by omitting a sign to the right of the savevalue number in the A-field or by inserting a + or a − sign.

QUESTIONS

1. What is a user-defined function? Explain the grammar of the function statement and its follower statement(s).
2. Why is a user-defined function commonly referenced in the A-field of a GENERATE or ADVANCE block while the general negative exponential function is commonly referenced in the B-field?
3. What is the difference, in theory and in grammar, between a discrete and a continuous function?

4. What is a GPSS parameter? How are parameters defined in GPSS? What purpose do they serve?

5. In what circumstances might a GATE block be used? What is the difference between a conditional and an unconditional GATE block?

6. How does a GPSS program which ends the run based on a transaction count differ from a GPSS program which ends the run based on the time on the simulator clock?

7. Explain the similarities and differences in the output of a GPSS run in which
   (a) there are several START statements which immediately follow each other
   (b) there are several START statements separated by CLEAR statements
   (c) there are several START statements separated by RESET statements.

8. What is a GPSS variable? What mathematical operators may be used in a GPSS variable?

9. Explain how a VARIABLE statement can cause the same event to occur every day at the same time.

10. Discuss the method for ending a simulation run in the minimum amount of time necessary to empty the system.

11. Suggest two possible uses for a PRINT block.

12. What is convenient and what is inconvenient about TRACE and UNTRACE blocks?

13. What is the format of output provided by
   (a) PRINT blocks?
   (b) blocks between a TRACE and an UNTRACE block?

14. Why is it necessary to specify all time values in a GPSS program in terms of the same unit of time?

15. What is a savevalue? Name two methods of giving values to savevalues. Explain why savevalues can be useful.

PROBLEMS

Assume that the data collected in the problem section of Chapter 2 and applied to the problem sections of Chapters 3 and 5 are applicable to the following problems. In each problem, run the model until 100 transactions have passed through the system unless instructed otherwise. The mean interarrival time was 421.75 seconds and mean service time was 316.769 seconds, both negative exponentially distributed.

1. Write a GPSS program in which interarrival time is as specified but service time is a continuous function which does not fit any common theoretical model. Service time is below 200 seconds for 30 percent of the time, between 200 and 300 seconds for 65 percent of the time, and between 300 and 350 seconds for 5 percent of the time.

2. Suppose that 30 percent of the customers have service times of 400 seconds and that the remainder have service times of 350 seconds. Use ASSIGN blocks to place service time in parameter 1 and use it to specify service time.

3. Suppose that the service facility is open between 8 A.M. and 4:30 P.M. daily. Customers begin to arrive at 7:45 A.M. but must wait in line until the facility opens. Any customers

in the queue at 4:30 P.M. will be served, but no additional customers will be permitted to join the queue. Write a GPSS program to model this situation in the most efficient manner possible. How many people arrived before 8 A.M.? At what time was the facility able to close for the day?

4. Run the simulation for 100 customers; then run it for a subsequent 100 customers, and compare the output. Run the simulation for 100 customers, clear the system, and then run it for another 100 customers and compare the output. Run the simulation for 100 customers reset the system, and then run it for another 100 customers and compare the output. Indicate the difference in interpretation among the three sets of runs.

5. Include a PRINT block in the basic model of the system which will cause the printing of queue and facility statistics after each 1200 sec. Comment on your results.

6. Include a TRACE and UNTRACE block in the basic model of the system beginning with the QUEUE block and ending with the RELEASE block. Comment on your results.

7. Write a GPSS program for the simple model which will cause transit time statistics to be printed for every fifth transaction. Comment on your results.

chapter seven

ADVANCED GPSS

VARYING INTERARRIVAL AND SERVICE TIMES

During the course of a normal workday, the distribution of interarrival and service times may change. It is important to understand the difference between a change in the distribution and variation in random draws from the same distribution. For example, interarrival time in the early morning hours may have a very different mean and distribution type from interarrival time in the late afternoon. Still, the effects of the distribution in the early phases of the system carry over to the later phases.

One way in which to alter the nature of the distribution of interarrival time at various points in the model is by the use of overlays. The GENERATE block is given a label and the model is run for a prescribed period of time during which the initial interarrival time condition prevails. Then the model is reset and a new GENERATE block with the same label but a different distribution is inserted after the RESET statement but prior to the next START statement. The new interarrival time information is substituted for the prior material until the count in the next START statement has been fully decremented and a new RESET, overlay and START statement, if any, have been encountered by the GPSS processor. Table 7-1 shows a GPSS program with several overlays of GENERATE blocks.

In this example the mean interarrival time, 101 seconds, happens to be the same for each of the three periods of the day which are modeled. The GENERATE block timer creates a dummy transaction every half-hour or 1800 seconds. The first period encompasses 9000 seconds or the first $2\frac{1}{2}$ hours in the morning. The second period occupies the middle 7200 seconds or 2 hours at lunchtime. The final 14,400 seconds or 4 hours represents afternoon activity. While the mean interarrival times

TABLE 7-1   USE OF OVERLAYS TO MODEL DIFFERENT
INTERARRIVAL TIME DISTRIBUTIONS

```
  JOB
EXPON FUNCTION RN1, C24
0, 0/. 1, . 104/. 2, . 222/. 3, . 355/. 4, . 509/. 5, . 69
. 6, . 915/. 7, 1. 2/. 75, 1. 38/. 8, 1. 6/. 84, 1. 83/. 88, 2. 12
. 9, 2. 3/. 92, 2. 52/. 94, 2. 81/. 95, 2. 99/. 96, 3. 2/. 97, 3. 5
. 98, 3. 9/. 99, 4. 6/. 995, 5. 3/. 998/, 6. 2/. 999, 7. 0/. 9997, 8. 0
OVLY GENERATE 101, FN$EXPON
  QUEUE TKTQ
  SEIZE TKTF
  DEPART TKTQ
  ADVANCE 84, 84
  RELEASE TKTF
  TERMINATE
  GENERATE 1800
  TERMINATE 1
  START 5
  RESET
OVLY GENERATE 101, 101
  START 4
  RESET
OVLY GENERATE 101
  START 8
  END
```

are hypothetically the same for this example, the interarrival time distribution is negative exponential in the first period, uniform in the second period, and constant in the third period. Although this pattern is unrealistic, it illustrates the flexibility of distribution modification with overlays. Any or all fields of the GENERATE block may be changed in an overlay. The GPSS compiler will issue an error warning diagnostic indicating that more than one block has the same label or tag when an overlay is present in the program. In the case of overlays, this duplication is intentional and the warning should be ignored.

It might be appropriate to change the distribution specified in an ADVANCE block when the individual performing the service changes during the course of the workday. Alteration of the mean and modifier of ADVANCE blocks during the course of the simulation run is somewhat more tricky than altering GENERATE blocks. The problem is that a block cannot be successfully overlaid if its current contents are nonzero. A GENERATE block will not have current contents if the block following the GENERATE block cannot deny entry to transactions, as should usually be the case. However, the principal purpose of an ADVANCE block is precisely to delay transactions. Methods comparable to those used for GENERATE blocks in Table 7-1 could be employed for ADVANCE blocks, but the RESET statements would have to be replaced by CLEAR statements to ensure that the

ADVANCE blocks are empty at the time of the overlay. This destroys the desirable statistical ''memory'' offered by the RESET block.

If the distribution type for the new server is the same as that for the previous server (e.g., both are uniform but with different means and/or modifiers or both are negative exponential but with different means), the problem is fairly simple. The arguments in field A and/or field B of the ADVANCE block are defined as savevalues rather than as constants, and their values are changed at appropriate times on the simulator clock or when other significant and noticeable events occur in the model.

For example, suppose that the lunchtime fill-in server is not as efficient as the regular server and processes customers according to a uniform distribution with a mean of 95 seconds and a modifier of 90 seconds. Table 7-2 shows a GPSS program which accomplishes this.

TABLE 7-2   USE OF SAVEVALUES
TO MODEL DIFFERENT SERVICE TIME
DISTRIBUTION PARAMETERS

```
        JOB
        INITIAL X1, 0
        INITIAL X2, 84
        INITIAL X3, 84
        GENERATE 101, 101
        QUEUE TKTQ
        SEIZE TKTF
        DEPART TKTQ
        ADVANCE X2, X3
        RELEASE TKTF
        TERMINATE
        GENERATE 1800
        SAVEVALUE 1+, K1
        TEST E X1, K6, NXT1
        SAVEVALUE 2, K95
        SAVEVALUE 3, K90
        TERMINATE 1
NXT1    SAVEVALUE 2, K84
        SAVEVALUE 3, K84
        TERMINATE 1
        START 16
        END
```

For the first six half-hours, the regular server processes customers in a time of 84, 84 seconds. During the one half-hour lunch break, the substitute server processes customers in a time of 95, 90. After the lunch break, regular service resumes at 84, 84.

If it is necessary to alter the nature of the service time distribution, the GPSS programming is more complex, and the topic is beyond the scope of this book. The interested reader is referred to any GPSS text or manual to study the subject head-ings FAVAIL, FUNAVAIL, and Logic Switches.

QUEUE-DEPENDENT SERVICE TIMES

Sometimes a server will feel compelled to work faster when a queue is relatively long. Suppose that the ticket seller's mean service time is 84 seconds when there is no one waiting, 82 seconds when there is one person waiting, 79 seconds when there are two people waiting, 75 seconds when there are three people waiting, and 73 seconds when there are four or more people waiting. To simulate this, it is necessary to reference the SNA for current queue length, Q or Q$. The function named WAIT will have as its X-coordinate queue length and as its Y-coordinate service time. Table 7-3 shows the GPSS program for this situation.

TABLE 7-3   GPSS PROGRAM
FOR TICKET SELLER MODEL WITH
QUEUE-DEPENDENT SERVICE TIMES

```
    JOB
WAIT FUNCTION Q$TKTQ, D5
0, 84/1, 82/2, 79/3, 75/4, 73
    GENERATE 101, 101
    QUEUE TKTQ
    SEIZE TKTF
    DEPART TKTQ
    ADVANCE FN$WAIT
    RELEASE TKTF
    TERMINATE 1
    START 100
    END
```

USER CHAINS

The QUEUE and DEPART blocks offer the modeler a method for measuring attrib-utes of waiting lines in the system under some popular, but rather restrictive, circumstances. The queue must be organized in FIFO order, and there must be no jockeying or reneging. Balking may be accomplished by having the transaction check the length of the queue prior to the QUEUE block via a TEST block and avoid the queue if its length is unacceptably long. Another disadvantage of the QUEUE-DEPART pair other than its relative inflexibility is that it requires a considerable amount of housekeeping on the part of the GPSS processor, which slows the execu-tion of the model and hence increases its cost.

A very attractive alternative but one which is totally different in grammar and concept is the user chain. By placing transactions on a user chain while they await the availability of a service facility or the existence of some condition, GPSS is able to "ignore" them until that availability or condition occurs.

Roughly analogous to the QUEUE-DEPART pair for standard queues are the LINK-UNLINK pair for user chains. The LINK block is positioned comparably to a QUEUE block in the model. Since some versions of GPSS which allow naming of queues demand numbering of user chains, we will give a number as the user-chain designator in the A-field of the LINK block. The B-field of the LINK block specifies the order in which the transactions are to be linked. Possible B-field arguments are "FIFO," "LIFO," or "P" followed by a parameter number.

If "FIFO" is chosen, the resulting sequence is the same as that for a QUEUE block. "LIFO" reverses the sequence so that newcomers are placed at the head of the chain. Sequencing by parameter places those transactions with a high value in the specified parameter at the back of the user chain.

The C-field of the user chain is optional. Any entry in the C-field would be the label of a block to which the transaction is to be routed next if the user chain is empty. This is typically a label attached to the SEIZE block for the facility for which the transaction is destined. If there is no entry in the C-field of the LINK block, the transaction will be attached unconditionally to the user chain and will remain there until it is detached by a different transaction unlinking it. This lack of a C-field argument can cause unintended errors in models written by a novice and should be applied with extreme caution.

While the LINK block commonly precedes a SEIZE block just as a QUEUE block does, an UNLINK block is located and behaves entirely differently from a DEPART block. Once a transaction is able to gain access to a SEIZE block following a QUEUE block, it proceeds to a DEPART block and removes *itself* from its position at the head of the queue. In the LINK-UNLINK system, it is some predecessor transaction which unlinks the next transaction from the user chain and directs it to its next destination in the model. The A-field of the UNLINK block gives the identifying number of the user chain from which a transaction is to be unlinked when some other transaction passes through the UNLINK block. The B-field of the UNLINK block gives the label of the block which is the destination of the *unlinked* transaction. In a fairly simple model, this label would be the same as the C-field label of the LINK block. The C-field of the UNLINK block specifies how many transactions are to be unlinked from the chain. While normally this number would be 1, any number or the word "ALL" are acceptable field arguments. "ALL" might apply when a school bus arrives at its destination and every occupant gets off the bus.

Fields D, E, and F of the UNLINK block are optional. The D-field may contain the number of a parameter. If there is no entry in the E-field of the UNLINK block, GPSS will attempt to unlink the first transaction in the user chain whose value in the specified parameter is the same as the value in the specified parameter of the unlinking transaction. If there is an entry in the E-field of the UNLINK block, GPSS will attempt to unlink the first transaction in the user chain whose value in the specified transaction is the same as that given in the E-field of the UNLINK block.

Another acceptable entry in the D-field of the UNLINK block is the word "BACK." When this is the case, transactions will be unlinked from the back of the user chain. Thus, a user chain onto which transactions are linked in LIFO order and which are unlinked from the back are effectively processed in FIFO order.

If the user chain is found to be empty or if the conditions for unlinking are not met (i.e., not enough transactions on the chain to permit unlinking of the specified number of transactions or no match between the desired parameter value of the unlinking transaction and others on the user chain), the F-field offers an alternative label address to which the *unlinking* transaction might be directed. This could be the label of another UNLINK block which is not so fussy about the attributes of the transactions which are to be unlinked, or it might be the label of a block whose count would reflect the number of unsuccessful matches, for example. If the D-, E-, and F-fields are unused or if there is a successful match, the destination of the unlinking transaction is assumed to be the next sequential block.

The summary output produced by a LINK-UNLINK pair is a bit more limited than the output of a QUEUE-DEPART pair. For each chain, the total number of entries, the average time per transaction, the current contents, the average contents, and the maximum contents are given.

Table 7-4 shows a GPSS program which replicates the conditions of a regular QUEUE-DEPART sequence with a LINK-UNLINK sequence.

TABLE 7-4   GPSS PROGRAM
FOR FIFO QUEUE EMPLOYING
USER CHAINS

```
        JOB
        GENERATE 101, 101
        LINK 1, FIFO, NXT
NXT SEIZE TKTF
        ADVANCE 84, 84
        RELEASE TKTF
        UNLINK 1, NXT, 1
        TERMINATE 1
        START 100
        END
```

Table 7-5 shows a portion of a somewhat more complex user-chain model. In this model, a bus en route to a football game picks up passengers at various stops (interarrival time distribution is not shown), and customers join the user chain (seats and standing positions) in FIFO order. We will assume that there are always sufficient spaces for all passengers. When the bus reaches the stadium, which is the destination of all passengers, passengers who are players are given priority in getting off the bus. When there are no more player-passengers left, the remainder of the passengers disembark in FIFO order.

TABLE 7-5   GPSS MODEL OF BUS
OPERATION EMPLOYING COMPLEX
USER CHAIN

```
JOB
GENERATE 101, 101
ASSIGN 1, 2
LINK 1, FIFO
GENERATE 30, 30
ASSIGN 1, 1
LINK 1, FIFO
GENERATE 1000
UNLINK 1, TERM, ALL, 1, 1
UNLINK 1, TERM, ALL
TERM TERMINATE 1
START 1
END
```

In this model, riders join the user chain at intervals of 101,101 seconds. They are not waiting for a facility to become free but rather for the bus ride, taking 1000 seconds, to be completed. The riders are linked unconditionally to the user chain and are unlinked only by a dummy transaction which signals the end of the ride. All those with a value of 1 in parameter 1 (player-passengers) disembark first, followed by the rest of the passengers.

ASSEMBLY SETS

Every transaction which is produced by a GENERATE block is "born" with a uniquely identifying sequence number. Transactions may, in turn, give birth to other transactions which possess many of their genetic characteristics. The group consisting of the parent transaction and any of its clones is called an assembly set. After the parent transaction is created by a GENERATE block, the clones are produced by one or more SPLIT blocks. The A-field of a SPLIT block gives the number of copies or clones to be made. The B-field states the label of the block to which the copies are to be routed. Often, the original and the copies proceed to the same destination.

The purpose of assembly sets and copy transactions is to permit the system and the model to work in parallel on different parts of a transaction and subsequently to synchronize their movements before they proceed to further processing. There are several block types which facilitate this. For example, pairs of MATCH blocks may be placed at different points in the model logic. Members of the same assembly set proceeding along different branches of the model will be halted at one of the paired MATCH blocks until both MATCH blocks have intercepted at least one member of the same assembly set. Both transactions will be simultaneously released to proceed with processing. An example of the usefulness of this feature would be to represent two canvassers who operate along different streets but agree to meet at certain points before proceeding further.

Just as a SPLIT block will create any desired number of copy transactions within the same assembly set, the ASSEMBLE block will destroy any desired number of copy transactions from the same assembly set. The parent transaction is not destroyed. The A-field of the ASSEMBLE block indicates the number of copy transactions to be destroyed.

The GATHER block shares some of the properties of the MATCH and ASSEMBLE blocks. As in the case of the MATCH block, it does not destroy transactions, yet as with the ASSEMBLE block, it attempts to cluster multiple units from the same assembly set in one spot in the model.

Imagine a meat rendering and packing plant in which beef carcasses proceed down a conveyor. A single carcass will be subdivided into ground meat sufficient for perhaps 10 dozen frankfurters. After the frankfurters have been placed in casings and sealed, they must be regrouped into units of a dozen for exterior wrapping and ultimate shipment. Table 7-6 shows a portion of a GPSS program that would accomplish this with assembly sets.

TABLE 7-6   GPSS PROGRAM TO
PREPARE AND PROCESS
FRANKFURTERS BY MEANS OF
ASSEMBLY SETS

```
          JOB
          GENERATE 101
          SEIZE DBONF
          ADVANCE 50
          RELEASE DBONF
          SEIZE CHOPF
          ADVANCE 30
          RELEASE CHOPF
          SPLIT 120, CASE
          TERMINATE
     CASE QUEUE CASEQ
          SEIZE CASEF
          DEPART CASEQ
          ADVANCE 3
          RELEASE CASEF
          ASSEMBLE 12
          TERMINATE 1
          START 10
          END
```

In Table 7-6, a single carcass or parent transaction is deboned and the meat is chopped, after which the bones are discarded at the TERMINATE block. The meat is partitioned into 120 equal copies or sections, each of which waits to be passed through the casing machine, a possible queuing situation in an otherwise paced environment. After the franks are cased, they are aggregated into collections of a dozen to be wrapped. After that, the relevant unit of measure of output is no longer the carcass or the frank but, rather, the package of a dozen. When the initially planned 10 dozen packages have been completed, the run ends.

EXTENSIONS OF EARLIER CONCEPTS

We have seen that savevalues allow the modeler to store constants whose values may change periodically. It is often useful to conceptualize not just a single savevalue or group of independent savevalues but an array of savevalues whose properties are interconnected in some way. For example, a medical clinic may consist of six physicians, each of whom works 8 hours a day. The receptionist may implement the scheduling heuristic that work should be distributed evenly among all doctors and that their appointments should be begun at 9 A.M. and continued contiguously until either the workday ends or the supply of patients is exhausted. Elements such as lunch and other breaks, start and stop time preferences of the various doctors, and length of appointments for various purposes may be added as embellishments to the model. The mechanism for holding the schedule is a matrix savevalue whose dimensions are 6 (doctors) by 8 (hours).

As mentioned already, most of the computations provided by GPSS are integer in nature to conserve computer time and space. This includes calculations in variable statements. While addition, subtraction, and multiplication of integers pose no threat of loss of accuracy in the result, division of one integer by another often produces a fractional component in the answer. A standard GPSS variable statement which includes this type of division will preserve the integer portion of the quotient and discard the decimal portion. A floating-point variable or FVARIABLE statement definition permits the full quotient to be preserved and used in the totality of calculations in the FVARIABLE statement. Any truncation of fractional parts is done only after all computations in the FVARIABLE statement have been completed rather than immediately after the division step or steps.

Integer and floating-point variable statements as described perform the common arithmetic operations of addition, subtraction, multiplication, division, and modulus division. Another type of variable statement employs logical operators rather than arithmetic operators. This type of variable is called a Boolean variable or BVARIABLE. It is frequently used in place of a series of TEST and/or GATE blocks. Each element in the BVARIABLE statement is evaluated as being true or false, having a value of 1 or 0, respectively, at the time the Boolean variable is referenced. Operators connecting the elements in the statement may be the same as the set used in test blocks (i.e., E, NE, L, G, LE, and GE) or the specifically Boolean operators ''and,'' symbolized by an asterisk, and ''or,'' symbolized by a plus sign. Thus, if a transaction were to proceed through a TEST block only if a particular function had a value within a specified range or if an arithmetic variable had a value less than some quantity and if a particular storage were not full, a Boolean variable combined with a single TEST block could establish whether or not the transaction might proceed.

We have discussed functions that are either discrete or continuous. In either case, the arguments are numeric. An attribute-valued function permits the Y-coordinates to be nonnumeric. Depending on the context within the model in which the function is called, the attribute values might be statement labels or SNAs.

The X-coordinate of a function is often a decimal fraction between 0 and 1 produced by one of the eight random number generators in GPSS. Sometimes it is desirable to control as well as to replicate the stream of values produced by the random number generators. One reason for this is to implement sophisticated methods of minimizing run length while keeping undesirable statistical properties of the output under control. This topic will be discussed in depth in Chapter 8. The RMULT statement enables the analyst to initialize one or more of the random number generators at some specifically desired values.

There are times when facilities or storages are prevented from serving customers even though they are not currently occupied by a transaction. These times might include lunch and other break periods. If the facility or storage were taken over by a high-priority transaction at the beginning of such a period and prevented from returning to regular activity until the prescribed period had elapsed, the desired effect on customers (having to wait, come back, or go away permanently) would be achieved. The problem with this approach is that the time during which the facility or storage was "busy" for these reasons would be included in the summary statistics for facility utilization and time per transaction, thus distorting perceptions of these statistics for regular transactions. One way in which to avoid these consequences while achieving the desired inactivity is to use the FAVAIL, FUNAVAIL, SAVAIL, and SUNAVAIL (facility or storage available or unavailable) blocks.

Sometimes it is necessary to change the priority of a transaction during its progress through the model. For example, after waiting a certain amount of time, a customer may become irate and demand immediate attention. A PRIORITY block at such a point can indeed change the customer's priority. The potential conflict lies in the fact that the GPSS processor attempts to advance any transaction as far as possible through the model once it begins to move that transaction in any particular unit of time. Possibly there are other customers whose priorities are as high as the new priority of the irate customer, and, since this irate customer might be construed as the last-in of the priority group, it is preferred that his or her progress be temporarily discontinued in favor of his predecessors at the same priority level after entering the PRIORITY block. This may be accomplished with a BUFFER block. When a transaction encounters a BUFFER block, it is held there while the GPSS processor returns to the string of other transactions eligible for movement in that time unit and causes them to progress appropriately. GPSS then returns to the transaction at the BUFFER block and moves it as far as is possible. It should be noted that this is not a delay in the sense that a transaction may be delayed in a queue or a user chain, since GPSS will try again to move the transaction in the same time unit at which it was buffered.

The relationship between members of the same assembly set is very specific—one transaction is the parent and the others are clones. Members of the set may be created, destroyed, synchronized, and batched. There are other commonalities of relationship other than parent to child which the modeler might want to identify and track. For example, customers who purchase at a wholesale price or employees who belong to a union may appear anywhere within a model, not necessarily created by the same GENERATE block and not necessarily sharing all of the same attributes. Aggregations of transactions which have at least one attribute in common other than a parent-child relationship are called GPSS groups. A transaction may "JOIN" or "REMOVE" itself from a group. Priorities or parameter values of one or more members of a group may be "ALTERed" *en masse*. A transaction may be "EXAMINED" to determine if it is a member of a particular group and routed accordingly to a destination block. The "SCAN" block augments the properties of the "EXAMINE" block to search for specific properties within the group or subject transaction.

The analyst may want to collect and tabulate transit time data for only a subset of the span of blocks between generation and termination of a transaction. To accomplish this, a MARK block is placed at the point where the subspan begins. The time at which the transaction passes this point is recorded in a parameter. A table statement in the housekeeping section at the end of the program is used in conjunction with the MARK block. The SNA MP followed by the parameter number in which the recorded MARK time has been stored is the argument of the table instead of M1, the total transit time.

A special type of table containing the frequency distribution of waiting times in a particular queue is called a QTABLE. The format of the QTABLE statement is similar to that of the TABLE statement except that the name or number of the queue must be added to the list of field arguments.

Most other high-level programming languages contain a mechanism by which some portion of the program will be repeated for a specified number of iterations. In FORTRAN, for example, this is called a "DO" loop, while in BASIC it is called a "FOR-NEXT" loop. The comparable GPSS block, whose name actually is LOOP, assumes that the number of times it is desired for a transaction to repeat the loop has been stored in a parameter whose number is found in the A-field of the LOOP block. After each pass through the loop, the A-field count is decremented by 1. If the resulting A-field count is not equal to 0, the transaction branches back to the block whose label is found in the B-field. When the count has been fully decremented, the transaction proceeds to the next sequential block.

The COUNT block allows the analyst to determine attributes of permanent entities such as storages and queues at any point in time. For example, the COUNT block will report the number of full storages or the number of queues whose contents are between 2 and 5.

The SELECT block behaves similarly to the COUNT block except for two characteristics. First, the transaction will be routed to the first facility or other permanent entity having the desired characteristic, or if no permanent entity meeting the specification is found, the transaction will be routed to an alternate exit. Second, the logical operators MAX and MIN may be used, for example, to send the transaction to the shortest queue.

OPERATING SYSTEM INTERACTIONS WITH GPSS

Two common functions of an operating system are to perform linking and to transform the object module into executable form. In the early stages of GPSS model building, the analyst may be more interested in finding and correcting compilation (grammatical) errors than in examining the standard simulation output. Compilation without execution saves computer time and cost when it is appropriate to avoid execution. Many GPSS compilers require the SIMULATE statement in the GPSS program to proceed to the execution phase. If the SIMULATE statement is absent, the program will compile but will not execute.

Every GPSS compiler includes a default maximum quantity of entities which the analyst may want to use, such as savevalues, storages, and facilities. Each of these entities requires a certain amount of computer work space. If the analyst needs more of one entity type and fewer of another than are offered by the default option, a REALLOCATE statement may be included in the program to specify the quantity of each entity to be provided in other than the default amount.

As will become evident in Chapter 8, it is difficult for the analyst to decide *a priori* just how long a simulation run will be necessary to derive the desired statistical conclusions. If the simulation run proves to be too short, extending it may involve repeating the portion of the run which has already been completed. The SAVE/READ feature avoids this unnecessary cost. When the model is initially run, a SAVE statement is placed at the end of the program. Assuming that appropriate operating system commands have been provided, the object program as it appears at the end of the run will be written onto a magnetic tape. If the analyst wants to continue the run at a later time, all that is needed is to define the tape via operating system commands and provide a few GPSS instructions: READ, to read the contents of what is now an input tape rather than an output tape; START, to specify the length of the additional run; SAVE, if the object program at the end of the extended run is also to be saved; and END.

Standard GPSS output offers both advantages and disadvantages. The advantages are the relative completeness of information provided and the lack of need for programming by the analyst to create this output. A REPORT feature of many GPSS compilers allows the user to define and obtain graphical output of information ordinarily stored by GPSS. However, this combination of graphical and tabular output of GPSS is insufficient for the analytical and statistical purposes of the user. Calculation ability within GPSS is limited by the essentially integer basis of the compiler. When more extensive mathematical operations are desired, a HELP block may be included in the program. The HELP block calls a subprogram in FORTRAN, PL/1, or Assembler to perform the requisite operations. Fields associated with the HELP block indicate which SNAs are to be passed to the subroutine in the other language. The GPSS program and the non-GPSS subprogram must be linked through operating system commands. Some of the valuable statistical tests of simulation output which might be carried out with a HELP block are shown in Chapter 8.

Sometimes the analyst may want to provide input transactions to the GPSS model in some manner other than by using GENERATE statements. One possible reason might be to ensure that subsequent variations of the model are subjected to precisely the same stream of input transactions as the original version of the model. The stream of input transactions may be written on a tape called a JOBTAPE either with the GPSS WRITE block or with some other programming language. The job-tape must be defined through operating system commands and will contain information for each transaction such as its interarrival time and the label of the block in the GPSS model for which the transaction is destined.

A summary of GPSS blocks and statements and their field requirements and options appears in Appendix Table VII.

THE STATE OF THE ART IN GPSS COMPILERS

The GPSS language was originally developed in the early 1960s by Geoffrey Gordon of IBM. By the late 1960s, GPSS II and GPSS III had been implemented. Soon to follow was GPSS/360, designed expressly for IBM's then-current System/360 series of computers. GPSS V appeared in the 1970s and remains IBM's most advanced version of the language. Within the past 10 years, other computer vendors such as Honeywell, Control Data, UNIVAC, and RCA,* to name a few, have developed GPSS compilers. Some proprietary software is offered by companies other than mainframe computer vendors. For example, a highly sophisticated version of GPSS, called GPSS/H, has been developed by James O. Henriksen of Wolverine Software in Annandale, Virginia, for use on IBM and IBM-compatible computers as well as on the Digital Equipment Corporation VAX series of super-minicomputers. Another version of GPSS which has been implemented on the VAX series was developed by faculty in the computer science department at the University of Western Ontario in London, Ontario, Canada.

As the cost of computer memories continues to fall and devices such as array processors, which expedite simultaneous arithmetic calculations, become more widely available, the cost of simulation execution will also drop and the demand for simulation modeling may be expected to increase. Concurrently, we should find GPSS software for smaller and smaller computers and GPSS software with more comprehensive capabilities being offered.

---

*The GPSS programs shown in this book were written for and compiled by the RCA flow simulator version of GPSS, now maintained by UNIVAC for its Series 90 computers.

SUMMARY

It is often desirable to alter the distribution of interarrival and/or service times during a simulation run. This may be accomplished by using overlays or savevalues to redefine distribution parameters. When service time is affected by queue length, the SNA for queue length may be used as the argument of a function defining service time.

For those situations in which waiting lines do not conform to FIFO queue discipline with no reneging or jockeying, user chains provide alternative queue disciplines which suit many cases. User chains also cause the simulation to run more efficiently, but their logic is much different from the logic of standard queues.

In production environments, raw materials are often subdivided into components which may be aggregated in other ways at other times to create final products. Assembly sets allow the analyst to subdivide, aggregate, and synchronize the movement of such items through the model.

There are a great many additional GPSS blocks and statements which are convenient for modeling specific aspects of real-world situations. There is reason to infer from the past history of the language that it will become even more powerful, efficient, and widely used in the future.

QUESTIONS

1. Why might the analyst want to redefine the distribution of interarrival times during a run?

2. What is an overlay? How is it programmed? Why does it generate an error message?

3. Why might the analyst want to redefine the distribution of service times during a run?

4. Why might it be impossible to effect the redefinition of service times in the same way that interarrival times can be redefined?

5. Suggest two ways of altering the distribution of service times during a run and explain the limitations of each method.

6. In what way might queue length affect service time? How might that effect be included in the simulation model?

7. Explain the difference in the programming of a standard QUEUE-DEPART sequence and the programming of a LINK-UNLINK sequence.

9. In what way are user chains more efficient than standard queues?

10. In what aspects are LINK blocks more flexible than QUEUE blocks?

11. In what ways are UNLINK blocks more flexible than DEPART blocks?

12. Briefly summarize the characteristics and usefulness of any three GPSS blocks discussed in general terms in the latter part of the chapter.

13. Briefly summarize the characteristics and usefulness of any three GPSS statements which involve interaction with the operating system which are discussed in general terms in the latter part of the chapter.

PROBLEMS

Assume that the data collected in the problem section of Chapter 2 and applied to the problem sections of Chapters 3, 6, and 7 are applicable to Problems 1 through 6. The mean interarrival time was 421.75 sec and mean service time was 316.769 sec, both negative exponentially distributed.

1. Suppose that customer traffic intensity increases after 3 P.M. so that the distribution of interarrival time after 3 P.M. is uniform with a mean of 400 seconds and a modifier of 300 seconds. Run this model for an 8-hour day beginning with 9 A.M. Briefly comment on your results.

2. Suppose that the prevailing service level given falls when a substitute server covers the lunch break from 12:15 P.M. to 1:00 P.M. The substitute server has a mean service time of 330 seconds, also negative exponentially distributed. Run this model for an 8-hour day beginning with 9 A.M. Briefly comment on your results.

3. Suppose that the model described is affected by the length of the waiting line. When no one is waiting, mean service time is approximately 317 seconds, as given. However, when one or two people are waiting, mean service time decreases to 310 seconds, and when more than 2 people are waiting, mean service time drops to 300 seconds. All times are negative exponentially distributed. Run this model for an 8-hour day beginning with 9 A.M. and briefly comment on your results. (Hint: A GPSS model may contain more than one type of function.)

4. Rewrite the basic program described so that a user chain is substituted for the queue. Assume that everything else remains the same, including the FIFO queue discipline and the destination of transactions. Run the model for an 8-hour day.

5. Suppose that in the basic model each customer requires two types of service which can be done in parallel by two different servers. Both servers exhibit the same distribution of service times for each type of work, but each server specializes in one portion of the work only. Arrange to make a copy transaction which can be processed by the second server and subsequently reunited with the original transaction before termination from the model. Run the system for a service period of 8 hours.

6. Choose any of the block types briefly discussed in the latter part of the chapter. Obtain the GPSS manual for your computer system, describe in detail the grammatical and field requirements for the block under the rules of your system, and write and run a small GPSS program which makes use of this block.

7. Choose any of the statement types requiring interaction with the operating system which are briefly discussed in the latter part of the chapter. Obtain the GPSS manual for your computer system and either the operating system manual or the advice of a system consultant. Describe in detail the grammatical and field requirements for the statement as well as the necessary operating system commands under the rules of your system. Write and run a small GPSS program which makes use of this statement.

chapter eight

TESTING FOR INDEPENDENCE AND STEADY STATE

PURPOSE OF THIS ANALYSIS

Previous chapters have enabled the analyst to design, program, and debug a simulation model so that no grammatical or obvious logical errors remain. There are subtle problems which may prevent the model from achieving the objectives for which it is intended. These objectives are to estimate system performance parameters and to determine which of several alternative configurations is superior. The statistical analysis to accomplish these objectives will be discussed in Chapter 9. In this chapter, we will test whether or not the model possesses the statistical properties necessary for further accurate analysis, and if it does not, some techniques for "sanitizing" the model will be given.

There are two properties which we will want the model to have. The first is that the model's output should be representative of the steady-state operation of the real system, unless the analyst has specifically decided that transient state conditions are of interest. The second is that the model's output should exhibt serial independence of successive performance observations. When these two properties are present, we may proceed to validate the model (i.e., test whether or not the output of the model is reasonable compared with original observations of the system, analytical model projections, or *a priori* guesstimates about system performance).

If the model is not in steady state at the time that performance measurements are taken, the analyst may make erroneous inferences about the system. For example, nearly every service establishment is closed for some portion of the day and night, only to reopen later. During the period immediately following reopening, there will be few or no customers in process, and performance measures such as transit time and queue length may be abnormally low until the system builds to more typical levels of operation.

If the performance observations of the model are autocorrelated, the mean value of these observations will be an unbiased estimate of the true mean system performance, but the standard deviation and variance of performance may be understated. That is, autocorrelated values may cluster together, and thus a sample of any given size is less likely to contain extreme values than is the case in the population at large. The impact of this on formal statistical testing and estimation is that confidence intervals will tend to be inappropriately narrow and hypothesis testing may encourage Type I errors, rejecting the null hypothesis of no difference in performance of alternative systems when there is, in fact, no significant difference.

Our tasks, then, are to examine the performance outputs of the model, to decide whether or not the system output reflects steady-state conditions, to decide whether or not consecutive performance observations are statistically independent, and if not, to take appropriate remedial actions. The remedial action for autocorrelation may be to delete certain observations from the sample of performance output or may be to substitute the mean values of adjacent observations for the original values of the observations. The remedial action for transients may be to truncate them or to include them in a longer simulation run to diffuse their impact. Any of these actions may necessitate running the simulation model for a longer period of time than was originally anticipated in order to boost the sample size to compensate for observations which have been deleted or averaged. We re-perform tests for steady state and statistical independence using the augmented sample and continue the cycle until the data successfully pass the tests.

VALIDATION OF THE TICKET SELLER SYSTEM

Table 8-1 shows the GPSS model of the ticket seller system with a PRINT block inserted just prior to termination of the transactions from the system.

TABLE 8-1 GPSS MODEL OF TICKET SELLER SYSTEM WITH PRINT BLOCK OUTPUT OF TRANSIT TIME

```
JOB
 INITIAL X1, 0
EXPON1 FUNCTION RN3, C24
0, 0/. 1, . 104/. 2, . 222/. 3, . 355/. 4, . 509/. 5, . 69
. 6, . 915/. 7, 1. 2/. 75, 1. 38/. 8, 1. 6/. 84, 1. 83/. 88, 2. 12
. 9, 2. 3/. 92, 2. 52/. 94, 2. 81/. 95, 2. 99/. 96, 3. 2/. 97, 3. 5
. 98, 3. 9/. 99, 4. 6/. 995, 5. 3/. 998, 6. 2/. 999, 7. 0/. 9997, 8. 0
```

TABLE 8-1 (cont.)

```
EXPON2 FUNCTION RN4, C24
0, 0/. 1, . 104/. 2, . 222/. 3, . 355/. 4, . 509/. 5, . 69
. 6, . 915/. 7, 1. 2/. 75, 1. 38/. 8, 1. 6/. 84, 1. 83/. 88, 2. 12
. 9, 2. 3/. 92, 2. 52/. 94, 2. 81/. 95, 2. 99/. 96, 3. 2/. 97, 3. 5
. 98, 3. 9/. 99, 4. 6/. 995, 5. 3/. 998, 6. 2/. 999, 7. 0/. 9997, 8. 0
  GENERATE 101, FN$EXPON1
  TABULATE INTARV
  QUEUE 1
  SEIZE 1
  DEPART 1
  ADVANCE 84, FN$EXPON2
  RELEASE 1
  SAVEVALUE 1, M1
  PRINT , , X
  TABULATE TRANSIT
  TERMINATE 1
INTARV TABLE IA, 0, 20, 20
TRANSIT TABLE M1, 50, 50, 20
  START 100
  RESET
  START 100
  END
```

To confirm the correct operation of the random number generator for interarrival time, the interarrival time for each transaction is captured and summarized in the GPSS table named INTARV, while transit time aggregate data is stored in the GPSS table named TRANSIT. A total of 200 transactions is requested. The first 100 might give some indication of transient-state conditions, since the model begins in the empty state. Some results of the GPSS run are shown in Table 8-2.

TABLE 8-2   SELECTED AGGREGATE OUTPUT STATISTICS
FROM TICKET SELLER GPSS RUN

	FIRST 100	SECOND 100
AVERAGE UTILIZATION	. 871	. 792
AVG. SVC. TIME/TRANS.	83. 460	90. 370
AVG. QUEUE CONTENTS	5. 034	2. 223
MAX. QUEUE CONTENTS	15	8
TL. QUEUE ENTRIES	108	104
ZERO QUEUE ENTRIES	14	24
AVG. Q TIME/TRANS.	446. 370	243. 875
NZ–AVG. Q TIME/TRANS.	512. 851	317. 037
CURRENT CONTENTS OF Q	8	4

TABLE 8-2 (cont.)

MEAN INTARV TIME	89.302	118.406
S.D. INTARV TIME	82.657	96.662
MEAN TRANSIT TIME	548.140	357.420
S.D. TRANSIT TIME	288.328	253.148
RELATIVE CLOCK TIME	9577	11410

There seems to be a substantial difference in the values of many of these statistics from one time period to the next. Comparing them with the original sample data collected to define the system in Chapter 2, we find that the summary statistics in Table 2-2 revealed average queue length to be 1.608, average transit time to be 213.304, average waiting time to be 128.804, and average server idle time percentage to be 17.17. These statistics are more comparable to those for the second group of 100 observations from the GPSS run than to those for the first group of 100. It is desirable to know whether the discrepancy is due to random variation or whether there is an inaccuracy in the GPSS model or in the original sample. Such inaccuracies lie in the GPSS program grammar or logic, in the random number generators, or in the biasedness of the original sample itself.

The original sample statistics are closer to the second GPSS grouping, suggesting that the system might not have been in steady state for the intitial 100 observations. This conjecture is flawed, however, by noting that the system was more congested in the first period than in the second. This result is inconsistent with the hypothesis of transients following the initialization of the model in an atypical empty state. A more plausible explanation is that mean interarrival time for the first period was 89 seconds, considerably below the stipulated mean of 101 seconds. Hence, system utilization will be higher than expected, and performance from the customer's point of view will suffer. The mean interarrival time for the second group is quite a bit higher than expected, however, so on the average, it seems fair to conclude that the random number generation process for interarrival times is operating correctly.

Since GPSS employs an integral clock and integer arithmetic, draws from continuous distributions such as the negative exponential may be distorted through truncation. When the mean values are rather large, as in this example, however, the effect of this distortion is probably not important.

We might also contrast these results to those of the analytical $M/M/1$ model in Chapter 3. This model projected average queue length to be 3.88, average transit time to be 476.55, average waiting time to be 393.02, and average server idle time percentage to be 17.528. All these estimates fall between the comparable values for the first and second batches of 100 simulated observations. This might lead us to the conclusion that the system has not yet stabilized after an initial transient period or that a larger, pooled sample is necessary to ensure representativeness.

At this point, we may consider it sufficient to say that the model appears to be generally valid; that is, the results are at least of the same order of magnitude as those expected based on the original sample and the analytical model.

Table 8-3 displays the 200 observations of transit time produced by the GPSS program in Table 8-2. Descriptive statistics calculated for these observations corroborate the GPSS summary statistics. Finally, the observations are shown in ascending rank order to facilitate validation with goodness-of-fit testing.

TABLE 8-3   DESCRIPTIVE STATISTICS FROM TICKET SELLER TRANSIT TIMES WITH SAMPLE SIZE OF 200 OBSERVATIONS

TRANSIT TIME RAW DATA

31.0	1.0	36.0	94.0	123.0	13.0	37.0	26.0	5.0	2.0
20.0	28.0	96.0	149.0	297.0	494.0	541.0	398.0	408.0	259.0
236.0	227.0	16.0	45.0	73.0	97.0	156.0	199.0	364.0	829.0
682.0	655.0	660.0	683.0	882.0	852.0	838.0	829.0	889.0	781.0
771.0	809.0	696.0	928.0	1047.0	1052.0	784.0	854.0	813.0	759.0
872.0	732.0	719.0	923.0	964.0	708.0	602.0	646.0	731.0	803.0
607.0	560.0	551.0	369.0	731.0	646.0	643.0	495.0	519.0	739.0
633.0	610.0	638.0	574.0	560.0	474.0	448.0	700.0	654.0	691.0
710.0	620.0	666.0	613.0	558.0	558.0	659.0	713.0	622.0	700.0
853.0	865.0	701.0	320.0	850.0	696.0	682.0	704.0	616.0	547.0
544.0	503.0	523.0	532.0	577.0	410.0	519.0	655.0	609.0	586.0
472.0	531.0	556.0	654.0	652.0	631.0	528.0	523.0	487.0	457.0
812.0	827.0	802.0	690.0	710.0	695.0	572.0	231.0	269.0	503.0
510.0	563.0	575.0	535.0	450.0	525.0	609.0	717.0	665.0	554.0
368.0	352.0	207.0	72.0	22.0	76.0	107.0	54.0	44.0	39.0
27.0	48.0	41.0	105.0	87.0	40.0	106.0	41.0	168.0	357.0
312.0	348.0	93.0	103.0	27.0	8.0	246.0	207.0	215.0	200.0
218.0	180.0	120.0	56.0	51.0	8.0	39.0	158.0	173.0	108.0
15.0	178.0	165.0	181.0	127.0	121.0	60.0	226.0	396.0	447.0
354.0	211.0	191.0	637.0	815.0	868.0	784.0	784.0	692.0	496.0

TRANSIT TIME SUMMARY STATISTICS

THE MEAN IS   452.7798
THE VARIANCE IS   83117.81
THE STANDARD DEVIATION IS   288.3015
THE SAMPLE SIZE IS   200

TRANSIT TIME SORTED DATA

1.0	2.0	5.0	8.0	8.0	13.0	15.0	16.0	20.0	22.0
26.0	27.0	27.0	28.0	31.0	36.0	37.0	39.0	39.0	40.0
41.0	41.0	44.0	45.0	48.0	51.0	54.0	56.0	60.0	72.0
73.0	76.0	87.0	93.0	94.0	96.0	97.0	103.0	105.0	106.0

TABLE 8-3 (cont.)

107.0	108.0	120.0	121.0	123.0	127.0	149.0	156.0	158.0	165.0
168.0	173.0	178.0	180.0	181.0	191.0	199.0	200.0	207.0	207.0
211.0	215.0	218.0	226.0	227.0	231.0	236.0	246.0	259.0	269.0
297.0	312.0	348.0	352.0	354.0	357.0	364.0	368.0	369.0	396.0
398.0	408.0	410.0	447.0	448.0	450.0	457.0	472.0	474.0	487.0
494.0	495.0	496.0	503.0	503.0	510.0	519.0	519.0	523.0	523.0
525.0	528.0	531.0	532.0	535.0	541.0	544.0	547.0	551.0	554.0
556.0	558.0	558.0	560.0	560.0	563.0	572.0	574.0	575.0	577.0
586.0	602.0	607.0	609.0	609.0	610.0	613.0	616.0	620.0	622.0
631.0	633.0	637.0	638.0	643.0	646.0	646.0	652.0	654.0	654.0
655.0	655.0	659.0	660.0	665.0	666.0	682.0	682.0	683.0	690.0
691.0	692.0	695.0	696.0	696.0	700.0	700.0	701.0	704.0	708.0
710.0	710.0	713.0	717.0	719.0	731.0	731.0	732.0	739.0	759.0
771.0	781.0	784.0	784.0	784.0	802.0	803.0	809.0	812.0	813.0
815.0	820.0	827.0	827.0	829.0	829.0	838.0	850.0	852.0	853.0
854.0	865.0	868.0	882.0	889.0	923.0	928.0	964.0	1047.0	1052.0

We note that there is a considerable difference between the mean and standard deviation of transit time. Would this difference be sufficient to contradict the hypothesis that transit times are negative exponentially distributed when interarrival times and service times are negative exponentially distributed? Table 8-4 shows the results of a goodness-of-fit test of the negative exponential distribution with a mean of 453, using only 10 classes to simplify calculations.

TABLE 8-4   GOODNESS-OF-FIT TEST OF THE NEGATIVE EXPONENTIAL DISTRIBUTION WITH MEAN 453 SEC TRANSIT TIME AND SAMPLE SIZE OF 200 OBSERVATIONS

CLASS BOUNDARY DEFINITIONS

CLASS IDENTIFIER	CUMULATIVE GREATER-THAN PROBABILITY	CLASS BOUNDARY	CLASS WIDTH
1	1.0000	0.00	47.73
2	0.9000	47.73	53.36
3	0.8000	101.08	60.49
4	0.7000	161.57	69.83
5	0.6000	231.40	82.59
6	0.5000	314.00	101.08
7	0.4000	415.08	130.32
8	0.3000	545.40	183.68
9	0.2000	729.08	314.00
10	0.1000	1043.07	999.99

TABLE 8-4 (cont.)

TEST CALCULATIONS

CLASS	CLASS IDENTIFIER	OBSERVED FREQUENCY	RELATIVE FREQUENCY	EXPECTED FREQUENCY	χ^2
1	47.7	24.00	0.100	20.00	0.80
2	101.1	13.00	0.100	20.00	2.45
3	161.6	12.00	0.100	20.00	3.20
4	231.4	17.00	0.100	20.00	0.45
5	314.0	6.00	0.100	20.00	9.80
6	415.1	11.00	0.100	20.00	4.05
7	545.4	24.00	0.100	20.00	0.80
8	729.1	58.00	0.100	20.00	72.20
9	1043.1	33.00	0.100	20.00	8.45
10	999.0	2.00	0.100	20.00	16.20
TOTALS	200.00	1.000	200.00	118.40	

Since "observed" data from the simulation model were used to establish the mean in the null hypothesis, degrees of freedom equals 8. Critical χ^2 at the 0.05 level of significance is 15.507, compared with a computed χ^2 value of 118.40. Thus, we must reject the null hypothesis that transit times are negative exponentially distributed with a mean of 453 seconds. Since we are only slightly interested in this aspect of the model, and since this result might be produced because of an inadequately large sample, we will not ponder the matter further and proceed to the evaluation of steady state and independence.

DETECTING THE PRESENCE OF TRANSIENTS

The procedures for assessing the presence or absence of steady state are heuristics rather than algorithms. That is, they are rules of thumb rather than formal statistical tests, although certain heuristics make use of statistics somewhere along the way. We shall consider three popular heuristics and apply them to the ticket seller output data.

The first steady-state heuristic we discuss was developed by R. W. Conway.* This simple decision rule examines each data value and compares it with all succeeding values in the sample. If the particular data value is either larger or smaller than all remaining values in the sample, it is said to be part of a transient state. If the particular data value is neither the largest nor the smallest of the remaining data items in the sample, it is presumed to be part of the steady-state output. Conway's rule is effective when there is a clear path of continuously ascending or descending tran-

*R. W. Conway et al., "Some Problems of Digital Machine Simulation," *Management Science*, Vol. 6, no. 1, October 1959.

sient values at the beginning of a simulation run or during some clearly separable interval in the middle of a run. Examples might be a newly opened service facility or a peak-period work load. Because of random variation, this is rarely the case. As soon as one observation in an otherwise transient period meets the criterion for steady state, no further inspection is done and steady state is defined to commence with that observation. According to Conway's rule, the first observation in the ticket seller data set serves as the beginning of steady state. It has a value of 31, which is neither the smallest (1) nor the largest (1052) of the remaining values in the data set. This conclusion essentially suggests that there is no transient period at the beginning of the ticket seller run.

Two additional heuristics for steady-state determination are attributable to Emshoff and Sisson.* Both decision rules by Emshoff and Sisson compare subsets of the data with each other. The size of the subset is intended to be statistically large, that is, at least 30 if a z test is to be used, or sufficient to yield expected class frequencies of 5 or more if a χ^2 test is to be used. Approximations can be made, however, using smaller subgroup sizes. The smaller the subgroup size, the more likely the advent of steady state may be pinpointed without "wasting" some steady-state observations in an otherwise transient subset.

The Emshoff and Sisson above-and-below rule suggests that a subgroup will be considered to belong to steady state if there are approximately equal numbers of observations in the subgroup above and below the cumulative average value of the previous subgroups. A χ^2 goodness-of-fit test is a simple way of testing this hypothesis. There will be two categories, above and below the previous cumulative mean. The expected class frequencies, respectively, for subgroup sizes of 10, 20, and 30 are $(10 \times 0.5) = 5$, $(20 \times 0.5) = 10$, and $(30 \times 0.5) = 15$. At the 0.05 level of significance with 1 degree of freedom, critical χ^2 is 3.84. Having established this, we may work backward to solve for observed values of the subgroup mean which are not significantly different from the expected value. If computed χ^2 falls below 3.84, there is no significant difference between observed and expected class frequencies and the subgroup is presumed to be part of steady state. Observed class frequencies which disqualify the subgroup for steady-state status, for subgroup sizes of 10, 20, and 30, respectively, are 0 and 10, for subgroup size = 10; 0, 1, 2, 3, 17, 18, 19, and 20, for subgroup size = 20; and 0, 1, 2, 3, 4, 5, 6, 7, 23, 24, 25, 26, 27, 28, 29, and 30, for subgroup size = 30.

Table 8-5 displays the calculations for the above-and-below decision rule as applied to the ticket seller example with subgroup size equal to 30.

The first subgroup of 30 observations is considered the standard for the initial comparison. Subgroup number 4 in Table 8-5 is the first instance in which the number above and below the previous cumulative subgroup average is within the bounds of statistical equality. Having noted this, the analyst might elect to truncate the first three subgroups, containing 90 observations, from the sample on the

*James R. Emshoff and Roger L. Sisson, *Design and Use of Computer Simulation Models* (New York: Macmillan, 1970).

TABLE 8-5   EMSHOFF AND SISSON ABOVE-AND-BELOW
CRITERION AS APPLIED TO TICKET SELLER DATA WITH
SAMPLE SIZE OF 200 OBSERVATIONS

SUBGP. NUMBER	PREVIOUS CUMULATIVE AVERAGE	NUMBER ABOVE	NUMBER BELOW	CHISQ
2	176.667	30	0	30.00
3	486.983	27	3	19.20
4	527.555	22	8	6.53
5	548.025	13	17	0.53
6	526.120	0	30	30.00
7	460.044	7	13	8.53

grounds that they contain transients. As we shall see, however, perfunctory application of heuristics is not always a prudent course of action.

Emshoff and Sisson's second heuristic suggests that steady state has been
achieved when there is no statistically significant difference between the average of
the previous subgroup and the average of the current subgroup. As before, the first
subgroup is used as the initial standard for comparison. The test produces a z statistic measuring the difference between means of independent samples. The format is

$$z = \frac{(XBARP - XBARC)}{SQRT(PSESQD + CSESQD)}$$

where XBARP is the mean of the previous subgroup, XBARC is the mean of the
current subgroup, PSESQD is the square of the standard error of the previous subgroup, and CSESQD is the square of the standard error of the current subgroup.
Application of the Emshoff and Sisson moving average criterion to the ticket seller
data is shown in Table 8-6. The first subgroup for which the computed value of z
falls below the critical level of 1.96 at the 0.05 level of significance is subgroup 4.
This seems to confirm the results of the other Emshoff and Sisson heuristic.

TABLE 8-6   EMSHOFF AND SISSON MOVING AVERAGE
CRITERION AS APPLIED TO TICKET SELLER DATA

SUBGP. NUMBER	AVERAGE PREVIOUS SBGP.	AVERAGE CURRENT SBGP.	ST. ERR. PREV. SQRD.	ST. ERR. CURR. SQRD.	z
2	76.667	797.300	1321.29	431.89	−14.82
3	97.300	608.700	431.89	259.25	7.17
4	08.700	609.433	259.25	489.74	−0.03
5	09.433	438.500	489.74	2327.30	3.22
6	38.500	129.667	2327.30	328.03	5.99
7	29.667	387.400	328.03	3990.31	−3.92

A glance at the χ^2 and z statistics for the various subgroups reveals the deficiencies of the Emshoff and Sisson heuristics—after all, they are only rules of thumb. The χ^2 and z values fall to acceptable levels for subgroup 4, only to rise again to significant levels immediately thereafter rather than signaling the end of a transient period. Like Conway's rule, the Emshoff and Sisson heuristics detected a point which was simply the end of a cycle but which was followed by values which violated the heuristics once again. Table 8-7 displays a plot of the 200 data points in which the cyclic pattern is more evident than it was in the data list.

TABLE 8-7   PLOT OF 200 OBSERVATIONS FROM TICKET SELLER SYSTEM

```
1052.000: -----------------------------------------------XX--------------------- 
1015.759: --------------------------------------------------------------------- 
 979.517: -----------------------------------------------------------X--------- 
 943.276: ---------------------------------------------------X----------------- 
 907.034: -------------------------------------------X-------------X----------- 
 870.793: -------------------------------------X-----------X------------------- 
 834.552: ---------------------------X-----XXX-----------X-------------------- 
 798.310: ------------------------------------X-X----X-X---------X---------- 
 762.069: -------------------------------------X--------X--------------------- 
 725.827: -------------------------------------------XX--X--X-----X----X 
 689.586: ----------------------------X--X--------X------------------------- 
 653.345: --------------------------XX-----------------------X-------XX--- 
 617.103: ----------------------------------------------------X---X--------- 
 580.862: --------------------------------------------------------------------- 
 544.621: --------------X-----------------------------------------------XX------- 
 508.379: --------------X-------------------------------------------------XX- 
 472.138: --------------------------------------------------------------------- 
 435.896: --------------------------------------------------------------------- 
 399.655: ----------------XX--------------------------------------------------- 
 363.414: -------------------------X-----------------------------------X------ 
 327.172: --------------------------------------------------------------------- 
 290.931: -------------X------------------------------------------------------- 
 254.690: ------------------X-------------------------------------------------- 
 218.448: -----------------XX-------------------------------------------------- 
 182.207: ------------------------X-------------------------------------------- 
 145.966: -------------X----------X------------------------------------------- 
 109.724: ---XX-------X----------X-------------------------------------------- 
  73.483: ----------------------X--------------------------------------------- 
  37.241: X-X---XX--XX----------X---------------------------------------------- 
   1.000: -X---X--XX-----------X----------------------------------------------- 
          0    5   10   15   20   25   30   35   40   45   50   55   60   65   70
```

```
1052.000: --------------------------------------------------------------------- 
1015.759: --------------------------------------------------------------------- 
 979.517: --------------------------------------------------------------------- 
```

TABLE 8-7 (cont.)

```
943.276: ------------------------------------------------------------------------------
907.034: ------------------------------------------------------------------------------
870.793: -------------------XX---------------------------------------------------------
834.552: ----------------------XX---------------------------------------X--------------
798.310: ------------------------------------------------------------X-X--------------
762.069: ----------------------------------------------------------------------------
725.827: ----------X------X-----------------------------------------X-----------X--
689.586: -------X-X---------X--X--XXX------------------------X-X--------------
653.345: --X-----X---X---X------------------X-----XX------------------------X-
617.103: XX---------X-X---X---------X---------X------X---------------------X---
580.862: ---X----------------------------X----X--------------X----XX-------
544.621: ----X---------XX-----------XX--X-------XX---X------------------X-----X
508.379: ----------------------------XX---X----------X-----------XX----X----
472.138: -----X--------------------------------X-------XX------------------
435.896: ------X-------------------------------------------------X-----
399.655: ----------------------------------X---------------------------
363.414: ----------------------------------------------------------------
327.172: ----------------------------------------------------------------
290.931: ----------------------------------------------------------------
254.690: -------------------------------------------------------X----------
218.448: -------------------------------------------------------X----------
182.207: ----------------------------------------------------------------
145.966: ----------------------------------------------------------------
109.724: ----------------------------------------------------------------
 73.483: ----------------------------------------------------------------
 37.241: ----------------------------------------------------------------
  1.000: ----------------------------------------------------------------
         70   75   80   85   90   95  100  105  110  115  120  125  130  135  140

1052.000: -----------------------------------------------------------
1015.759: -----------------------------------------------------------
 979.517: -----------------------------------------------------------
 943.276: -----------------------------------------------------------
 907.034: -----------------------------------------------------------
 870.793: -----------------------------------------------------X----
 834.552: -----------------------------------------------------------
 798.310: ------------------------------------------------X-XX--
 762.069: -----------------------------------------------------------
 725.827: -----------------------------------------------------------
 689.586: -------------------------------------------------X-
 653.345: -----------------------------------------------X------
 617.103: -----------------------------------------------------------
 580.862: -----------------------------------------------------------
 544.621: -----------------------------------------------------------
 508.379: ----------------------------------------------------------X
 472.138: -----------------------------------------------------------
```

TABLE 8-7 (cont.)

```
435.896: -------------------------------------------------X----------
399.655: -------------------------------------------------X----------
363.414: XX---------------X-X-----------------------------X---------
327.172: ------------------X----------------------------------------
290.931: ------------------------------------------------------------
254.690: ------------------------X-----------------------------------
218.448: --X----------------------XX-X----------------X---X---------
182.207: -----------------X---------X-X------X--XXX-------X------
145.966: ----------------------------------------X-------------------
109.724: ------X------X--X-----XX--------X------X----XX--------------
 73.483: ---X-X--------X-----------------X-----------X------------
 37.241: ----X--XXXXXX--X-X------X---------X-X---------------------
  1.000: ----------------------X---------X----X------------------
          140   145   150   155   160   165   170   175   180   185   190   195   200
```

TESTING FOR AUTOCORRELATION

It appears that the presence of cycles in the data decreases the accuracy of the steady-state heuristics. Since we also want to remove serial dependence as a precondition for the statistical analysis in Chapter 9, we now use the runs-up-and-down test and the Spearman rank correlation test introduced in Chapter 4 to measure the extent of autocorrelation in the data. Table 8-8 shows the results of the runs-up-and-down test.

TABLE 8-8   RESULTS OF RUNS-UP-AND-DOWN
TEST FOR TICKET SELLER DATA WITH SAMPLE
SIZE OF 200 OBSERVATIONS

THE ORIGINAL SAMPLE SIZE IS	200
THE EFFECTIVE SAMPLE SIZE IS	198
THE NUMBER OF RUNS IS 91	
THE EXPECTED NUMBER OF RUNS IS	131.67
THE STANDARD DEVIATION IS 5.9057	
THE VALUE OF Z IS −6.8013	

The z value of the runs-up-and-down test indicates a large amount of autocorrelation. There are far fewer runs (too much clustering, too little variance) than would probably occur by sheer chance. One hopeful sign is that there are not many tie values in the transit time data, as evidenced by the closeness of the values of the original and effective sample sizes. The Spearman test results are summarized in Table 8-9.

The z value of the Spearman test confirms the extensive presence of autocorrelation in the data. There are two possible ways in which to decrease the degree of

TABLE 8-9 RESULTS OF SPEARMAN
TEST FOR TICKET SELLER DATA WITH
SAMPLE SIZE OF 200 OBSERVATIONS

THE ORIGINAL SAMPLE SIZE IS	200
THE SUM OF D SQUARED IS	130445.50
SUMU=	12.50
SUMV=	12.50
R= 0.90	
Z= 12.67	

autocorrelation in simulation output data. One way is to average each set of M successive observations in the data set and substitute the subgroup means as a new data set. Another way is to select every Nth observation to be placed in a new data set and discard the intervening observations. The advantage of the averaging method is that it does not entirely waste the rather expensive-to-generate simulation output because it incorporates the flavor of the individual data values in the average. However, it may require longer to remove the autocorrelation from the sets of means than from the sets of every Nth observation.

It is difficult to fix a value for M or N in advance. On the one hand, it is efficient to average or throw away as few observations as possible. On the other hand, the time of the analyst and the computing facility are valuable, too, and it is efficient to minimize their iterative use for averaging or selecting, runs testing, and Spearman testing. Since the z values in Tables 8-8 and 8-9 are very robust, we will try averaging successive subgroups of 5 from our data set of 200 observations. The result will be 200/5 or 40 means for the new data set. These means are shown along with summary statistics in Table 8-10.

TABLE 8-10 FORTY MEANS AND DESCRIPTIVE STATISTICS FOR
SUBGROUPS OF SIZE 5 FROM TICKET SELLER SYSTEM WITH 200
OBSERVATIONS OF TICKET SELLER SYSTEM

SUBGROUP MEANS

57.0	16.6	118.0	420.0	119.4	329.0	712.4	837.8	850.2	852.4
833.0	698.0	563.6	608.4	603.0	593.4	633.4	650.4	817.8	649.0
535.8	555.8	573.0	525.2	768.2	454.0	526.6	614.0	204.2	64.0
61.6	142.4	176.6	175.2	125.0	97.2	133.2	250.0	441.6	724.8

SUMMARY STATISTICS

THE MEAN IS 452.7788
THE VARIANCE IS   73890.75
THE STANDARD DEVIATION IS 271.8284
THE SAMPLE SIZE IS   40

Comparing the magnitudes of the summary statistics for the original 200 observations in Table 8-3, the means are, or course, the same but the variance has decreased. The proportionate change in the variance is far less than the proportionate change in the sample size, so we doubt that the reduced sample will pass the runs and Spearman tests. These are shown, respectively, in Tables 8-11 and 8-12.

TABLE 8-11   RESULTS OF RUNS-UP-AND-DOWN TEST FOR 40 MEANS OF SUBGROUPS OF 5 FOR TICKET SELLER DATA WITH 200 OBSERVATIONS

THE ORIGINAL SAMPLE SIZE IS 40
THE EFFECTIVE SAMPLE SIZE IS 40
THE NUMBER OF RUNS IS 18
THE EXPECTED NUMBER OF RUNS IS 26.33
THE STANDARD DEVIATION IS 2.6055
THE VALUE OF Z IS -3.0064

TABLE 8-12   RESULTS OF SPEARMAN TEST FOR 40 MEANS OF SUBGROUPS OF 5 FOR TICKET SELLER DATA WITH 200 OBSERVATIONS

THE ORIGINAL SAMPLE SIZE IS 40
THE SUM OF D SQUARED IS 2186.00
SUMU= 0.00
SUMV= 0.00
R= 0.78
Z= 4.80

While the degree of serial correlation as indicated by the z statistics in the runs test and the Spearman test has fallen, it is still statistically significant. Further averaging would deplete the sample size to 20 or fewer observations, well below the recommended levels for a z test. While Appendix Table VIII gives critical levels of Spearman's R for small sample sizes and while comparable tables are available for the runs-up-and-down test for small sample sizes,* it is preferable to make inferences based on larger samples if possible.

What would have happened if we had selected every fifth observation from the 200 values in the original data set and discarded all intervening values? Table 8-13 shows the 40 observations (not means) which remain, along with statistics summarizing them.

The mean of the selected observations in Table 8-13 differs somewhat from the common mean of the 200 observations and 40 averages because some of the original data items are no longer represented in the sample. The statistics for the standard

*Jean Dickinson Gibbons, *Nonparametric Methods for Quantitative Analysis* (New York: Holt, Rinehart and Winston, 1976).

TABLE 8-13   EVERY FIFTH OBSERVATION BEGINNING WITH THE FIRST AND
DESCRIPTIVE STATISTICS FOR THESE OBSERVATIONS SELECTED FROM THE
SAMPLE OF 200 OBSERVATIONS OF TICKET SELLER SYSTEM

31.0	13.0	20.0	494.0	236.0	97.0	682.0	852.0	771.0	1052.0
827.0	708.0	607.0	646.0	633.0	474.0	710.0	558.0	853.0	696.0
544.0	410.0	472.0	631.0	812.0	695.0	510.0	525.0	368.0	76.0
27.0	40.0	312.0	8.0	218.0	8.0	15.0	121.0	354.0	868.0

SUMMARY STATISTICS

THE MEAN IS 449.3499
THE VARIANCE IS   94749.38
THE STANDARD DEVIATION IS 307.8137
THE SAMPLE SIZE IS   40

deviation and variance also differ. The runs test and the Spearman test results for
this data set are shown, respectively, in Tables 8-14 and 8-15.

TABLE 8-14   RESULTS OF RUNS-UP-AND-DOWN
TEST FOR SET OF EVERY FIFTH OBSERVATION
SELECTED FROM ORIGINAL 200

THE ORIGINAL SAMPLE SIZE IS   40
THE EFFECTIVE SAMPLE SIZE IS   40
THE NUMBER OF RUNS IS   22
THE EXPECTED NUMBER OF RUNS IS   26.33
THE STANDARD DEVIATION IS   2.6055
THE VALUE OF Z IS -1.4712

TABLE 8-15   RESULTS OF SPEARMAN TEST
FOR SET OF EVERY FIFTH OBSERVATION
SELECTED FROM ORIGINAL 200

THE ORIGINAL SAMPLE SIZE IS   40
THE SUM OF D SQUARED IS   2870.00
SUMU=   0.50
SUMV=   0.50
R=   0.71
Z=   4.37

EXTENDING THE SIMULATION RUN

While the z statistic in the runs test is not significant for the reduced sample of 40
data items, the z statistic for the Spearman test is still too high for accurate use of the
statistics in Chapter 9. The Spearman test is considered more reliable than the runs

test, so we shall resign ourselves to extending the length of the simulation run. There is no scientific way in which to determine how long the run should be: it is a trial-and-error process just as the averaging or selection of every Nth was. Suppose that we choose a run length of 1000 observations. If we have made use of the GPSS READ/SAVE feature to store the status of the model at the end of the prior execution phase, an overlay of the START block will produce the next 800 transactions without having to re-create the first 200. If we have not used READ/SAVE, it will be necessary to begin the run again, continuing for the full 1000 transactions. The latter type of program is shown in Table 8-16.

TABLE 8-16 GPSS MODEL TO CREATE 1000 TRANSACTIONS FOR THE TICKET SELLER SYSTEM

```
JOB
INITIAL X1,0
EXPON1 FUNCTION RN3,C24
0,0/.1,.104/.2,.222/.3,.355/.4,.509/.5,.69
.6,.915/.7,1.2/.75,1.38/.8,1.6/.84,1.83/.88,2.12
.9,2.3/.92,2.52/.94,2.81/.95,2.99/.96,3.2/.97,3.5
.98,3.9/.99,4.6/.995,5.3/.998,6.2/.999,7.0/.9997,8.0
EXPON2 FUNCTION RN4,C24
0,0/.1,.014/.2,.222/.3,.355/.4,.509/.5,.69
.6,.915/.7,1.2/.75,1.38/.8,1.6/.84,1.83/.88,2.12
.9,2.3/.92,2.52/.94,2.81/.95,2.99/.96,3.2/.97,3.5
.98,3.9/.99,4.6/.995,5.3/.998,6.2/.999,7.0/.9997,8.0
GENERATE 101,FN$EXPON1
TABULATE INTARV
QUEUE 1
SEIZE 1
DEPART 1
ADVANCE 84,FN$EXPON2
RELEASE 1
SAVEVALUE 1,M1
PRINT ,,X
TABULATE TRANSIT
TERMINATE 1
INTARV TABLE IA,0,20,20
TRANSIT TABLE M1,20,20,20
START 1000
END
```

The relative clock time at the end of this simulation run is 102,737 seconds. No matter what type of computer is used, this length of run will be quite costly, and the analyst should consider whether it might be possible to evaluate the system less expensively, if less accurately. If READ/SAVE is not available, however, it will be cheaper to plan and execute a rather long run instead of having to repeat the initial segments of shorter runs which prove to be statistically unsatisfactory.

Aggregate GPSS output for the run of 1000 transactions shows average system utilization as 0.805, average service time per transaction as 82.666, maximum queue length as 16, average queue length as 2.428, zero queue entries as 216 out of 1006, average queue time per transaction as 247.939, nonzero average queue time per transaction as 315.730, average interarrival time as 102.229, and average transit time as 330.832. The standard deviation of interarrival time is 95.162, and the standard deviation of transit time is 316.053, suggesting that the negative exponential distribution might fit the expanded model output rather well. Average interarrival time and average service time are much closer to the stipulated parameters than they were in the run with 200 transactions. Table 8-17 shows the 1000 observations of transit time, summary statistics, and a sorted data list.

TABLE 8-17   DESCRIPTIVE STATISTICS FOR 1000 OBSERVATIONS FROM TICKET SELLER SYSTEM

RAW DATA

31.0	1.0	36.0	94.0	123.0	13.0	37.0	26.0	5.0	2.0
20.0	28.0	96.0	149.0	297.0	494.0	541.0	398.0	408.0	259.0
236.0	227.0	16.0	45.0	73.0	97.0	156.0	199.0	364.0	829.0
682.0	655.0	660.0	683.0	882.0	852.0	838.0	829.0	889.0	781.0
771.0	809.0	696.0	928.0	1047.0	1052.0	784.0	854.0	813.0	759.0
827.0	732.0	719.0	923.0	964.0	708.0	602.0	646.0	731.0	803.0
607.0	560.0	551.0	369.0	731.0	646.0	643.0	495.0	519.0	739.0
633.0	610.0	638.0	574.0	560.0	474.0	448.0	700.0	654.0	691.0
710.0	620.0	666.0	613.0	558.0	558.0	659.0	713.0	622.0	700.0
853.0	865.0	701.0	820.0	850.0	696.0	682.0	704.0	616.0	547.0
544.0	503.0	523.0	532.0	577.0	410.0	519.0	655.0	609.0	586.0
472.0	531.0	556.0	654.0	652.0	631.0	528.0	523.0	487.0	457.0
812.0	827.0	802.0	690.0	710.0	695.0	572.0	231.0	269.0	503.0
510.0	563.0	575.0	535.0	450.0	525.0	609.0	717.0	665.0	554.0
368.0	352.0	207.0	72.0	22.0	76.0	107.0	54.0	44.0	39.0
27.0	48.0	41.0	105.0	87.0	40.0	106.0	41.0	168.0	357.0
312.0	348.0	93.0	103.0	27.0	8.0	246.0	207.0	215.0	200.0
218.0	180.0	120.0	56.0	51.0	8.0	39.0	158.0	173.0	108.0
15.0	178.0	165.0	181.0	127.0	121.0	60.0	226.0	396.0	447.0
354.0	211.0	191.0	637.0	815.0	868.0	784.0	784.0	692.0	496.0
427.0	350.0	282.0	252.0	318.0	262.0	130.0	174.0	115.0	176.0
268.0	267.0	395.0	460.0	662.0	598.0	512.0	541.0	563.0	521.0
541.0	528.0	257.0	70.0	51.0	13.0	99.0	402.0	387.0	292.0
182.0	142.0	219.0	223.0	185.0	44.0	143.0	87.0	22.0	39.0
14.0	58.0	37.0	373.0	429.0	505.0	734.0	693.0	634.0	639.0
282.0	110.0	211.0	218.0	31.0	3.0	64.0	99.0	122.0	85.0
106.0	169.0	84.0	106.0	25.0	35.0	21.0	9.0	352.0	416.0

TABLE 8-17 (cont.)

367.0	327.0	193.0	190.0	96.0	100.0	89.0	161.0	53.0	85.0
40.0	160.0	79.0	10.0	96.0	246.0	220.0	182.0	237.0	166.0
412.0	408.0	179.0	98.0	20.0	106.0	192.0	297.0	347.0	350.0
356.0	295.0	235.0	94.0	121.0	49.0	129.0	101.0	84.0	78.0
139.0	157.0	189.0	206.0	178.0	303.0	31.0	24.0	49.0	38.0
120.0	196.0	103.0	26.0	81.0	259.0	266.0	224.0	387.0	285.0
645.0	837.0	805.0	801.0	719.0	703.0	712.0	595.0	836.0	657.0
216.0	275.0	373.0	327.0	339.0	337.0	310.0	324.0	376.0	327.0
151.0	154.0	155.0	5.0	109.0	48.0	74.0	279.0	357.0	286.0
248.0	70.0	131.0	28.0	88.0	92.0	102.0	248.0	54.0	7.0
42.0	25.0	47.0	43.0	74.0	221.0	160.0	193.0	186.0	289.0
500.0	130.0	79.0	52.0	552.0	576.0	605.0	536.0	631.0	606.0
644.0	638.0	632.0	628.0	441.0	282.0	356.0	308.0	352.0	327.0
225.0	166.0	163.0	205.0	211.0	191.0	186.0	389.0	460.0	413.0
180.0	113.0	80.0	11.0	177.0	214.0	59.0	94.0	7.0	120.0
71.0	284.0	301.0	387.0	408.0	168.0	98.0	146.0	5.0	405.0
466.0	539.0	427.0	389.0	589.0	592.0	579.0	661.0	622.0	519.0
597.0	604.0	523.0	462.0	337.0	160.0	99.0	5.0	11.0	121.0
121.0	152.0	66.0	214.0	344.0	435.0	428.0	528.0	541.0	474.0
384.0	344.0	278.0	106.0	199.0	59.0	131.0	173.0	36.0	74.0
225.0	131.0	65.0	53.0	55.0	207.0	246.0	158.0	194.0	60.0
48.0	121.0	4.0	39.0	46.0	127.0	118.0	170.0	210.0	376.0
567.0	983.0	897.0	1236.0	1218.0	1465.0	1719.0	1510.0	1708.0	1787.0
1738.0	1786.0	1751.0	1870.0	1871.0	1911.0	1879.0	1793.0	1563.0	1704.0
1625.0	1757.0	1832.0	1488.0	1482.0	1386.0	1348.0	1275.0	1120.0	1036.0
969.0	708.0	795.0	539.0	652.0	749.0	753.0	972.0	970.0	964.0
929.0	771.0	778.0	835.0	911.0	903.0	703.0	672.0	745.0	617.0
478.0	485.0	334.0	300.0	374.0	341.0	384.0	322.0	277.0	221.0
239.0	185.0	374.0	321.0	358.0	549.0	546.0	524.0	503.0	369.0
46.0	82.0	79.0	95.0	22.0	20.0	61.0	352.0	300.0	279.0
213.0	262.0	452.0	420.0	183.0	21.0	66.0	80.0	47.0	31.0
42.0	179.0	167.0	174.0	93.0	46.0	54.0	88.0	55.0	444.0
402.0	313.0	365.0	312.0	372.0	355.0	387.0	491.0	437.0	441.0
437.0	455.0	548.0	566.0	372.0	172.0	174.0	162.0	154.0	103.0
133.0	148.0	169.0	562.0	390.0	421.0	719.0	702.0	458.0	456.0
507.0	466.0	575.0	516.0	493.0	452.0	493.0	398.0	306.0	152.0
339.0	338.0	352.0	362.0	300.0	289.0	287.0	236.0	82.0	43.0
14.0	482.0	529.0	611.0	549.0	437.0	104.0	107.0	103.0	9.0
97.0	155.0	180.0	37.0	103.0	47.0	25.0	27.0	116.0	68.0
106.0	95.0	60.0	96.0	161.0	104.0	32.0	54.0	17.0	10.0
57.0	101.0	110.0	310.0	249.0	40.0	276.0	286.0	295.0	309.0
101.0	19.0	54.0	54.0	8.0	32.0	71.0	68.0	272.0	24.0
23.0	130.0	57.0	129.0	195.0	203.0	217.0	152.0	40.0	33.0
41.0	256.0	231.0	60.0	172.0	298.0	191.0	209.0	193.0	52.0
196.0	159.0	196.0	275.0	411.0	148.0	191.0	22.0	26.0	58.0

TABLE 8-17 (cont.)

117.0	26.0	149.0	323.0	41.0	51.0	85.0	47.0	109.0	83.0
60.0	58.0	22.0	22.0	262.0	268.0	357.0	403.0	372.0	440.0
421.0	354.0	524.0	310.0	253.0	204.0	55.0	67.0	11.0	20.0
12.0	197.0	174.0	247.0	198.0	23.0	179.0	266.0	355.0	165.0
179.0	128.0	50.0	6.0	24.0	75.0	235.0	418.0	420.0	440.0
520.0	509.0	491.0	486.0	511.0	415.0	435.0	331.0	643.0	500.0
439.0	453.0	433.0	343.0	433.0	428.0	434.0	127.0	140.0	219.0
340.0	339.0	439.0	259.0	456.0	466.0	313.0	540.0	326.0	242.0
253.0	237.0	288.0	142.0	141.0	155.0	215.0	185.0	110.0	106.0
26.0	139.0	83.0	107.0	464.0	478.0	398.0	382.0	448.0	271.0
269.0	30.0	48.0	116.0	146.0	92.0	76.0	136.0	249.0	181.0
137.0	113.0	104.0	53.0	230.0	264.0	396.0	380.0	421.0	334.0
342.0	321.0	321.0	243.0	64.0	78.0	179.0	282.0	97.0	343.0
336.0	281.0	295.0	114.0	279.0	252.0	336.0	388.0	324.0	455.0
440.0	348.0	307.0	115.0	121.0	86.0	78.0	66.0	179.0	84.0
5.0	233.0	41.0	2.0	94.0	29.0	6.0	92.0	44.0	29.0
10.0	251.0	215.0	59.0	99.0	85.0	156.0	69.0	162.0	50.0
16.0	44.0	115.0	30.0	87.0	17.0	15.0	83.0	5.0	190.0
39.0	67.0	47.0	157.0	112.0	128.0	114.0	13.0	26.0	191.0
301.0	244.0	114.0	78.0	462.0	476.0	374.0	310.0	120.0	131.0
208.0	325.0	654.0	543.0	516.0	506.0	246.0	199.0	355.0	255.0
276.0	202.0	99.0	54.0	34.0	122.0	127.0	299.0	269.0	156.0
144.0	171.0	72.0	87.0	35.0	62.0	110.0	30.0	10.0	19.0
2.0	133.0	193.0	192.0	192.0	174.0	308.0	351.0	353.0	360.0
399.0	405.0	453.0	225.0	213.0	222.0	331.0	268.0	324.0	444.0
204.0	203.0	121.0	4.0	24.0	26.0	193.0	250.0	292.0	264.0
396.0	395.0	281.0	314.0	429.0	437.0	500.0	533.0	488.0	583.0
275.0	235.0	257.0	311.0	439.0	384.0	660.0	662.0	508.0	353.0

SUMMARY STATISTICS

THE MEAN IS 331.6929
THE VARIANCE IS   99798.25
THE STANDARD DEVIATION IS 315.9084
THE SAMPLE SIZE IS   1000

SORTED DATA LIST

1.0	2.0	2.0	2.0	3.0	4.0	4.0	5.0	5.0	5.0
5.0	5.0	5.0	6.0	6.0	7.0	7.0	8.0	8.0	8.0
9.0	9.0	10.0	10.0	10.0	10.0	11.0	11.0	11.0	12.0
13.0	13.0	13.0	14.0	14.0	15.0	15.0	16.0	16.0	17.0
17.0	19.0	19.0	20.0	20.0	20.0	20.0	21.0	21.0	22.0
22.0	22.0	22.0	22.0	22.0	23.0	23.0	24.0	24.0	24.0
24.0	25.0	25.0	25.0	26.0	26.0	26.0	26.0	26.0	26.0

TABLE 8-17 (cont.)

26. 0	27. 0	27. 0	27. 0	28. 0	28. 0	29. 0	29. 0	30. 0	30. 0
30. 0	31. 0	31. 0	31. 0	31. 0	32. 0	32. 0	33. 0	34. 0	35. 0
35. 0	36. 0	36. 0	37. 0	37. 0	37. 0	38. 0	39. 0	39. 0	39. 0
39. 0	39. 0	40. 0	40. 0	40. 0	40. 0	41. 0	41. 0	41. 0	41. 0
41. 0	42. 0	42. 0	43. 0	43. 0	44. 0	44. 0	44. 0	44. 0	45. 0
46. 0	46. 0	46. 0	47. 0	47. 0	47. 0	47. 0	47. 0	48. 0	48. 0
48. 0	48. 0	49. 0	49. 0	50. 0	50. 0	51. 0	51. 0	51. 0	52. 0
52. 0	53. 0	53. 0	53. 0	54. 0	54. 0	54. 0	54. 0	54. 0	54. 0
54. 0	55. 0	55. 0	55. 0	56. 0	57. 0	57. 0	58. 0	58. 0	58. 0
59. 0	59. 0	59. 0	60. 0	60. 0	60. 0	60. 0	60. 0	61. 0	62. 0
64. 0	64. 0	65. 0	66. 0	66. 0	66. 0	67. 0	67. 0	68. 0	68. 0
69. 0	70. 0	70. 0	71. 0	71. 0	72. 0	72. 0	73. 0	74. 0	74. 0
74. 0	75. 0	76. 0	76. 0	78. 0	78. 0	78. 0	78. 0	79. 0	79. 0
79. 0	80. 0	80. 0	81. 0	82. 0	82. 0	83. 0	83. 0	83. 0	84. 0
84. 0	84. 0	85. 0	85. 0	85. 0	85. 0	86. 0	87. 0	87. 0	87. 0
87. 0	88. 0	88. 0	89. 0	92. 0	92. 0	92. 0	93. 0	93. 0	94. 0
94. 0	94. 0	94. 0	95. 0	95. 0	96. 0	96. 0	96. 0	96. 0	97. 0
97. 0	97. 0	98. 0	98. 0	99. 0	99. 0	99. 0	99. 0	99. 0	100. 0
101. 0	101. 0	101. 0	102. 0	103. 0	103. 0	103. 0	103. 0	103. 0	104. 0
104. 0	104. 0	105. 0	106. 0	106. 0	106. 0	106. 0	106. 0	106. 0	106. 0
107. 0	107. 0	107. 0	108. 0	109. 0	109. 0	110. 0	110. 0	110. 0	110. 0
112. 0	113. 0	113. 0	114. 0	114. 0	114. 0	115. 0	115. 0	115. 0	116. 0
116. 0	117. 0	118. 0	120. 0	120. 0	120. 0	120. 0	121. 0	121. 0	121. 0
121. 0	121. 0	121. 0	121. 0	122. 0	122. 0	123. 0	127. 0	127. 0	127. 0
127. 0	128. 0	128. 0	129. 0	129. 0	130. 0	130. 0	130. 0	131. 0	131. 0
131. 0	131. 0	133. 0	133. 0	136. 0	137. 0	139. 0	139. 0	140. 0	141. 0
142. 0	142. 0	143. 0	144. 0	146. 0	146. 0	148. 0	148. 0	149. 0	149. 0
151. 0	152. 0	152. 0	152. 0	154. 0	154. 0	155. 0	155. 0	155. 0	156. 0
156. 0	156. 0	157. 0	157. 0	158. 0	158. 0	159. 0	160. 0	160. 0	160. 0
161. 0	161. 0	162. 0	162. 0	163. 0	165. 0	165. 0	166. 0	166. 0	167. 0
168. 0	168. 0	169. 0	169. 0	170. 0	171. 0	172. 0	172. 0	173. 0	173. 0
174. 0	174. 0	174. 0	174. 0	174. 0	176. 0	177. 0	178. 0	178. 0	179. 0
179. 0	179. 0	179. 0	179. 0	179. 0	180. 0	180. 0	180. 0	181. 0	181. 0
182. 0	182. 0	183. 0	185. 0	185. 0	185. 0	186. 0	186. 0	189. 0	190. 0
190. 0	191. 0	191. 0	191. 0	191. 0	191. 0	192. 0	192. 0	192. 0	193. 0
193. 0	193. 0	193. 0	193. 0	194. 0	195. 0	196. 0	196. 0	196. 0	197. 0
198. 0	199. 0	199. 0	199. 0	200. 0	202. 0	203. 0	203. 0	204. 0	204. 0
205. 0	206. 0	207. 0	207. 0	207. 0	208. 0	209. 0	210. 0	211. 0	211. 0
211. 0	213. 0	213. 0	214. 0	214. 0	215. 0	215. 0	215. 0	216. 0	217. 0
218. 0	218. 0	219. 0	219. 0	220. 0	221. 0	221. 0	222. 0	223. 0	224. 0
225. 0	225. 0	225. 0	226. 0	227. 0	230. 0	231. 0	231. 0	233. 0	235. 0
235. 0	235. 0	236. 0	236. 0	237. 0	237. 0	239. 0	242. 0	243. 0	244. 0
246. 0	246. 0	246. 0	246. 0	247. 0	248. 0	248. 0	249. 0	249. 0	250. 0
251. 0	252. 0	252. 0	253. 0	253. 0	255. 0	256. 0	257. 0	257. 0	259. 0

TABLE 8-17   (cont.)

259.0	259.0	262.0	262.0	262.0	264.0	264.0	266.0	266.0	267.0
268.0	268.0	268.0	269.0	269.0	269.0	271.0	272.0	275.0	275.0
275.0	276.0	276.0	277.0	278.0	279.0	279.0	279.0	281.0	281.0
282.0	282.0	282.0	282.0	284.0	285.0	286.0	286.0	287.0	288.0
289.0	289.0	292.0	292.0	295.0	295.0	295.0	297.0	297.0	298.0
299.0	300.0	300.0	300.0	301.0	301.0	303.0	306.0	307.0	308.0
308.0	309.0	310.0	310.0	310.0	310.0	311.0	312.0	312.0	313.0
313.0	314.0	318.0	321.0	321.0	321.0	322.0	323.0	324.0	324.0
324.0	325.0	326.0	327.0	327.0	327.0	327.0	331.0	331.0	334.0
334.0	336.0	336.0	337.0	337.0	338.0	339.0	339.0	339.0	340.0
341.0	342.0	343.0	343.0	344.0	344.0	347.0	348.0	348.0	350.0
350.0	351.0	352.0	352.0	352.0	352.0	352.0	353.0	353.0	354.0
354.0	355.0	355.0	355.0	356.0	356.0	357.0	357.0	357.0	358.0
360.0	362.0	364.0	365.0	367.0	368.0	369.0	369.0	372.0	372.0
372.0	373.0	373.0	374.0	374.0	374.0	376.0	376.0	380.0	382.0
384.0	384.0	384.0	387.0	387.0	387.0	387.0	388.0	389.0	389.0
390.0	395.0	395.0	396.0	396.0	396.0	398.0	398.0	398.0	399.0
402.0	402.0	403.0	405.0	405.0	408.0	408.0	408.0	410.0	411.0
412.0	413.0	415.0	416.0	418.0	420.0	420.0	421.0	421.0	421.0
427.0	427.0	428.0	428.0	429.0	429.0	433.0	433.0	434.0	435.0
435.0	437.0	437.0	437.0	437.0	439.0	439.0	439.0	440.0	440.0
440.0	441.0	441.0	444.0	444.0	447.0	448.0	448.0	450.0	452.0
452.0	453.0	453.0	455.0	455.0	456.0	456.0	457.0	458.0	460.0
460.0	462.0	462.0	464.0	466.0	466.0	466.0	472.0	474.0	474.0
476.0	478.0	478.0	482.0	485.0	486.0	487.0	488.0	491.0	491.0
493.0	493.0	494.0	495.0	496.0	500.0	500.0	500.0	503.0	503.0
503.0	505.0	506.0	507.0	508.0	509.0	510.0	511.0	512.0	516.0
516.0	519.0	519.0	519.0	520.0	521.0	523.0	523.0	523.0	524.0
524.0	525.0	528.0	528.0	528.0	529.0	531.0	532.0	533.0	535.0
536.0	539.0	539.0	540.0	541.0	541.0	541.0	541.0	543.0	544.0
546.0	547.0	548.0	549.0	549.0	551.0	552.0	554.0	556.0	558.0
558.0	560.0	560.0	562.0	563.0	563.0	566.0	567.0	572.0	574.0
575.0	575.0	576.0	577.0	579.0	583.0	586.0	589.0	592.0	595.0
597.0	598.0	602.0	604.0	605.0	606.0	607.0	609.0	609.0	610.0
611.0	613.0	616.0	617.0	620.0	622.0	622.0	628.0	631.0	631.0
632.0	633.0	634.0	637.0	638.0	638.0	639.0	643.0	643.0	644.0
645.0	646.0	646.0	652.0	652.0	654.0	654.0	654.0	655.0	655.0
657.0	659.0	660.0	660.0	661.0	662.0	662.0	665.0	666.0	672.0
682.0	682.0	683.0	690.0	691.0	692.0	693.0	695.0	696.0	696.0
700.0	700.0	701.0	702.0	703.0	703.0	704.0	708.0	708.0	710.0
710.0	712.0	713.0	717.0	719.0	719.0	719.0	731.0	731.0	732.0
734.0	739.0	745.0	749.0	753.0	759.0	771.0	771.0	778.0	781.0
784.0	784.0	784.0	795.0	801.0	802.0	803.0	805.0	809.0	812.0
813.0	815.0	820.0	827.0	827.0	829.0	829.0	835.0	836.0	837.0
838.0	850.0	852.0	853.0	854.0	865.0	868.0	882.0	889.0	897.0

TABLE 8-17 (cont.)

903.0	911.0	923.0	928.0	929.0	964.0	964.0	969.0	970.0	972.0
983.0	1036.0	1047.0	1052.0	1120.0	1218.0	1236.0	1275.0	1348.0	1386.0
1465.0	1482.0	1488.0	1510.0	1563.0	1625.0	1704.0	1708.0	1719.0	1738.0
1751.0	1757.0	1786.0	1787.0	1793.0	1832.0	1870.0	1871.0	1879.0	1911.0

A goodness-of-fit test of the negative exponential distribution with a mean of 332 is given in Table 8-18. Ten classes are used because optimizing 200 classes would be unwieldy, and with such a robust sample size, the test with 10 classes should be quite accurate.

TABLE 8-18 GOODNESS OF FIT OF THE NEGATIVE EXPONENTIAL DISTRIBUTION WITH A MEAN OF 332 AND A SAMPLE SIZE OF 1000 TRANSACTIONS FROM TICKET SELLER SYSTEM

CLASS BOUNDARIES

CLASS IDENTIFIER	CUMULATIVE GREATER-THAN PROBABILITY	CLASS BOUNDARY	CLASS WIDTH
1	1.0000	0.00	34.98
2	0.9000	34.98	39.10
3	0.8000	74.08	44.33
4	0.7000	118.42	51.18
5	0.6000	169.59	60.53
6	0.5000	230.12	74.08
7	0.4000	304.21	95.51
8	0.3000	399.72	134.61
9	0.2000	534.33	230.12
10	0.1000	764.46	999.99

GOODNESS-OF-FIT CALCULATIONS

CLASS	CLASS IDENTIFIER	OBSERVED FREQUENCY	RELATIVE FREQUENCY	EXPECTED FREQUENCY	χ^2
1	35.0	89.00	0.100	100.00	1.21
2	74.1	102.00	0.100	100.00	0.04
3	118.4	102.00	0.100	100.00	0.04
4	169.6	81.00	0.100	100.00	3.61
5	230.1	102.00	0.100	100.00	0.04
6	304.2	91.00	0.100	100.00	0.81
7	399.7	113.00	0.100	100.00	1.69

TABLE 8-18 (cont.)

8	534.3	119.00	0.100	100.00	3.61
9	764.5	127.00	0.100	100.00	7.29
10	999.0	74.00	0.100	100.00	6.76
TOTALS	1000.00	1.000	1000.00	25.10

We used the GPSS output sample mean transit time of 332 as our best estimate of the unknown population mean transit time. Hence, degrees of freedom are (10 − 2) = 8, and critical χ^2 at the 0.05 level of significance is 15.507. While the distribution pattern in the sample of 1000 is still not close enough to negative exponential with a mean of 332 to permit us to accept the null hypothesis, it is certainly closer than was the pattern in the sample of 200. Since the goodness-of-fit test is not crucial to further statistical analysis, we will note the discrepancy and proceed.

Our next step will be either to average subgroups of a certain size, say, 25, or to select every Nth, say, twenty-fifth value for evaluation with the runs-up-and-down test and the Spearman rank correlation test. The 40 means of subgroups of size 25 in the sample of 1000 transactions are shown in Table 8-19.

TABLE 8-19   MEANS OF SUBGROUPS OF SIZE 25 FROM SAMPLE OF 1000 TRANSACTIONS

SUBGROUP MEANS

146.2	716.4	661.2	668.8	591.6	372.6	136.2	329.4	348.8	263.8
148.8	173.3	135.3	477.4	116.7	397.0	204.6	365.0	224.6	608.6
1435.1	631.5	273.4	218.5	394.3	295.0	96.7	123.5	161.6	194.3
252.3	379.7	209.5	217.4	210.4	78.3	230.0	157.6	243.9	378.6

SUMMARY STATISTICS

THE MEAN IS 331.6914
THE VARIANCE IS   62372.18
THE STANDARD DEVIATION IS 249.7442
THE SAMPLE SIZE IS   40

Tables 8-20 and 8-21, respectively, display the results of the runs-up-and-down test and the Spearman rank correlation test for the 40 means of subgroups of size 25.

The z statistics for both the runs-up-and-down test and the Spearman test are well within the range of acceptance of the null hypothesis that the data are independent. Therefore, no further consolidation is needed. We now repeat the heuristics for steady-state determination in Table 8-22. Since the sample size has been reduced to 40, further subgrouping for the Emshoff and Sisson tests leaves subgroup sizes which are theoretically inappropriate for a z test. We will cautiously apply the heuristics using subgroups of size 10.

TABLE 8-20   RUNS-UP-AND-DOWN TEST FOR MEANS OF SUBGROUPS OF SIZE 25 FROM SAMPLE OF 1000 TRANSACTIONS

THE ORIGINAL SAMPLE SIZE IS	40
THE EFFECTIVE SAMPLE SIZE IS	40
THE NUMBER OF RUNS IS 25	
THE EXPECTED NUMBER OF RUNS IS 26.33	
THE STANDARD DEVIATION IS 2.6055	
THE VALUE OF Z IS −0.3198	

TABLE 8-21. SPEARMAN RANK CORRELATION TEST FOR MEANS OF SUBGROUPS OF SIZE 25 FROM SAMPLE OF 1000 TRANSACTIONS

THE ORIGINAL SAMPLE SIZE IS	40
THE SUM OF D SQUARED IS	7506.00
SUMU=	0.00
SUMV=	0.00
R=	0.24
Z=	1.48

TABLE 8-22   RESULTS OF STEADY-STATE HEURISTICS APPLIED TO MEANS OF SUBGROUPS OF SIZE 25 FROM SAMPLE OF 1000 TRANSACTIONS

THE FIRST STEADY STATE OBSERVATION ACCORDING TO CONWAY'S RULE IS NUMBER 1.

THE VALUE OF THAT OBSERVATION IS	146.20
THE MINIMUM VALUE IN THE REMAINING SET IS	78.32
THE MAXIMUM VALUE IN THE REMAINING SET IS	1435.12

EMSHOFF AND SISSON
ABOVE-AND-BELOW TEST

SUBGP. NUMBER	PREVIOUS CUMULATIVE AVERAGE	NUMBER ABOVE	NUMBER BELOW	CHISQ
2	423.484	2	8	3.60
3	354.304	3	7	1.60
4	363.667	2	8	3.60

TABLE 8-22 (cont.)

EMSHOFF AND SISSON
MOVING AVERAGE TEST

AVERAGE PREVIOUS SBGP.	AVERAGE CURRENT SBGP.	ST. ERR. PREVIOUS SQRD.	ST. ERR. CURRENT SQRD.	Z
423.484	285.124	4807.75	2805.61	1.59
285.124	382.392	2805.61	16106.53	−0.71
382.392	235.772	16106.53	823.07	1.13

TABLE 8-23   PLOT OF 40 MEANS OF SUBGROUPS OF SIZE 25 FROM SAMPLE OF 1000 TRANSACTIONS

```
1435.120: -----------------------------------------
1388.333: -----------------------------------------
1341.547: -----------------------------------------
1294.761: -----------------------------------------
1247.975: -----------------------------------------
1201.189: -----------------------------------------
1154.402: -----------------------------------------
1107.616: -----------------------------------------
1060.830: -----------------------------------------
1014.044: -----------------------------------------
 967.258: -----------------------------------------
 920.471: -----------------------------------------
 873.685: -----------------------------------------
 826.899: -----------------------------------------
 780.113: -----------------------------------------
 733.327: -X---------------------------------------
 686.540: ---X-------------------------------------
 639.754: --X---------------X----------------------
 592.968: ----X-------------X--------------------
 546.182: -----------------------------------------
 499.396: -------------X-------------------------
 452.609: -----------------------------------------
 405.823: ---------------X--------X---------------
 359.037: -----X--X--------X-------------X-------X
 312.251: -------X-----------------X-------------
 265.465: ---------X-----------X-------X-------X-
 218.679: ---------------X-X----X---------XXX-X---
 171.892: ----------XX----------------XX-------X--
 125.106:X-----X-----X-X------------X-----------
  78.320: ------------------------X--------X----
          0    5    10   15   20   25   30   35   40
```

Critical χ^2 at the 0.05 level of significance with 1 degree of freedom is 3.84. According to the Emshoff and Sisson above-and below criterion, the first subgroup for which the computed χ^2 is below 3.84 is subgroup 2. The critical value of z at the 0.05 level of significance is 1.96. According to the E & S moving-average criterion, the first subgroup for which the computed z is below 1.96 is subgroup 2. The results of all three steady-state heuristics indicate that the sample of 40 means exhibits steady-state characteristics. A plot of the 40 means is given in Table 8-23.

Suppose that we had elected to extract 40 individual observations from the GPSS sample of 1000 transactions. Table 8-24 shows the data set of every twenty-fifth observation, beginning with the first observation, from the sample of 1000.

TABLE 8-24   LIST OF FORTY SUBGROUP MEANS WITH SUMMARY STATISTICS

SUBGROUP MEANS

31.0	97.0	827.0	474.0	544.0	695.0	27.0	8.0	427.0	13.0
282.0	100.0	356.0	259.0	151.0	221.0	225.0	168.0	121.0	207.0
1738.0	749.0	239.0	21.0	437.0	452.0	97.0	40.0	41.0	51.0
12.0	415.0	253.0	92.0	336.0	29.0	39.0	506.0	2.0	26.0

SUMMARY STATISTICS

THE MEAN IS 270.2000
THE VARIANCE IS   105610.30
THE STANDARD DEVIATION IS 324.9773
THE SAMPLE SIZE IS   40

Table 8-25 shows the steady-state heuristics as applied to the sample if every twenty-fifth observation, beginning with the first, from the augmented sample of 1000 observations from the ticket seller system.

TABLE 8-25   RESULTS OF STEADY-STATE HEURISTICS APPLIED TO EVERY TWENTY-FIFTH VALUE IN A SAMPLE OF 1000

THE FIRST STEADY STATE OBSERVATION ACCORDING TO CONWAY'S RULE IS NUMBER 1.
THE VALUE OF THAT OBSERVATION IS   31.00
THE MINIMUM VALUE IN THE REMAINING SET IS   2.00
THE MAXIMUM VALUE IN THE REMAINING SET IS   1738.00

TABLE 8-25 (cont.)

EMSHOFF AND SISSON
ABOVE–AND–BELOW TEST

SUBGP. NUMBER	PREVIOUS CUMULATIVE AVERAGE	NUMBER ABOVE	NUMBER BELOW	CHISQ
2	314.300	1	9	6.40
3	261.650	4	6	0.40
4	303.267	3	7	1.60

EMSHOFF AND SISSON
MOVING AVERAGE TEST

AVERAGE PREVIOUS SBGP.	AVERAGE CURRENT SBGP.	ST. ERR. PREVIOUS SQRD.	ST. ERR. CURRENT SQRD.	Z
314.300	209.000	9926.46	605.69	1.03
209.000	386.500	605.69	28399.43	−1.04
386.500	171.000	28399.43	3604.51	1.20

The critical value of χ^2 at the 0.05 level of significance is 3.84. According to the Emshoff and Sisson above-and-below criterion, the first subgroup for which the computed χ^2 is below 3.84 is subgroup 3. The critical value of z at the 0.05 level of significance is 1.96. According to the E & S moving-average criterion, the first subgroup for which the computed z is below 1.96 is subgroup 2.

The conclusions of the three heuristics about the point of onset of steady state are mixed. The Conway rule and the Emshoff and Sisson moving-average rule suggest that steady state commences with subgroup 2, while the Emshoff and Sisson above-and-below rule indicates that steady state starts with subgroup 3. Although it would be most conservative to truncate the first subgroup, that action would leave either a sample of size 30 for the further analysis in Chapter 9 or the alternative of extending the simulation run once again and repeating the steps. In the interest of minimizing computational costs, we will keep the sample of 40 without truncation if the sample successfully passes the runs and Spearman tests which follow. Table 8-26 shows a plot of the 40 values currently under analysis.

Tables 8-27 and 8-28 confirm that there is little trace of serial correlation in the set of every twenty-fifth observation.

TABLE 8-26 PLOT OF EVERY TWENTY-FIFTH VALUE
IN A SAMPLE OF 1000

```
1738.000: -------------------X------------------
1678.138: --------------------------------------
1618.276: --------------------------------------
1558.414: --------------------------------------
1498.552: --------------------------------------
1438.689: --------------------------------------
1378.827: --------------------------------------
1318.965: --------------------------------------
1259.103: --------------------------------------
1199.241: --------------------------------------
1139.379: --------------------------------------
1079.517: --------------------------------------
1019.655: --------------------------------------
 959.793: --------------------------------------
 899.931: --------------------------------------
 840.069: --X-----------------------------------
 780.207: --------------------------------------
 720.345: -----X--------------X-----------------
 660.483: --------------------------------------
 600.621: --------------------------------------
 540.759: ----X---------------------------------
 480.896: ---X--------------------X-----------X--
 421.034: --------X--------------X------X--------
 361.172: -----------X--------------------X-----
 301.310: ----------X---------------------------
 241.448: -------------X-XX-----X--------X-------
 181.586: ----------------X-X-------------------
 121.724: -X---------X--X---X-------X------X-----
  61.862: -----------------------XXX------X---
   2.000: X-----XX-X-----------X------X----X--XX
          0    5   10   15   20   25   30   35   40
```

TABLE 8-27 RUNS-UP-AND-DOWN TEST FOR EVERY
TWENTY-FIFTH VALUE IN A SAMPLE OF 1000

THE ORIGINAL SAMPLE SIZE IS   40
THE EFFECTIVE SAMPLE SIZE IS   40
THE NUMBER OF RUNS IS   25
THE EXPECTED NUMBER OF RUNS IS   26.33
THE STANDARD DEVIATION IS   2.6055
THE VALUE OF Z IS -0.3198

TABLE 8-28   SPEARMAN RANK CORRELATION
TEST FOR EVERY TWENTY-FIFTH VALUE IN A
SAMPLE OF 1000

THE ORIGINAL SAMPLE SIZE IS	40
THE SUM OF D SQUARED IS	8853.00
SUMU=	0.50
SUMV=	0.50
R= 0.10	
Z= 0.64	

The data set containing every twenty-fifth observation could be produced directly by the GPSS program shown in Table 8-29.*

TABLE 8-29   GPSS MODEL TO CREATE 1000 TRANSACTIONS
AND PRINT EVERY TWENTY-FIFTH TRANSIT TIME VALUE

```
JOB
INITIAL X1, 0
INITIAL X2, 0
EXPON1 FUNCTION RN3, C24
0, 0/. 1, . 104/. 2, . 222/. 3, . 355/. 4, . 509/. 5, . 69
. 6, . 915/. 7, 1. 2/. 75, 1. 38/. 8, 1. 6/. 84, 1. 83/. 88, 2. 12
. 9, 2. 3/. 92, 2. 52/. 94, 2. 81/. 95, 2. 99/. 96, 3. 2/. 97, 3. 5
. 98, 3. 9/. 99, 4. 6/. 995, 5. 3/. 998, 6. 2/. 999, 7. 0/. 9997, 8. 0
EXPON2 FUNCTION RN4, C24
0, 0/. 1, . 104/. 2, . 222/. 3, . 355/. 4, . 509/. 5, . 69
. 6, . 915/. 7, 1. 2/. 75, 1. 38/. 8, 1. 6/. 84, 1. 83/. 88, 2. 12
. 9, 2. 3/. 92, 2. 52/. 94, 2. 81/. 95, 2. 99/. 96, 3. 2/. 97, 3. 5
. 98, 3. 9/. 99, 4. 6/. 995, 5. 3/. 998, 6. 2/. 999, 7. 0/. 9997, 8. 0
GENERATE 101, FN$EXPON1
TABULATE INTARV
QUEUE 1
SEIZE 1
DEPART 1
ADVANCE 84, FN$EXPON2
RELEASE 1
TEST E X2, K24, TERM1
SAVEVALUE 2, K0
SAVEVALUE 1, M1
PRINT , , X
TABULATE TRANSIT
TERMINATE 1
```

*The GPSS program will select every twenty-fifth item, beginning with the twenty-fifth value rather than with the first value, as given in the data set in Table 8-24.

TABLE 8-29 (cont.)

```
TERM1 SAVEVALUE 2+, K1
    TERMINATE 1
    INTARV TABLE IA, 0, 20, 20
    TRANSIT TABLE M1, 20, 20, 20
      START 1000
      END
```

It appears that either the set of 40 means of adjacent subgroups of 25 observations or the set of 40 values resulting from a choice of every twenty-fifth observation from the simulation run of 1000 transactions will serve the purpose of further statistical analysis in Chapter 9 adequately. There is no evidence of significant serial correlation in either set, and there is no strong evidence of the presence of transients. If transients were present and the rest of the data seemed free of serial correlation, the transients might be isolated roughly from the plot and truncated from the sample.

SUMMARY

The ultimate objective of simulation may be to obtain interval estimates of performance parameters of a system or to enable selection of the system with the best performance among a set of alternative configurations. Whichever of these may be the objective, some preliminary statistical analysis must be done to ensure that the conclusions will be accurate. The data values from the model which are used to meet the objective must be characteristic of steady-state operation of the system, unless it is specifically desired to evaluate transient-state conditions and must be reasonably free of serial correlation. Serial correlation may be detected using the runs-up-and-down test or the Spearman rank correlation test. If significant serial correlation is present, an unbiased sample with reduced serial correlation may be chosen in one of two ways. First, the means of adjacent M data value may be substituted for the original data values themselves. Second, every Nth observation may be chosen and the intervening observations deleted. The sample resulting from either of these two procedures is then reevaluated for independence and the cycle continues until either the level of serial correlation is acceptable or the sample size dwindles below the level at which the tests can be performed. When the sample size becomes too small, the original simulation run must be extended to create additional observations.

There is no formal statistical way in which to establish the onset of steady state. Heuristics or rules of thumb have been proposed by Conway and by Emshoff and Sisson to assist the analyst in deciding whether and where transients occur in the output of a simulation model. Transients are most likely to be found when a system begins in the empty state or when a peak-load situation is imposed. For evaluation of the steady-state performance of a system, transients should be truncated or averaged into an extended data stream.

QUESTIONS

1. What are the goals of model building? What statistical impediments may stand in the way of the achievement of these goals?
2. Define serial correlation. Why might it pose a problem to the modeler?
3. By what methods might one detect the presence of serial correlation? Explain intuitively how each method works.
4. If serial correlation is found to be present in simulation output, what two remedies might be applied? What are the advantages and disadvantages of each remedy?
5. What is a heuristic? How does it differ from an algorithm? In what ways are heuristics preferable or not preferable to algorithms?
6. Define steady state. Why is a modeler interested in establishing the presence or absence of transients?
7. Suggest three heuristics for evaluating the presence of transients in a model. Explain briefly the logic underlying each heuristic. What statistical requirements, if any, should be met by the data set under evaluation to justify use of the heuristic?
8. What two strategies might be adopted if transients are found in the output of a model?
9. Suppose that the sample size is inadequate to perform tests for serial correlation and transients? What might the modeler do? What factors must be considered in planning the next move?
10. In what cases might the steady-state heuristics lead the modeler to be too cautious (i.e., truncate more of a data set than is necessary) or too rash (i.e., truncate too little of a data set)? Discuss.

PROBLEMS

1. Consider the output of a GPSS program to model the $M/M/1$ situation which prevailed in problems for Chapters 2 and 3. The model is as follows:

```
JOB
INITIAL X1, 0
EXPON 1 FUNCTION RN3, C24
0, 0/.1, .104/.2, .222/.3, .355/.4, .509/.5, .69
0.6, .915/.7, 1.2/.75, 1.38/.8, 1.6/.84, 1.83/.88, 2.12
0.9, 2.3/.92, 2.52/.94, 2.81/.95, 2.99/.96, 3.2/.97, 3.5
0.98, 3.9/.99, 4.6/.995, 5.3/.998, 6.2/.999, 7.0/.9997, 8.0
EXPON2 FUNCTION RN4, C24
0, 0/.1, .104/.2, .222/.3, .355/.4, .509/.5, .69
0.6, .915/.7, 1.2/.75, 1.38/.8, 1.6/.84, 1.83/.88, 2.12
0.9, 2.3/.92, 2.52/.94, 2.81/.95, 2.99/.96, 3.2/.97, 3.5
0.98, 3.9/.99, 4.6/.995, 5.3/.998, 6.2/.999, 7.0/.9997, 8.0
GENERATE 422, FN$EXPON1
TABULATE INTARV
QUEUE 1
```

```
                    SEIZE 1
                    DEPART 1
                    ADVANCE 317, FN$EXPON2
                    RELEASE 1
                    SAVEVALUE 1, M1
                    PRINT , , X
                    TABULATE TRANSIT
                    TERMINATE 1
                 INTARV TABLE IA, 0, 20, 20
                 TRANSIT TABLE M1, 20, 20, 20
                    START 100
                    RESET
                    START 100
                    END
```

A selected group of GPSS summary statistics is as follows:

	FIRST 100	SECOND 100
AVG. UTILIZATION	0.803	0.641
AVG. SVC. TIME/TRANS.	352.00	282.49
AVG. QUEUE CONTENTS	2.176	0.696
MAX. QUEUE CONTENTS	10	5
TL. QUEUE ENTRIES	100	100
ZERO QUEUE ENTRIES	25	40
AVG. Q TIME/TRANS.	954.31	306.61
NZ-AVG. Q TIME/TRANS.	1272.413	511.017
CURRENT CONTENTS OF Q	0	0
MEAN INTARV TIME	436.50	445.21
S.D. INTARV TIME	423.977	441.775
MEAN TRANSIT TIME	1306.31	589.10
S.D. TRANSIT TIME	1152.804	508.703
RELATIVE CLOCK TIME	43861	44058
ABSOLUTE CLOCK TIME	43861	87919

Make some tentative comments about
- **(a)** the reliability of the random number generators
- **(b)** the accuracy of the original sample information
- **(c)** the presence of start-up transients
- **(d)** the consistency with predictions of the $M/M/1$ analytical models in Chapter 3

*2. Suppose that the following 200 data values are the output produced by the PRINT block shown in Problem 1.

```
 115.0     3.0   135.0   354.0   458.0    49.0   138.0    98.0    20.0     7.0
  75.0   101.0   361.0   547.0   295.0  1009.0  1893.0  2130.0  2545.0  2063.0
2306.0  3548.0  3234.0  4095.0  3525.0  3091.0  2990.0  2667.0  2592.0  4234.0
4504.0  3916.0  3570.0  3399.0  2701.0  2574.0  2345.0  2644.0  2929.0  2940.0
```

*Answers to problems preceded by * can be found in the Selected Numerical Answers.

2798.0	1828.0	1016.0	1087.0	1218.0	319.0	374.0	75.0	386.0	398.0
607.0	657.0	545.0	88.0	465.0	1148.0	773.0	141.0	311.0	703.0
968.0	813.0	1128.0	1159.0	1179.0	459.0	210.0	238.0	1139.0	954.0
745.0	813.0	875.0	47.0	63.0	125.0	267.0	147.0	699.0	468.0
736.0	876.0	192.0	987.0	1559.0	1655.0	1777.0	1623.0	1554.0	1095.0
1311.0	1329.0	1346.0	1967.0	2126.0	2113.0	1648.0	1348.0	1050.0	706.0
241.0	254.0	226.0	22.0	447.0	741.0	715.0	765.0	164.0	236.0
786.0	69.0	5.0	162.0	1123.0	518.0	855.0	655.0	701.0	59.0
359.0	1085.0	405.0	175.0	161.0	397.0	112.0	96.0	680.0	616.0
853.0	1496.0	1331.0	1525.0	1337.0	277.0	92.0	355.0	196.0	48.0
87.0	96.0	500.0	279.0	260.0	159.0	219.0	356.0	686.0	372.0
421.0	951.0	1107.0	545.0	535.0	385.0	360.0	502.0	844.0	343.0
398.0	112.0	75.0	135.0	173.0	274.0	982.0	1703.0	1614.0	1079.0
854.0	797.0	873.0	195.0	1139.0	1429.0	1459.0	1007.0	1174.0	2186.0
2285.0	2185.0	1041.0	905.0	1044.0	660.0	486.0	359.0	93.0	495.0
61.0	244.0	161.0	564.0	319.0	238.0	222.0	35.0	265.0	243.0

The mean is 947.7048, the variance is 927104.10, and the standard deviation is 962.8623.

(a)   Examine the summary statistics and comment about the possible distribution of the transit time data. Explain the rationale for your choice of distribution.

(b)   The 200 data points in sorted order are as follows:

3.0	5.0	7.0	20.0	22.0	35.0	47.0	48.0	49.0	59.0
61.0	63.0	69.0	75.0	75.0	75.0	87.0	88.0	92.0	93.0
96.0	96.0	98.0	101.0	112.0	112.0	115.0	125.0	135.0	135.0
138.0	141.0	147.0	159.0	161.0	161.0	162.0	164.0	173.0	175.0
192.0	195.0	196.0	210.0	219.0	222.0	226.0	236.0	238.0	238.0
241.0	243.0	244.0	254.0	260.0	265.0	267.0	274.0	277.0	279.0
295.0	311.0	319.0	319.0	343.0	354.0	355.0	356.0	359.0	359.0
360.0	361.0	372.0	374.0	385.0	386.0	397.0	398.0	398.0	405.0
421.0	447.0	458.0	459.0	465.0	468.0	486.0	495.0	500.0	502.0
518.0	535.0	545.0	545.0	547.0	564.0	607.0	616.0	655.0	657.0
660.0	680.0	686.0	699.0	701.0	703.0	706.0	715.0	736.0	741.0
745.0	765.0	773.0	786.0	797.0	813.0	813.0	844.0	853.0	854.0
855.0	873.0	875.0	876.0	905.0	951.0	954.0	968.0	982.0	987.0
1007.0	1009.0	1016.0	1041.0	1044.0	1050.0	1079.0	1085.0	1087.0	1095.0
1107.0	1123.0	1128.0	1139.0	1139.0	1148.0	1159.0	1174.0	1179.0	1218.0
1311.0	1329.0	1331.0	1337.0	1346.0	1348.0	1429.0	1459.0	1496.0	1525.0
1554.0	1559.0	1614.0	1623.0	1648.0	1655.0	1703.0	1777.0	1828.0	1893.0
1967.0	2063.0	2113.0	2126.0	2130.0	2185.0	2186.0	2285.0	2306.0	2345.0
2545.0	2574.0	2592.0	2644.0	2667.0	2701.0	2798.0	2929.0	2940.0	2990.0
3091.0	3234.0	3399.0	3525.0	3548.0	3570.0	3916.0	4095.0	4234.0	4504.0

Test your hypothesis about the nature of the distribution of transit times using a Chi-square test with 10 classes. Show your calculations and comment on your results.

3. The results of the runs-up-and-down test as applied to the data in Problem 2 are as follows:

```
THE ORIGINAL SAMPLE SIZE IS 200
THE EFFECTIVE SAMPLE SIZE IS 200
THE NUMBER OF RUNS IS 97
THE EXPECTED NUMBER OF RUNS IS 133.00
THE STANDARD DEVIATION IS 5.9358
THE VALUE OF Z IS -5.9807
```

What would you conclude from these results? Why?

4. The results of the Spearman rank correlation test as applied to the data in Problem 2 are shown as follows:

```
R=    0.78
Z=   10.99
```

What would you conclude from these results? Why?

5. The results of the Conway and Emshoff and Sisson steady-state heuristics as applied to the data in Problem 2 are as follows:

```
THE FIRST STEADY STATE OBSERVATION ACCORDING TO CONWAY'S RULE IS NUMBER 1.
THE VALUE OF THAT OBSERVATION IS            115.00
THE MINIMUM VALUE IN THE REMAINING SET IS     3.00
THE MAXIMUM VALUE IN THE REMAINING SET IS  4504.00
```

*A*B*O*V*E*-*A*N*D*-*B*E*L*O*W*--*M*O*V*I*N*G*-*A*V*E*R*A*G*E*

SUBGP. NUMBER	PREVIOUS CUMULATIVE AVERAGE	NUMBER ABOVE	NUMBER BELOW	CHISQ	AVERAGE PREVIOUS SBGP.	AVERAGE CURRENT SBGP.	ST.ERR. PREVIOUS SQRD.	ST.ERR. CURRENT SQRD.	Z
2	1489.267	12	18	1.20	1489.267	1548.633	71351.88	58476.03	-0.16
3	1518.950	5	25	13.33	1548.633	818.333	58476.03	8887.17	2.81
4	1285.411	8	22	6.53	818.333	789.600	8887.17	13236.21	0.19
5	1161.458	4	26	16.13	789.600	487.000	13236.21	6616.47	2.15
6	1026.567	9	21	4.80	487.000	788.366	6616.47	9588.16	-2.37
7	986.866	4	16	16.13	788.366	595.250	9588.16	20266.98	1.12

What do you conclude? What would you do next? Why?

*6. Select every fifth value, beginning with the first item, from the data set in Problem 2.
   (a) Calculate the mean and standard deviation for this set of 40 observations. Compare them with the mean and standard deviation of the original data set and comment on any differences.
   (b) Apply the runs-up-and-down test to the 40 data items. Comment on your results.
   (c) Apply the Spearman rank correlation test to the 40 data items. Comment on your results.

        (d)   Apply the Conway and Emshoff and Sisson heuristics to the 40 data items. Use subgroups of size 10. Comment on your results.

        (e)   What would you do next? Why?

    7. Solve Problems 1 through 5 using the output of a simulation model which you have constructed.

APPENDIX

Six programs from the SIMSTAT package are used in Chapter 8. The reader is already acquainted with four of these—DESCST (descriptive statistics), RUNUPD (Runs-up-and-down test), SPEAR (Spearman rank correlation test), and GOFFIT (goodness-of-fit tests). GOFFIT was previously used to test the goodness of fit of the model input data in Chapter 2 and to test the goodness of fit of random number streams in Chapter 4. RUNUPD and SPEAR were previously used in the analysis of random numbers in Chapter 4. DESCST was used in Chapters 2 and 4. However, the second option of DESCST, which allows the user to average M adjacent values or select every Nth value from a data set, is invoked for the first time in this chapter.

    Two programs are introduced in this chapter. One is PLOT, which produces a grid to offer a visual representation of the data set. The second is STEADY, which applies the Conway rule and the two Emshoff and Sisson rules for detecting the onset of steady state.*

    Since GOFFIT, RUNUPD, and SPEAR are already familiar to the reader, the tables which are based on those programs will not be replicated in this appendix. Likewise, the tables which use only option 1 of DESCST to create descriptive statistics and a sorted data list will not be included. An occurrence of tables which employ PLOT, STEADY, or option 2 of DESCST will be displayed, including the conversation which generated the table output.

    Table 8A-1 shows the conversation with STEADY that produced the output given in Table 8-6. Table 8A-2 indicates that no real conversation is required to use the PLOT program, but a few preliminary remarks are made by the program itself prior to creating the graph. The graph in Table 8A-2 is identical with the one in Table 8-7.

    Table 8A-3 utilizes option 2 of DESCST to condense the original 200 observations of the ticket seller simulation into a set of 40 means of subgroups of size 5. This table shows the conversation which generated the output given in Table 8-10 as well as the output itself.

---

*A third program is used to extract the transit time values produced by the PRINT block from the image of GPSS output stored on disk. This program, SIGN, searches the lines of GPSS output for a plus sign in the space allocated to print the contents of SAVEVALUE 1. When it finds such a plus sign, it removes the sign and writes the value following the sign as a floating-point integer on disk. The values are then available for processing with RUNUPD, SPEAR, DESCST, and other FORTRAN programs which demand floating-point values for free-format input. The reader is cautioned that the location of the PRINT block output of SAVEVALUE 1 is system dependent.

Table 8A-4 offers the alternative in option 2 of DESCST to choose every *n*th observation from the original data set and place it in a new data set, along with summary statistics about the new data set. Again, the source is the original 200 ticket seller observations. The output is identical with that in Table 8-13.

TABLE 8A-1   STEADY-STATE HEURISTICS BY CONWAY AND EMSHOFF AND SISSON
FOR SAMPLE SIZE OF 200 OBSERVATIONS

THIS PROGRAM PERFORMS THREE HEURISTICS FOR STEADY STATE DETECTION.

IF YOU WANT INFORMATION, TYPE A 1. OTHERWISE, TYPE A 0.
• 1

THIS PROGRAM COMPUTES THE POINT OF STEADY STATE BY THESE CRITERIA:
1. CONWAY'S RULE OF THE FIRST BEING NEITHER THE MAX NOR THE MIN OF THE REST;
2. EMSHOFF AND SISSON'S RULE OF THE FIRST SUBGROUP HAVING AS MANY ABOVE
    THE PREVIOUS MEAN AS BELOW IT;
3. EMSHOFF AND SISSON'S RULE OF THE FIRST SUBGROUP HAVING
    A MOVING AVERAGE SIMILAR TO THAT OF THE PREVIOUS SUBGROUP.

ACCEPTABLE SUBGROUP SIZES ARE 10, 20 OR 30.
TYPE THE DESIRED SUBGROUP SIZE.
• 30

IF DATA ARE ENTERED FROM DISK, TYPE A 1. OTHERWISE, TYPE A 0.
• 1

THE FIRST STEADY STATE OBSERVATION ACCORDING TO CONWAY'S RULE IS NUMBER 1.
THE VALUE OF THAT OBSERVATION IS   31.00
THE MINIMUM VALUE IN THE REMAINING SET IS   1.00
THE MAXIMUM VALUE IN THE REMAINING SET IS   1052.00

IF YOU WANT TO SEE DETAILS OF THE EMSHOFF AND SISSON TESTS,
TYPE A 1. OTHERWISE, TYPE A 0.
• 1

```
        *A*B*O*V*E*-*A*N*D*--*B*E*L*O*W*---*M*O*V*I*N*G*-*A*V*E*R*A*G*E*
        PREVIOUS                        AVERAGE AVERAGE  ST.ERR.  ST.ERR.
SUBGP.  CUMULATIVE NUMBER NUMBER        PREVIOUS CURRENT PREVIOUS CURRENT
NUMBER   AVERAGE   ABOVE  BELOW  CHISQ   SBGP   SBGP.    SQRD.    SQRD.     Z
  2      176.667    30      0   30.00  176.667  797.300 1321.29  431.89 -14.82
  3      531.254    25      5   13.33  797.300  608.700  431.89  259.25   7.17
  4      558.588    16     14    0.13  608.700  609.433  259.25  489.74  -0.03
  5      571.852    11     19    2.13  609.433  438.500  489.74 2327.30   3.22
  6      544.262     0     30   30.00  438.500  129.667 2327.30  328.03   5.99
  7      473.188     7     13    8.53  129.667  387.400  328.03 3990.31  -3.92
```

TABLE 8A-1 (cont.)

THE CRITICAL VALUE OF CHI-SQUARE AT THE .05 LEVEL OF SIGNIFICANCE IS 3.84.
ACCORDING TO THE E&S ABOVE-AND-BELOW CRITERION, THE
FIRST SUBGROUP FOR WHICH THE COMPUTED CHI-SQUARE IS
BELOW 3.84 IS SUBGROUP NUMBER 4.

THE CRITICAL VALUE OF Z AT THE .05 LEVEL OF SIGNIFICANCE IS 1.96.
ACCORDING TO THE E&S MOVING-AVERAGE CRITERION, THE FIRST SUBGROUP
FOR WHICH THE COMPUTED Z IS BELOW 1.96 IS SUBGROUP NUMBER 4.

TABLE 8A-2   PLOT OF 200 OBSERVATIONS FROM TICKET SELLER SYSTEM

THIS PROGRAM PLOTS A GRAPH OF DATA POINTS. THE Y-AXIS CONSISTS OF THE RANGE
OF VALUES OF THE SUCCESSIVE OBSERVATIONS, X, DIVIDED INTO 29 EQUAL INCREMENTS.
THE X-AXIS GIVES A SEQUENCE NUMBER FOR EACH OBSERVATION, UP TO THE SAMPLE
SIZE, N. EACH X IS PLOTTED AT THE Y-AXIS POINT CLOSEST TO ITS TRUE VALUE.

IF INPUT IS FROM DISK, TYPE A 1. OTHERWISE, TYPE A 0.
• 1

```
1052.000:--------------------------------------------XX-----------------------
1015.759:-----------------------------------------------------------------------
 979.517:-------------------------------------------------------X--------------
 943.276:-------------------------------------------------X--------------------
 907.034:---------------------------------------X-------------X----------------
 870.793:--------------------------------X-----------------X-------------------
 834.552:------------------------------X-----XXX-------------X-----------------
 798.310:----------------------------------------X-X----X-X---------X----------
 762.069:----------------------------------------X--------X-------------------
 725.827:------------------------------------------------XX--X--X-----X----X
 689.586:---------------------------X--X---------X----------------------------
 653.345:----------------------------XX-------------------------X-------XX---
 617.103:-----------------------------------------------------X---X---------
 580.862:-----------------------------------------------------------------------
 544.621:--------------X-------------------------------------------XX-------
 508.379:-------------X---------------------------------------------------XX-
 472.138:-----------------------------------------------------------------------
 435.896:-----------------------------------------------------------------------
 399.655:---------------XX-----------------------------------------------------
 363.414:-----------------------X------------------------------------X------
 327.172:-----------------------------------------------------------------------
 290.931:------------X--------------------------------------------------------
 254.690:-----------------X---------------------------------------------------
 218.448:-------------------XX------------------------------------------------
 182.207:------------------------X--------------------------------------------
 145.966:------------X-----------X--------------------------------------------
```

TABLE 8A-2 (cont.)

```
 109.724: ---XX-------X-----------X---------------------------------------------
  73.483: ----------------------X-----------------------------------------------
  37.241: X-X---XX--XX----------X------------------------------------------------
   1.000: -X---X--XX-----------X-------------------------------------------------
          0    5   10   15   20   25   30   35   40   45   50   55   60   65   70

1052.000: ----------------------------------------------------------------------
1015.759: ----------------------------------------------------------------------
 979.517: ----------------------------------------------------------------------
 943.276: ----------------------------------------------------------------------
 907.034: ----------------------------------------------------------------------
 870.793: -----------------XX---------------------------------------------------
 834.552: -----------------------XX-------------------------------X-------------
 798.310: -------------------------------------------------------X-X------------
 762.069: ----------------------------------------------------------------------
 725.827: ---------X------X-----------------------------------X-----------X--
 689.586: -------X-X---------X--X--XXX-------------------------X-X-----------
 653.345: --X-----X---X---X--------------------X-----XX------------------X-
 617.103: XX---------X-X----X---------X--------X------X----------------X---
 580.862: ---X-----------------------------X---X----------X----XX------
 544.621: ----X---------XX------------XX--X-------XX---X--------------X-----X
 508.379: -----------------------XX---X----------X---------XX----X----
 472.138: -----X-------------------------------X------XX-------------------
 435.896: ------X------------------------------------------------------X-----
 399.655: --------------------------------X-------------------------------------
 363.414: ----------------------------------------------------------------------
 327.172: ----------------------------------------------------------------------
 290.931: ----------------------------------------------------------------------
 254.690: -------------------------------------------------X------------
 218.448: -------------------------------------------------X------------
 182.207: ----------------------------------------------------------------------
 145.966: ----------------------------------------------------------------------
 109.724: ----------------------------------------------------------------------
  73.483: ----------------------------------------------------------------------
  37.241: ----------------------------------------------------------------------
   1.000: ----------------------------------------------------------------------
          70   75   80   85   90   95  100  105  110  115  120  125  130  135  140

1052.000: -------------------------------------------------------------
1015.759: -------------------------------------------------------------
 979.517: -------------------------------------------------------------
 943.276: -------------------------------------------------------------
 907.034: -------------------------------------------------------------
 870.793: -----------------------------------------------------X----
 834.552: -------------------------------------------------------------
```

```
798.310: -------------------------------------------------------X-XX--
762.069: ----------------------------------------------------------------
725.827: ----------------------------------------------------------------
689.586: -------------------------------------------------------------X-
653.345: -----------------------------------------------------X------
617.103: ----------------------------------------------------------------
580.862: ----------------------------------------------------------------
544.621: ----------------------------------------------------------------
508.379: -------------------------------------------------------------X
472.138: ----------------------------------------------------------------
435.896: --------------------------------------------X----------
399.655: ---------------------------------------X----------
363.414:XX---------------X-X--------------------X---------
327.172: --------------------X----------------------------
290.931: ----------------------------------------------------------------
254.690: -----------------------X------------------------
218.448: --X----------------------XX-X----------------X---X--------
182.207: ------------------X----------X-X------X--XXX-------X-------
145.966: -----------------------------------X---------------------
109.724: ------X------X--X-----XX--------X------X----XX-------------
 73.483: ---X-X--------X------------------X-----------X---------
 37.241: ----X--XXXXXX--X-X------X---------X-X-------------------
  1.000: --------------------X---------X----X------------------
         140   145   150   155   160   165   170   175   180   185   190   195   200
```

TABLE 8A-3   MEANS OF SUCCESSIVE SUBGROUPS OF SIZE 5 FROM TICKET SELLER DATA WITH 200 OBSERVATIONS

DESCRIPTIVE STATISTICS

DO YOU WANT INSTRUCTIONS? TYPE 1 FOR YES, 0 FOR NO.
• 0

WHICH OPTION? TYPE 1 OR 2.
• 2

IF INPUT IS TO BE PROVIDED FROM A DISK
FILE, TYPE A 1. OTHERWISE TYPE A 0.
• 1

IF YOU WANT TO AVERAGE ADJACENT OBSERVATIONS, TYPE A 1.
IF YOU WANT TO SELECT EVERY N–TH OBSERVATION, TYPE A 0.
• 1

TABLE 8A-3 (cont.)

HOW MANY ADJACENT OBSERVATIONS DO YOU WANT TO AVERAGE?
- 5

IF YOU WANT YOUR OUTPUT WRITTEN ON PAPER ONLY, TYPE A 1.
IF YOU WANT YOUR OUTPUT WRITTEN ON DISK ONLY, TYPE A 2.
IF YOU WANT YOUR OUTPUT WRITTEN BOTH ON PAPER AND ON DISK, TYPE A 3.
- 3

57.0	16.6	118.0	420.0	119.4	329.0	712.4	837.8	850.2	852.4
833.0	698.0	563.6	608.4	603.0	593.4	633.4	650.4	817.8	649.0
535.8	555.8	573.0	525.2	768.2	454.0	526.6	614.0	204.2	64.0
61.6	142.4	176.6	175.2	125.0	97.2	133.2	250.0	441.6	724.8

TABLE 8A-4   EVERY FIFTH OBSERVATION FROM TICKET SELLER DATA WITH 200 OBSERVATIONS

DESCRIPTIVE STATISTICS

DO YOU WANT INSTRUCTIONS? TYPE 1 FOR YES, 0 FOR NO.
- 0

WHICH OPTION? TYPE 1 OR 2.
- 2

IF INPUT IS TO BE PROVIDED FROM A DISK
FILE, TYPE A 1. OTHERWISE TYPE A 0.
- 1

IF YOU WANT TO AVERAGE ADJACENT OBSERVATIONS, TYPE A 1.
IF YOU WANT TO SELECT EVERY N-TH OBSERVATION, TYPE A 0.
- 0

YOU CHOSE SELECTION OF EVERY N-TH OBSERVATION.
TYPE THE VALUE OF N.
- 5

IF YOU WANT YOUR OUTPUT WRITTEN ON PAPER ONLY, TYPE A 1.
IF YOU WANT YOUR OUTPUT WRITTEN ON DISK ONLY, TYPE A 2.
IF YOU WANT YOUR OUTPUT WRITTEN BOTH ON PAPER AND ON DISK, TYPE A 3.
- 3

31.0	13.0	20.0	494.0	236.0	97.0	682.0	852.0	771.0	1052.0
827.0	708.0	607.0	646.0	633.0	474.0	710.0	558.0	853.0	696.0
544.0	410.0	472.0	631.0	812.0	695.0	510.0	525.0	368.0	76.0
27.0	40.0	312.0	8.0	218.0	8.0	15.0	121.0	354.0	868.0

chapter nine

ESTIMATING SYSTEM PARAMETERS AND EVALUATING ALTERNATIVE SYSTEMS

EVALUATING ALTERNATIVE SYSTEMS

Now that the simulation output data may be assumed to be free of transients and autocorrelation, we may estimate system performance parameters and evaluate alternative systems in a statistically meaningful way. We say "statistically meaningful" because the point estimates (means) provided in the GPSS summary report are not usually sufficient to establish accurate ranges within which the system is likely to perform. A reliable estimate of the population standard deviation and variance based on serially independent data is needed to accomplish interval estimation of population performance parameters and testing of hypotheses about differences among alternative systems.

Classical statistics offers interval estimation and hypothesis testing for means, proportions, differences in means, and differences in proportions. We shall confine our discussion to interval estimation of means for several reasons. First, mean transit time is the performance parameter most often desired by management. Second, the principles and formulas for interval estimates of proportions, differences in means, and differences in proportions are quite analogous to those for means and proportions and may be found in most introductory textbooks on applied statistics. Finally, interval estimates of differences are limited to pairwise differences between two systems and are actually a cumbersome way of performing hypothesis tests. We shall consider more straightforward and powerful hypothesis testing techniques later in this chapter.

ESTIMATING CONFIDENCE INTERVALS

If the sample size is at least 25 or 30, the central limit theorem provides that sample means will be normally distributed around their population mean. The mean of the sample means is the population mean. The standard deviation of the distribution of sample means is called the standard error. It is important to distinguish the standard deviation of the individual items in a sample or population from the standard error, which is the standard deviation of the distribution of sample means and which decreases in proportion to the square root of the sample size. That is, the larger the sample size, the smaller the standard error. If the entire population were to be sampled, there would be no error at all in estimating its mean, but ordinarily an entire population cannot feasibly be sampled. For distributions of sample means, the standard error is equal to the population standard deviation divided by the square root of the sample size. Expressed as a formula, then,

$$\text{standard error} = \frac{\text{S.D.}}{\sqrt{n}}$$

where S.D. is the sample standard deviation and n is the sample size. Thus, it is intuitively clear that the larger the sample size, the lower the standard error. However, it is also painfully evident that the standard error is not reduced in direct proportion to an increase in the sample size but rather in proportion to the square root of the increase in the sample size. It can be prohibitively expensive to increase sampling to the point where estimates of system performance are exceedingly accurate. The analyst and management must determine a satisfactory trade-off between the cost of sampling and the cost of making somewhat inaccurate predictions of system performance.

INTERVAL ESTIMATION OF TRANSIT TIME

In Chapter 8 we obtained a sample of 40 observations which were chosen from a set of 1000 transit time values at intervals of 25. These observations were found to exhibit minimal levels of serial correlation. Those 40 observations appeared in Table 8-24, with a mean of 270.2 seconds, a variance of 105610.3 seconds, and a standard deviation of 324.9773 seconds. Suppose that we want to define a confidence interval for the true mean transit time for the ticket seller system. Our best single estimate is the sample mean, but creating an interval estimate permits us to estimate a range of values which is highly likely to include the true mean, whereas the point estimate is virtually guaranteed to deviate from the true population mean by an undetermined amount. The standard error of the mean will be

$$\text{standard error} = \frac{270.2}{\sqrt{40}} = 42.72 \text{ seconds}$$

A confidence interval must be defined by the level of confidence. Commonly used levels of confidence are 90, 95, and 99 percent. These are roughly analogous to levels of significance of 0.10, 0.05, and 0.01 in hypothesis testing. While it is technically incorrect to say that a 95 percent confidence interval has a 0.95 probability of containing the true population mean,* that statement offers a fairly good basis for choosing levels of confidence and interpreting a specific confidence interval.

The larger the level of confidence, the more likely it is that the confidence interval will accurately forecast the range containing the true population mean. However, the larger the level of confidence, the wider the confidence interval and the more vague the notion of where the true population mean may lie within the interval. To illustrate, we might state with 99.99 percent confidence that the true mean height of an American adult male is between 3 feet and 8.5 feet, but this interval is not at all useful in assisting manufacturers of men's clothing to decide how long to cut pants for ready-to-wear suits. If a very high level of confidence is desired, the sample size should be increased concomitantly to decrease the standard error and avoid impractical statements such as the one about average heights.

The facts necessary to compute a confidence interval are the sample point estimate (mean), the standard error of the sample point estimator, and the level of confidence desired. The level of confidence may be translated into a z statistic—1.645 for 90 percent confidence, 1.96 for 95 percent confidence, and 2.575 for 99 percent confidence† The lower and upper limits of the confidence interval are

$$\text{L.L.} = \text{P.E.} - (z \times \text{standard error})$$

$$\text{U.L.} = \text{P.E.} + (z \times \text{standard error})$$

where P.E. is the sample point estimate of the mean, proportion, or difference and z is the value of z corresponding to the desired level of confidence.

A 90 percent confidence interval for the true mean transit time for the ticket seller system would be $270.2 \pm (1.645 \times 42.72) = 270.2 \pm 70.27$ seconds, and the upper limit is $270.2 + 70.27 = 340.47$ seconds. A 95 percent confidence interval for the true mean transit time for the ticket seller system would be $270.2 \pm (1.96 \times 42.72 = 270.2 +$ or $- 83.73$ seconds. The lower limit of this

---

*Since the true population mean, which is an unknown constant, either falls within a particular confidence interval or it does not, the probability that the specific confidence interval contains the population mean is either 0 or 1. A statistically correct statement would be that if a large number of samples of the same size were taken and if confidence intervals were computed from the data given by each sample, 95 percent of the computed confidence intervals would contain the true population mean and 5 percent would not include it.

†Other levels of confidence are available, although 90, 95, and 99 percent are the most commonly used.

confidence interval is $270.2 - 83.73 = 186.47$ seconds, and the upper limit is $270.2 + 8373 = 353.93$ seconds. A 99 percent confidence interval for the true mean transit time would be $270.2 \pm (2.575 \times 42.72) = 270.2 \pm 110.00$ seconds. The lower limit of this confidence interval is $270.2 - 110.00 = 160.2$ seconds, and the upper limit is $270.2 + 110.00 = 380.2$ seconds. Table 9-1 displays the results we have just calculated in compact form.

TABLE 9-1 RELATIONSHIP AMONG LEVEL OF
CONFIDENCE, Z VALUE, AND CONFIDENCE LIMITS
FOR TICKET SELLER SYSTEM TRANSIT TIME DATA

CONFIDENCE LEVEL	Z VALUE	LOWER LIMIT	UPPER LIMIT	INTERVAL WIDTH
90	1.645	227.5	312.9	85.4
95	1.96	186.5	353.9	167.4
99	2.575	160.2	380.2	220.0

If management wants to be extremely sure that the interval it quotes will indeed include the true population mean, it will have to be content with a range of 220 seconds between the possible lower limit and upper limit of mean transit time. In exchange for slightly greater risk of misstatement of the true mean, a substantial drop in the range or width of the confidence interval may be achieved. Again, if management wants to be extremely sure that the quoted interval will encompass the true mean transit time and also wants that interval estimate to be narrow, it must pay the cost of additional sampling to decrease the standard error.

Why might such precision matter to management? If the transaction being processed were perishable after some amount of processing time had elapsed, it would be important to establish that the mean processing time for the system under consideration fell well within the bounds of survivability for the product.

Another example might be the weighing of cost of time in the system against the cost of servers. If there were a substantial possibility that transit time was very large, as indicated by a high upper limit on the confidence interval, management might deem it worthwhile to hire a faster server or an additional server rather than incur the wrath of customers and cost of lost sales. On the other hand, management might be loath to take such a rash step without solid evidence that transit time is indeed unacceptably high.

TESTING DIFFERENCES AMONG ALTERNATIVES

Often, the objective of statistical analysis of simulation output is to determine whether or not there is a significant difference in the performance of two or more alternative system configurations. If the difference in sample simulation results is not statistically significant, the best decision will be to select the system whose costs are lowest. If there is a significant difference in system performance, the manager must decide whether the enhanced system performance is worth the additional cost, if any. For a test comparing two alternatives (the original and one other), the Mann-Whitney-Wilcoxon test will be used. For a test comparing three or more alternatives, the Kruskal-Wallis test will be used. It should be noted here that since we are using nonparametric hypothesis tests, the population parameter for which we will be testing differences is the median, not the mean. Since we are simply trying to establish whether or not one system is preferable, the substitution of median for mean in the tests should not cause any real problems. Tables for the Mann-Whitney-Wilcoxon and Kruskal-Wallis tests using small samples are available in nonparametric statistics tests.

Suppose that we want to decide whether there is a statistically significant difference between the mean transit time in our standard ticket seller model and the faster-server model described in Chapter 3. We retain the 40 observations which were chosen as every twenty-fifth value in the run of 1000 transactions. We must now perform the tests and associated operations described in Chapter 8 for the alternative system. The GPSS program for the alternative system is shown in Table 9-2.

TABLE 9-2   GPSS PROGRAM FOR FASTER-SERVER SYSTEM

```
JOB
INITIAL X1, 0
EXPON1 FUNCTION RN3, C24
0, 0/. 1, . 104/. 2, . 222/. 3, . 355/. 4, . 509/. 5, . 69
0. 6, . 915/. 7, 1. 2/. 75, 1. 38/. 8, 1. 6/. 84, 1. 83/. 88, 2. 12
0. 9, 2. 3/. 92, 2. 52/. 94, 2. 81/. 95, 2. 99/. 96, 3. 2/. 97, 3. 5
0. 98, 3. 9/. 99, 4. 6/. 995, 5. 3/. 998, 6. 2/. 999, 7. 0/. 9997, 8. 0
EXPON2 FUNCTION RN4, C24
0, 0/. 1, . 104/. 2, . 222/. 3, . 355/. 4, . 509/. 5, . 69
0. 6, . 915/. 7, 1. 2/. 75, 1. 38/. 8, 1. 6/. 84, 1. 83/. 88, 2. 12
0. 9, 2. 3/. 92, 2. 52/. 94, 2. 81/. 95, 2. 99/. 96, 3. 2/. 97, 3. 5
0. 98, 3. 9/. 99, 4. 6/. 995, 5. 3/. 998, 6. 2/. 999, 7. 0/. 9997, 8. 0
GENERATE 101, FN$EXPON1
TABULATE INTARV
QUEUE 1
SEIZE 1
DEPART 1
ADVANCE 70, FN$EXPON2
RELEASE 1
SAVEVALUE 1, M1
PRINT , , X
```

TABLE 9-2 (cont.)

```
TABULATE TRANSIT
TERMINATE 1
INTARV TABLE IA, 0, 20, 20
TRANSIT TABLE M1, 20, 20, 20
START 1000
END
```

By following the procedures for detecting transients and serial correlation in Chapter 8, we eventually choose every twentieth value from the output of 1000 observations produced by the faster-server program in Table 9-2. This results in a data set consisting of 50 observations shown in Table 9-3.

TABLE 9-3   FIFTY TRANSIT TIME VALUES FROM FASTER-SERVER SYSTEM

25.0	393.0	104.0	61.0	181.0	124.0	268.0	374.0	61.0	317.0
95.0	251.0	133.0	6.0	224.0	23.0	403.0	280.0	382.0	497.0
19.0	45.0	904.0	1396.0	602.0	118.0	736.0	1139.0	497.0	106.0
240.0	30.0	420.0	278.0	11.0	79.0	378.0	399.0	116.0	37.0
320.0	223.0	107.0	58.0	257.0	316.0	88.0	454.0	165.0	337.0

THE MEAN IS 281.5398
THE VARIANCE IS   79787.13
THE STANDARD DEVIATION IS 282.4661
THE SAMPLE SIZE IS   50

When this data set is concatenated with the data set of 40 observations from the original ticket seller system, the result is a pooled data set consisting of 90 observations—the first 40 followed by the second 50. This combined data set is shown in Table 9-4.

TABLE 9-4   FORTY TRANSIT TIME VALUES FROM STANDARD-SERVER SYSTEM
AND FIFTY TRANSIT TIME VALUES FROM FASTER-SERVER SYSTEM

31.0	97.0	827.0	474.0	544.0	695.0	27.0	8.0	427.0	13.0
282.0	100.0	356.0	259.0	151.0	221.0	225.0	168.0	121.0	207.0
1738.0	749.0	239.0	21.0	437.0	452.0	97.0	40.0	41.0	51.0
12.0	415.0	253.0	92.0	336.0	29.0	39.0	506.0	2.0	26.0
25.0	393.0	104.0	61.0	181.0	124.0	268.0	374.0	61.0	317.0
95.0	251.0	133.0	6.0	224.0	23.0	403.0	280.0	382.0	497.0

TABLE 9-4 (cont.)

19.0	45.0	904.0	1396.0	602.0	118.0	736.0	1139.0	497.0	106.0
240.0	30.0	420.0	278.0	11.0	79.0	378.0	399.0	116.0	37.0
320.0	223.0	107.0	58.0	257.0	316.0	88.0	454.0	165.0	337.0

THE MEAN IS 276.5000

THE VARIANCE IS   90238.56

THE STANDARD DEVIATION IS 300.3972

THE SAMPLE SIZE IS   90

We are now ready to perform the Mann-Whitney-Wilcoxon test of differences between two alternative configurations. The null hypothesis is that there is no statistically significant difference between the two, and the alternative hypothesis is two-tailed. The procedure is to tag each data value with the system alternative which produced it and then sort the combined data set into ascending order. The resulting sorted list is assigned ranks beginning with 1 for the lowest-valued item. Tied values receive a common average rank. Then the sum of the ranks of the data items from configuration 1 is calculated, as well as the sum of the ranks of the data items from configuration 2. If the null hypothesis is true, there should be no substantial difference between the sums of the ranks of the two groups. The mean of the sampling distribution of rank sums for one of the samples is $m(n + 1)/2$, where m is the number of observations in the one group and n is the total number of observations in both groups. The standard deviation of the distribution is

$$\text{S.D.} = \sqrt{\frac{(l \times m) \times (n + 1)}{12}}$$

where l is the number of observations in the other group.

Either sample rank sum may be used for the test. A continuity correction factor of 0.5 in the direction of making the z value less significant is often applied. Also, a tie correction factor is available, but it tends to make the test less conservative and should be used only if there are numerous ties in ranks.

Table 9-5 shows the Mann-Whitney-Wilcoxon test results for the two alternative systems.

Using the first sample as a base, the z value corrected for continuity and ties is -0.678. The z value corrected for continuity only is also -0.678, since there are no tied values. The normal approximation may be considered valid for this test for sample sizes of 10 or more in each sample.

Evidently there is no statistically significant difference in mean transit times for the two systems. Assuming that the cost of employing the two servers is equal, we would perhaps choose the faster one to try to minimize any cost of lost sales, however insignificant. However, if there is a premium wage paid to the faster server, management might well decide that the nonsignificant finding supports the retention of the slower server.

TABLE 9-5   MANN-WHITNEY-WILCOXON TEST OF
DIFFERENCES BETWEEN THE STANDARD-SERVER AND
FASTER-SERVER SYSTEM TRANSIT TIMES

GROUP NUMBER	OBSERVATION NUMBER	VALUE	INITIAL RANK	FINAL RANK
1	1	31.00	1.00	15.00
1	2	97.00	2.00	29.50
1	3	827.00	3.00	86.00
1	4	474.00	4.00	77.00
1	5	544.00	5.00	81.00
1	6	695.00	6.00	83.00
1	7	27.00	7.00	12.00
1	8	8.00	8.00	3.00
1	9	427.00	9.00	73.00
1	10	13.00	10.00	6.00
1	11	282.00	11.00	58.00
1	12	100.00	12.00	31.00
1	13	356.00	13.00	64.00
1	14	259.00	14.00	54.00
1	15	151.00	15.00	40.00
1	16	221.00	16.00	45.00
1	17	225.00	17.00	48.00
1	18	168.00	18.00	42.00
1	19	121.00	19.00	37.00
1	20	207.00	20.00	44.00
1	21	1738.00	21.00	90.00
1	22	749.00	22.00	85.00
1	23	239.00	23.00	49.00
1	24	21.00	24.00	8.00
1	25	437.00	25.00	74.00
1	26	452.00	26.00	75.00
1	27	97.00	27.00	29.50
1	28	40.00	28.00	18.00
1	29	41.00	29.00	19.00
1	30	51.00	30.00	21.00
1	31	12.00	31.00	5.00
1	32	415.00	32.00	71.00
1	33	253.00	33.00	52.00
1	34	92.00	34.00	27.00
1	35	336.00	35.00	62.00
1	36	29.00	36.00	13.00
1	37	39.00	37.00	17.00

TABLE 9-5 (cont.)

GROUP NUMBER	OBSERVATION NUMBER	VALUE	INITIAL RANK	FINAL RANK
1	38	506.00	38.00	80.00
1	39	2.00	39.00	1.00
1	40	26.00	40.00	11.00
2	1	25.00	41.00	10.00
2	2	393.00	42.00	68.00
2	3	104.00	43.00	32.00
2	4	61.00	44.00	23.50
2	5	181.00	45.00	43.00
2	6	124.00	46.00	38.00
2	7	268.00	47.00	55.00
2	8	374.00	48.00	65.00
2	9	61.00	49.00	23.50
2	10	317.00	50.00	60.00
2	11	95.00	51.00	28.00
2	12	251.00	52.00	51.00
2	13	133.00	53.00	39.00
2	14	6.00	54.00	2.00
2	15	224.00	55.00	47.00
2	16	23.00	56.00	9.00
2	17	403.00	57.00	70.00
2	18	280.00	58.00	57.00
2	19	382.00	59.00	67.00
2	20	497.00	60.00	78.50
2	21	19.00	61.00	7.00
2	22	45.00	62.00	20.00
2	23	904.00	63.00	87.00
2	24	1396.00	64.00	89.00
2	25	602.00	65.00	82.00
2	26	118.00	66.00	36.00
2	27	736.00	67.00	84.00
2	28	1139.00	68.00	88.00
2	29	497.00	69.00	78.50
2	30	106.00	70.00	33.00
2	31	240.00	71.00	50.00
2	32	30.00	72.00	14.00
2	33	420.00	73.00	72.00
2	34	278.00	74.00	56.00
2	35	11.00	75.00	4.00
2	36	79.00	76.00	25.00
2	37	378.00	77.00	66.00
2	38	399.00	78.00	69.00

TABLE 9-5 (cont.)

GROUP NUMBER	OBSERVATION NUMBER	VALUE	INITIAL RANK	FINAL RANK
2	39	116.00	79.00	35.00
2	40	37.00	80.00	16.00
2	41	320.00	81.00	61.00
2	42	223.00	82.00	46.00
2	43	107.00	83.00	34.00
2	44	58.00	84.00	22.00
2	45	257.00	85.00	53.00
2	46	316.00	86.00	59.00
2	47	88.00	87.00	26.00
2	48	454.00	88.00	76.00
2	49	165.00	89.00	41.00
2	50	337.00	90.00	63.00

DIFFERENCES AMONG MORE THAN TWO ALTERNATIVES

The procedure for comparing more than two alternatives is completely analogous to that for testing two alternatives, except that the test statistic for three or more is generally χ^2 distributed whereas the test statistic for two was generally normally distributed. The Kruskal-Wallis H test has a null hypothesis of no significant differences among the several treatment groups. The alternative hypothesis is that some difference exists. Data sets from the various treatment groups are pooled, sorted, and ranked, and the rank sum for each group is computed. The Kruskal-Wallis H statistic is computed as

$$H = \frac{12 - \text{DEVSQR}}{n(n+1)}$$

where n is the total number of observations and DEVSQR is a measure of the deviation of each group average rank sum from the overall average rank sum. DEVSQR is calculated as

$$\text{DEVSQR} = \sum \text{ for } j \text{ groups} = n(j) \times [r(j) - r]^2$$

where $n(j)$ is the size of the jth treatment group, $r(j)$ is the average rank in group j, and r is the average rank in the entire pooled sample. Table 9-6 shows a GPSS program describing a third alternative, a two-server ticket seller system.

TABLE 9-6 GPSS PROGRAM FOR TWO-SERVER SYSTEM

```
JOB
 INITIAL X1, 0
EXPON1 FUNCTION RN3, C24
0, 0/. 1, . 104/. 2, . 222/. 3, . 355/. 4, . 509/. 5, . 69
0. 6, . 915/. 7, 1. 2/. 75, 1. 38/. 8, 1. 6/. 84, 1. 83/. 88, 2. 12
```

TABLE 9-6 (cont.)

0.9,2.3/.92,2.52/.94,2.81/.95,2.99/.96,3.2/.97,3.5

0.98,3.9/.99,4.6/.995,5.3/.998,6.2/.999,7.0/.9997,8.0

EXPON2 FUNCTION RN4, C24

0,0/.1,.104/.2,.222/.3,.355/.4,.509/.5,.69

0.6,.915/.7,1.2/.75,1.38/.8,1.6/.84,1.83/.88,2.12

0.9,2.3/.92,2.52/.94,2.81/.95,2.99/.96,3.2/.97,3.5

0.98,3.9/.99,4.6/.995,5.3/.998,6.2/.999,7.0/.9997,8.0

```
GENERATE 101, FN$EXPON1
TABULATE INTARV
QUEUE 1
ENTER 1
DEPART 1
ADVANCE 70, FN$EXPON2
LEAVE 1
SAVEVALUE 1, M1
PRINT , , X
TABULATE TRANSIT
TERMINATE 1
1 STORAGE 2
INTARV TABLE IA, 0, 20, 20
TRANSIT TABLE M1, 20, 20, 20
START 1000
END
```

It happens that a data set consisting of every tenth value from the 1000 produced by this program passes the tests for transients and serial correlation. These 100 observations are shown in Table 9-7.

TABLE 9-7   ONE HUNDRED TRANSIT TIME VALUES FROM TWO-SERVER SYSTEM

25.0	30.0	22.0	24.0	166.0	61.0	12.0	53.0	5.0	59.0
166.0	112.0	148.0	13.0	6.0	61.0	2.0	142.0	36.0	33.0
10.0	9.0	99.0	121.0	5.0	4.0	80.0	235.0	208.0	13.0
35.0	152.0	51.0	132.0	38.0	83.0	5.0	20.0	88.0	30.0
13.0	124.0	7.0	85.0	199.0	129.0	33.0	10.0	140.0	8.0
10.0	4.0	75.0	100.0	14.0	104.0	74.0	97.0	83.0	28.0
2.0	61.0	3.0	23.0	5.0	111.0	193.0	83.0	7.0	56.0
54.0	72.0	160.0	25.0	81.0	57.0	311.0	243.0	27.0	14.0
15.0	12.0	31.0	40.0	41.0	369.0	50.0	92.0	77.0	91.0
174.0	19.0	29.0	2.0	140.0	172.0	95.0	124.0	52.0	52.0

TABLE 9-7 (cont.)

THE MEAN IS 71.9600
THE VARIANCE IS   4978.41
THE STANDARD DEVIATION IS 70.5579
THE SAMPLE SIZE IS   100

Table 9-8 gives the results of the Kruskal-Wallis test of the three alternative system configurations.

TABLE 9-8   KRUSKAL-WALLIS TEST OF DIFFERENCES AMONG THE STANDARD-SERVER, FASTER-SERVER, AND TWO-SERVER SYSTEMS

GROUP NUMBER	OBSERVATION NUMBER	VALUE	INITIAL RANK	FINAL RANK
1	1	31.00	1.00	53.50
1	2	97.00	2.00	102.00
1	3	827.00	3.00	186.00
1	4	474.00	4.00	177.00
1	5	544.00	5.00	181.00
1	6	695.00	6.00	183.00
1	7	27.00	7.00	45.50
1	8	8.00	8.00	16.50
1	9	427.00	9.00	173.00
1	10	13.00	10.00	27.50
1	11	282.00	11.00	156.00
1	12	100.00	12.00	105.50
1	13	356.00	13.00	163.00
1	14	259.00	14.00	152.00
1	15	151.00	15.00	127.00
1	16	221.00	16.00	141.00
1	17	225.00	17.00	144.00
1	18	168.00	18.00	133.00
1	19	121.00	19.00	115.50
1	20	207.00	20.00	139.00
1	21	1738.00	21.00	190.00
1	22	749.00	22.00	185.00
1	23	239.00	23.00	146.00
1	24	21.00	24.00	36.00
1	25	437.00	25.00	174.00

TABLE 9-8 (cont.)

1	26	452.00	26.00	175.00
1	27	97.00	27.00	102.00
1	28	40.00	28.00	62.50
1	29	41.00	29.00	64.50
1	30	51.00	30.00	68.50
1	31	12.00	31.00	24.00
1	32	415.00	32.00	171.00
1	33	253.00	33.00	150.00
1	34	92.00	34.00	97.50
1	35	336.00	35.00	161.00
1	36	29.00	36.00	48.50
1	37	39.00	37.00	61.00
1	38	506.00	38.00	180.00
1	39	2.00	39.00	2.50
1	40	26.00	40.00	44.00
2	1	25.00	41.00	42.00
2	2	393.00	42.00	168.00
2	3	104.00	43.00	107.50
2	4	61.00	44.00	80.00
2	5	181.00	45.00	136.00
2	6	124.00	46.00	118.00
2	7	268.00	47.00	153.00
2	8	374.00	48.00	165.00
2	9	61.00	49.00	80.00
2	10	317.00	50.00	159.00
2	11	95.00	51.00	99.50
2	12	251.00	52.00	149.00
2	13	133.00	53.00	122.00
2	14	6.00	54.00	12.50
2	15	224.00	55.00	143.00
2	16	23.00	56.00	38.50
2	17	403.00	57.00	170.00
2	18	280.00	58.00	155.00
2	19	382.00	59.00	167.00
2	20	497.00	60.00	178.50
2	21	19.00	61.00	33.50
2	22	45.00	62.00	66.00
2	23	904.00	63.00	187.00
2	24	1396.00	64.00	189.00
2	25	602.00	65.00	182.00
2	26	118.00	66.00	114.00
2	27	736.00	67.00	184.00

TABLE 9-8 (cont.)

2	28	1139.00	68.00	188.00
2	29	497.00	69.00	178.50
2	30	106.00	70.00	109.00
2	31	240.00	71.00	147.00
2	32	30.00	72.00	51.00
2	33	420.00	73.00	172.00
2	34	278.00	74.00	154.00
2	35	11.00	75.00	22.00
2	36	79.00	76.00	87.00
2	37	378.00	77.00	166.00
2	38	399.00	78.00	169.00
2	39	116.00	79.00	113.00
2	40	37.00	80.00	59.00
2	41	320.00	81.00	160.00
2	42	223.00	82.00	142.00
2	43	107.00	83.00	110.00
2	44	58.00	84.00	76.00
2	45	257.00	85.00	151.00
2	46	316.00	86.00	158.00
2	47	88.00	87.00	94.50
2	48	454.00	88.00	176.00
2	49	165.00	89.00	130.00
2	50	337.00	90.00	162.00
3	1	25.00	91.00	42.00
3	2	30.00	92.00	51.00
3	3	22.00	93.00	37.00
3	4	24.00	94.00	40.00
3	5	166.00	95.00	131.50
3	6	61.00	96.00	80.00
3	7	12.00	97.00	24.00
3	8	53.00	98.00	72.00
3	9	5.00	99.00	9.50
3	10	59.00	100.00	77.00
3	11	166.00	101.00	131.50
3	12	112.00	102.00	112.00
3	13	148.00	103.00	126.00

TABLE 9-8 (cont.)

3	14	13.00	104.00	27.50
3	15	6.00	105.00	12.50
3	16	61.00	106.00	80.00
3	17	2.00	107.00	2.50
3	18	142.00	108.00	125.00
3	19	36.00	109.00	58.00
3	20	33.00	110.00	55.50
3	21	10.00	111.00	20.00
3	22	9.00	112.00	18.00
3	23	99.00	113.00	104.00
3	24	121.00	114.00	115.50
3	25	5.00	115.00	9.50
3	26	4.00	116.00	6.50
3	27	80.00	117.00	88.00
3	28	235.00	118.00	145.00
3	29	208.00	119.00	140.00
3	30	13.00	120.00	27.50
3	31	35.00	121.00	57.00
3	32	152.00	122.00	128.00
3	33	51.00	123.00	68.50
3	34	132.00	124.00	121.00
3	35	38.00	125.00	60.00
3	36	83.00	126.00	91.00
3	37	5.00	127.00	9.50
3	38	20.00	128.00	35.00
3	39	88.00	129.00	94.50
3	40	30.00	130.00	51.00
3	41	13.00	131.00	27.50
3	42	124.00	132.00	118.00
3	43	7.00	133.00	14.50
3	44	85.00	134.00	93.00
3	45	199.00	135.00	138.00
3	46	129.00	136.00	120.00
3	47	33.00	137.00	55.50
3	48	10.00	138.00	20.00
3	49	140.00	139.00	123.50
3	50	8.00	140.00	16.50
3	51	10.00	141.00	20.00
3	52	4.00	142.00	6.50
3	53	75.00	143.00	85.00
3	54	100.00	144.00	105.50
3	55	14.00	145.00	30.50

TABLE 9-8 (cont.)

3	56	104.00	146.00	107.50
3	57	74.00	147.00	84.00
3	58	97.00	148.00	102.00
3	59	83.00	149.00	91.00
3	60	28.00	150.00	47.00
3	61	2.00	151.00	2.50
3	62	61.00	152.00	80.00
3	63	3.00	153.00	5.00
3	64	23.00	154.00	38.50
3	65	5.00	155.00	9.50
3	66	111.00	156.00	111.00
3	67	193.00	157.00	137.00
3	68	83.00	158.00	91.00
3	69	7.00	159.00	14.50
3	70	56.00	160.00	74.00
3	71	54.00	161.00	73.00
3	72	72.00	162.00	83.00
3	73	160.00	163.00	129.00
3	74	25.00	164.00	42.00
3	75	81.00	165.00	89.00
3	76	57.00	166.00	75.00
3	77	311.00	167.00	157.00
3	78	243.00	168.00	148.00
3	79	27.00	169.00	45.50
3	80	14.00	170.00	30.50
3	81	15.00	171.00	32.00
3	82	12.00	172.00	24.00
3	83	31.00	173.00	53.50
3	84	40.00	174.00	62.50
3	85	41.00	175.00	64.50
3	86	369.00	176.00	164.00
3	87	50.00	177.00	67.00
3	88	92.00	178.00	97.50
3	89	77.00	179.00	86.00
3	90	91.00	180.00	96.00
3	91	174.00	181.00	135.00
3	92	19.00	182.00	33.50
3	93	29.00	183.00	48.50
3	94	2.00	184.00	2.50
3	95	140.00	185.00	123.50
3	96	172.00	186.00	134.00
3	97	95.00	187.00	99.50

TABLE 9-8 (cont.)

3	98	124.00	188.00	118.00
3	99	52.00	189.00	70.50
3	100	52.00	190.00	70.50

GROUP NUMBER	MEAN RANK	SAMPLE SIZE	WEIGHTED DIFFERENCE
1	116.60	40	17808.40
2	127.48	50	51136.02
3	71.07	100	59682.49

THE TIE CORRECTION FACTOR = 0.9999.

H = 42.54.

IF THE SAMPLE SIZES ARE > 5, THIS STATISTIC IS χ^2 DISTRIBUTED WITH 2 DEGREES OF FREEDOM.

We conclude that the differences among the three groups are highly significant, since critical χ^2 with 2 degrees of freedom is only 5.99. It is likely after examining the average ranks that group 3 (two-server system) is superior to groups 1 and 2. Often, when statistical significance is indicated by the Kruskal-Wallis test or other types of analysis of variance, it is not so obvious which groups have caused the statistic to be significant. If we were to perform three Mann-Whitney-Wilcoxon tests, respectively comparing groups 1 and 2, groups 2 and 3, and groups 1 and 3, we would be able to establish a hierarchy of significant differences. However, we would pay a potential price for this precision. Each test is performed at the 0.05 level of significance. Thus, the probability of *not* committing a type I error (rejecting the null hypothesis when it is true) for *each* test is 0.95. The probability of not committing a type I error on any of three tests is, thus, $0.95 \times 0.95 \times 0.95 = 0.86$. The probability of committing a type I error on one or more of the tests is $1 - 0.86 = 0.14$. This is considerably higher than the 0.05 probability which prevails in an analysis of variance, no matter how many different treatment groups are compared.

MULTIPLE COMPARISONS TECHNIQUES

It would be preferable to decide which groups are significantly different from which other groups by more sophisticated statistical methods. Dunn's test of multiple comparisons offers just such a vehicle. For each pair of groups, I and J, to be compared, the absolute value of the difference in their average ranks is calculated. This absolute value is compared with the following criterion statistic:

$$\text{criterion} = z \sqrt{(A - B) \times (C/D)}$$

$$A = n(n^2 - 1)$$

$$B = \sum u^3 - \sum u$$

$$C = \frac{1}{n_i} + \frac{1}{n_j}$$

$$D = 12(n - 1)$$

n is the combined number of observations in all treatment groups, n_i is the number of observations in the ith treatment group, n_j is the number of observations in the jth treatment group, and sum $u3$ and sum u are terms which correct for ties. u is the number of observations in all samples combined which are tied at a given rank; these are then summed over all tied ranks. While the tie correction generally has little effect, it tends to make the test more conservative.

Let us apply Dunn's test of multiple comparisons to the results of the Kruskal-Wallis test of differences among the three ticket seller systems. Table 9-9 shows the partial calculations.

TABLE 9-9   DUNN'S TEST OF MULTIPLE COMPARISONS
AS APPLIED TO THE THREE TICKET SELLER SYSTEMS

GROUP I	GROUP J	ABSOLUTE DIFFERENCE	CRITERION VALUE AT 0.05	CRITERION VALUE AT 0.10
1	2	10.880	27.926	24.823
1	3	45.530	24.629	21.892
2	3	56.410	22.802	20.268

COST OF SYSTEM ALTERNATIVES

Dunn's test corroborates our conjecture that the significant differences lie between group 3 and the other two groups. The difference between group 1 and group 2 is not statistically significant. This presents management with a more subjective problem. The cost of employing two ticket sellers will almost certainly exceed the cost of employing even the faster single ticket seller. Is this additional expenditure for service justifiable in terms of displacement of costs of lost sales?

Let us investigate how high the expected cost of lost sales would have to be to leave management exactly indifferent between implementing the two-server system and not implementing it. Suppose, as we assumed in Chapter 3, that each ticket seller is paid $4.00 per hour for an 8-hour workday. The average customer arrival rate, the reciprocal of interarrival time on an hourly basis, is 35.545 customers per hour. The point of indifference between employing the two servers and not employing them is set where expected cost of lost sales per unit time is just equal to cost of wages per unit time. The cost of wages per day is $4.00 × 8 hours × 2 servers = $64.00. What figure shall we use as the mean customer transit time? If we were to construct a confidence interval estimate of the true mean transit time based on the sample mean and standard deviation of the group of 100 observations of the two-server system, we would be providing considerably more information than by simply using the mean as a point estimate.

A 95 percent confidence interval for the true mean transit time in the two-server system is

$$\text{L.L.} = 71.96 - 1.96(70.56/\sqrt{100})$$

$$\text{U.L.} = 71.96 + 1.96(70.56/\sqrt{100})$$

Thus, the 95 percent confidence interval for the true mean transit time has a lower limit of 58.14 and an upper limit of 85.78 seconds. Expressed as fractions of an hour, the lower and upper confidence limits are, respectively, 0.01615 and 0.0238277 hours. First, we will estimate the expected cost of lost sales per day using the lower confidence limit for mean transit time. This is 35.545 × 8 × 0.01615 × chh, where chh is the upper limit on cost of lost sales per hour which would justify the two-server system. Setting this expression equal to the daily wage cost and solving for chh we have:

$$chh = \frac{\$\,64.00}{4.59} = \$\,13.94$$

Applying the same methodology to the lower limit on expected cost of lost sales, we obtain a daily equation of 35.545 × 8 × 0.0238277 × chl, where chl is the low estimate of cost of lost sales per hour:

$$chl = \frac{\$\,64.00}{6.78} = \$\,9.44$$

Our original guesstimate of the cost of lost sales per hour in Chapter 3 was $10.00, a figure which falls within this range. Hence, we might say that the use of the two-server system appears to be warranted. If the true cost of lost sales were even higher, of course, a fourth alternative system with even better performance characteristics could be cost justified.

ESTIMATING SYSTEM UTILIZATION

A proportion is defined to be a value between 0 and 1, inclusive. The proportion of most obvious interest in simulation is the system utilization factor. Since the formula for the standard error of a proportion is based on the average proportion and not a separate estimate of variance, we would not need to take the trouble to collect individual observations of system utilization and test them for serial correlation, since serial correlation does not affect averages, only variances and standard deviations. It is fortunate that individual observations of proportions are not necessary, since they would pose additional problems in collection and analysis. The GPSS SNA called FR contains a cumulative utilization statistic rather than a snapshot of whether or not the facility is in use at the time of printing; cumulative statistics are undesirable for detecting autocorrelation. The SNA called FV returns a value of 1 if the facility is busy or 0 if it is idle at the time the transaction passes the SAVEVALUE and PRINT blocks. The resulting data stream consists only of 0's and 1's, discrete data with many tie values which invalidate the use of the runs-up-and-down and Spearman rank correlation tests. However, the analyst must find some method of ensuring that the system was in steady state at the time the statistics were gathered. The use of one or more GPSS RESET blocks would probably be sufficient to enable the analyst to determine whether or not steady state was achieved during the various run segments.

The system utilization factor is in one respect unlike the proportions which are commonly the subject of confidence interval estimation and hypothesis testing. A typical proportion case might involve a sample of 36 people who are examined to determine whether or not they are more than 6 feet tall. If 9 of the 36 individuals are more than 6 feet tall and our sample is drawn randomly, 9/36 or 0.25 is our sample point estimate of the true population proportion which exceeds 6 feet in height. The standard error of the proportion is related to the sample size:

$$\text{S.E.P.} = \sqrt{\frac{p - (1 - p)}{n}}$$

where p is the sample proportion and n is the sample size. It is clear that the proportion in the height example is the proportion *of people sampled*. In dealing with transit time, it is clear that average transit time is the mean *of all people sampled*. System utilization, however, is not a proportion of the transactions sampled but rather a proportion of the simulation run time. n, then, is the length of the simulation run in simulated clock units, and p is the proportion of the run for which the facility was busy.

SUMMARY

When the data for the simulated system are free of autocorrelation and transients, it is possible to estimate system performance parameters using confidence intervals. These performance data may also be used to test the hypothesis that two or more alternative systems are statistically equivalent in performance. When there are only two alternatives to be compared, the Mann-Whitney-Wilcoxon test is employed; when there are three or more alternatives, the Kruskal-Wallis test is applied.

It is helpful to add Dunn's test of multiple comparisons when the Kruskal-Wallis test is used, to clarify which, if any, alternatives are significantly different without increasing the probability of a type I error.

An optimal system will balance wage costs per unit time with expected cost of lost sales per unit time. Whether or not this ideal balance is achieved, the best system of those evaluated will have the lowest sum of wage costs plus expected costs of lost sales.

QUESTIONS

1. What are the advantages and disadvantages of using point estimates of system parameters as compared with interval estimates? Give an intuitive and a statistically valid definition of a 95 percent confidence interval for true mean transit time.
2. What is the difference between a standard error and a standard deviation? What is the relationship between the standard error and the sample size? What trade-offs must management make because of this relationship?
3. What are the major differences between the Kruskal-Wallis test and the Mann-Whitney-Wilcoxon test? What are the main similarities between the two tests?
4. Suppose that the hypothesis tests indicate that there is no significant difference in the performance of two systems you are considering. Which system would you choose? On what basis?
5. Explain the purpose of Dunn's test of multiple comparisons. If one system alternative were shown to be significantly superior to the others in performance, would you implement it without any further consideration? Explain.
6. What are the similarities and differences in working with confidence intervals and hypothesis tests concerning facility utilization as compared with transit time?

PROBLEMS

1. The GPSS program which was shown in Chapter 8, Problem 1, has been run for a total of 1000 observations of transit time. Using the methods in Chapter 8, these data were eventually reduced to a sample of every twenty-fifth value. A set of 40 observations which passed the runs-up-and-down test, the Spearman test, and the steady-state heuristics resulted. These 40 values are as follows:

115.0	3091.0	607.0	125.0	241.0	397.0	421.0	1429.0	229.0	1186.0
15.0	315.0	1652.0	1114.0	456.0	317.0	503.0	1269.0	1629.0	994.0
777.0	1123.0	1263.0	427.0	1646.0	4483.0	4505.0	1693.0	418.0	1538.0
502.0	414.0	523.0	2504.0	2576.0	147.0	6.0	2248.0	560.0	5265.0

THE MEAN IS   1218.0740
THE VARIANCE IS   1621598.00
THE STANDARD DEVIATION IS   1273.4190
THE SAMPLE SIZE IS   40

Construct a 95 percent and a 99 percent confidence interval for the true mean transit time in this system. Compare your results and interpret them. What might you do if you wanted the confidence interval to be narrower?

*2. The following data are from a GPSS model of the same system as in Problem 1, except that the server operates with an average service time of 340 seconds instead of 317 seconds. Perform a Mann-Whitney-Wilcoxon test of the difference between the system whose data are found in Problem 1 and the system whose data (every twenty-fifth observation from a set of 1000) are as follows:

124.0	2888.0	195.0	524.0	402.0	922.0	1917.0	2162.0	1266.0	584.0
855.0	3411.0	3368.0	2414.0	3045.0	82.0	479.0	1938.0	3307.0	2809.0
1864.0	2412.0	925.0	936.0	1306.0	518.0	3365.0	194.0	1145.0	760.0
319.0	145.0	1030.0	586.0	6065.0	1876.0	1364.0	742.0	375.0	985.0

THE MEAN IS   1490.0990
THE VARIANCE IS   1650885.00
THE STANDARD DEVIATION IS   1284.8670
THE SAMPLE SIZE IS   40

Explain your results and decide what steps you might take next.

3. Suppose that a third alternative system were added for consideration. This system has two servers, each with a mean service time of 150 seconds. Forty observations from this system (every twenty-fifth observation from a set of 1000) are as follows:

55.0	639.0	121.0	62.0	747.0	208.0	67.0	80.0	292.0	250.0
150.0	457.0	200.0	76.0	32.0	330.0	40.0	5.0	12.0	61.0
11.0	291.0	92.0	310.0	289.0	41.0	76.0	35.0	274.0	51.0
47.0	40.0	9.0	61.0	215.0	315.0	62.0	80.0	4.0	290.0

THE MEAN IS   161.9250
THE VARIANCE IS   29235.13
THE STANDARD DEVIATION IS   170.9828
THE SAMPLE SIZE IS   40

Explain your results and decide what steps you might take next.

4. Apply Dunn's test of multiple comparisons to the data used for Problem 3. (Remember that the data sets from Problems 1 and 2 will also be needed.) Explain your results.

*Answers to problems preceded by * can be found in the Selected Numerical Answers.

5. Suppose that the cost of lost sales is $10.00 per hour of time spent in the system, average interarrival time is 422 seconds, and the wage cost of servers is as follows:
   a. original single server—$4.00 per hour
   b. slower single server—$3.50 per hour
   c. each of the two faster servers—$10.00 per hour

   Compute the wage cost of each system and the expected cost of lost sales per hour using the transit time summary statistics given for Problems 1, 2, and 3. Which system would you choose? Why?

APPENDIX

The only new program used in Chapter 9 is NANOVA, a routine which performs either the Mann-Whitney-Wilcoxon test of two alternative systems or the Kruskal-Wallis test of three or more alternative systems. When the Kruskal-Wallis test is used, Dunn's test of multiple comparisons is also shown.

The program asks whether data are to come from disk or via key input. It then requests the number of treatment groups to decide which test is to be performed. The size of the groups is requested next. The user may ask for a detail listing of the individual data items and their original and final rank ordering, or the user may bypass this lengthy printout to reach the summary results.

Table 9A-1 shows the interaction with NANOVA which generated Table 9-5, the Mann-Whitney-Wilcoxon example. Table 9A-2 shows the conversation with NANOVA which produced Table 9-8, the Kruskal-Wallis example, and Table 9-9, the example of Dunn's test.

TABLE 9A-1   CONVERSATION WITH NANOVA FOR MANN-WHITNEY-WILCOXON TEST

THIS PROGRAM PERFORMS NONPARAMETRIC ONE-WAY ANALYSIS
OF VARIANCE USING THE KRUSKAL-WALLIS "H" PROCEDURE IF THE NUMBER
OF TREATMENT GROUPS IS THREE OR MORE.
DUNN'S TEST OF MULTIPLE COMPARISONS IS ALSO PERFORMED.
IF THERE ARE ONLY TWO TREATMENT GROUPS, THE MANN-WHITNEY-WILCOXON
TEST IS PERFORMED INSTEAD.

IF INPUT FROM A DISK FILE IS DESIRED, TYPE A 1
IF DATA ARE TO BE DIRECTLY ENTERED, TYPE A 0.
• 1

ENTER THE NUMBER OF TREATMENTS (GROUPS).
• 2

TABLE 9A-1 (cont.)
_____

IF ALL GROUPS ARE THE SAME SIZE, ENTER
THAT SAMPLE SIZE. OTHERWISE ENTER 999.
• 999

ENTER THE SAMPLE SIZE FOR GROUP   1
• 40

ENTER THE SAMPLE SIZE FOR GROUP   2
• 50

DO YOU WANT TO SEE THE RAW DATA? IF YES, TYPE A 1. IF NO, TYPE A 0.
• 1

GROUP NUMBER	OBSERVATION NUMBER	VALUE	INITIAL RANK	FINAL RANK
1	1	31.00	1.00	15.00
1	2	97.00	2.00	29.50
1	3	827.00	3.00	86.00
1	4	474.00	4.00	77.00
1	5	544.00	5.00	81.00
1	6	695.00	6.00	83.00
1	7	27.00	7.00	12.00
1	8	8.00	8.00	3.00
1	9	427.00	9.00	73.00
1	10	13.00	10.00	6.00
.				
.				
.				
1	31	12.00	31.00	5.00
1	32	415.00	32.00	71.00
1	33	253.00	33.00	52.00
1	34	92.00	34.00	27.00
1	35	336.00	35.00	62.00
1	36	29.00	36.00	13.00
1	37	39.00	37.00	17.00
1	38	506.00	38.00	80.00
1	39	2.00	39.00	1.00
1	40	26.00	40.00	11.00
2	1	25.00	41.00	10.00
2	2	393.00	42.00	68.00

TABLE 9A-1 (cont.)

GROUP NUMBER	OBSERVATION NUMBER	VALUE	INITIAL RANK	FINAL RANK
2	3	104.00	43.00	32.00
2	4	61.00	44.00	23.50
2	5	181.00	45.00	43.00
2	6	124.00	46.00	38.00
2	7	268.00	47.00	55.00
2	8	374.00	48.00	65.00
2	9	61.00	49.00	23.50
2	10	317.00	50.00	60.00
.				
.				
.				
2	41	320.00	81.00	61.00
2	42	223.00	82.00	46.00
2	43	107.00	83.00	34.00
2	44	58.00	84.00	22.00
2	45	257.00	85.00	53.00
2	46	316.00	86.00	59.00
2	47	88.00	87.00	26.00
2	48	454.00	88.00	76.00
2	49	165.00	89.00	41.00
2	50	337.00	90.00	63.00

USING THE FIRST SAMPLE AS A BASE,
THE Z VALUE CORRECTED FOR CONTINUITY AND TIES IS -0.678
THE Z VALUE CORRECTED FOR CONTINUITY ONLY IS -0.678

THE NORMAL APPROXIMATION IS VALID FOR SAMPLE SIZES
OF TEN OR MORE IN EACH SAMPLE.

DO YOU WANT TO SOLVE ANOTHER PROBLEM?
IF YES, TYPE 1. IF NO, TYPE 0.
• 0

TABLE 9A-2   CONVERSATION WITH NANOVA FOR KRUSKAL-WALLIS TEST AND DUNN'S TEST

THIS PROGRAM PERFORMS NONPARAMETRIC ONE-WAY ANALYSIS
OF VARIANCE USING THE KRUSKAL-WALLIS "H" PROCEDURE IF THE NUMBER
OF TREATMENT GROUPS IS THREE OR MORE.
DUNN'S TEST OF MULTIPLE COMPARISONS IS ALSO PERFORMED.
IF THERE ARE ONLY TWO TREATMENT GROUPS, THE MANN-WHITNEY-WILCOXON
TEST IS PERFORMED INSTEAD.

TABLE 9A-2 (cont.)

IF INPUT FROM A DISK FILE IS DESIRED, TYPE A 1.
IF DATA ARE TO BE DIRECTLY ENTERED, TYPE A 0.
• 1

ENTER THE NUMBER OF TREATMENTS (GROUPS).
• 3

IF ALL GROUPS ARE THE SAME SIZE, ENTER
THAT SAMPLE SIZE. OTHERWISE ENTER 999.
• 999

ENTER THE SAMPLE SIZE FOR GROUP   1
• 40

ENTER THE SAMPLE SIZE FOR GROUP   2
• 50

ENTER THE SAMPLE SIZE FOR GROUP   3
• 100

DO YOU WANT TO SEE THE RAW DATA? IF YES, TYPE A 1. IF NO, TYPE A 0.
• 1

GROUP NUMBER	OBSERVATION NUMBER	VALUE	INITIAL RANK	FINAL RANK
1	1	31.00	1.00	53.50
1	2	97.00	2.00	102.00
1	3	827.00	3.00	186.00
1	4	474.00	4.00	177.00
1	5	544.00	5.00	181.00
1	6	695.00	6.00	183.00
1	7	27.00	7.00	45.50
1	8	8.00	8.00	16.50
1	9	427.00	9.00	173.00
1	10	13.00	10.00	27.50
.				
.				
.				
1	31	12.00	31.00	24.00
1	32	415.00	32.00	171.00
1	33	253.00	33.00	150.00

TABLE 9A-2 (cont.)

GROUP NUMBER	OBSERVATION NUMBER	VALUE	INITIAL RANK	FINAL RANK
1	34	92.00	34.00	97.50
1	35	336.00	35.00	161.00
1	36	29.00	36.00	48.50
1	37	39.00	37.00	61.00
1	38	506.00	38.00	180.00
1	39	2.00	39.00	2.50
1	40	26.00	40.00	44.00
2	1	25.00	41.00	42.00
2	2	393.00	42.00	168.00
2	3	104.00	43.00	107.50
2	4	61.00	44.00	80.00
2	5	181.00	45.00	136.00
2	6	124.00	46.00	118.00
2	7	268.00	47.00	153.00
2	8	374.00	48.00	165.00
2	9	61.00	49.00	80.00
2	10	317.00	50.00	159.00
⋮				
2	41	320.00	81.00	160.00
2	42	223.00	82.00	142.00
2	43	107.00	83.00	110.00
2	44	58.00	84.00	76.00
2	45	257.00	85.00	151.00
2	46	316.00	86.00	158.00
2	47	88.00	87.00	94.50
2	48	454.00	88.00	176.00
2	49	165.00	89.00	130.00
2	50	337.00	90.00	162.00
3	1	25.00	91.00	42.00
3	2	30.00	92.00	51.00
3	3	22.00	93.00	37.00
3	4	24.00	94.00	40.00
3	5	166.00	95.00	131.50
3	6	61.00	96.00	80.00
3	7	12.00	97.00	24.00
3	8	53.00	98.00	72.00
3	9	5.00	99.00	9.50
3	10	59.00	100.00	77.00

TABLE 9A-2 (cont.)

GROUP NUMBER	OBSERVATION NUMBER	VALUE	INITIAL RANK	FINAL RANK
.				
.				
.				
3	91	174.00	181.00	135.00
3	92	19.00	182.00	33.50
3	93	29.00	183.00	48.50
3	94	2.00	184.00	2.50
3	95	140.00	185.00	123.50
3	96	172.00	186.00	134.00
3	97	95.00	187.00	99.50
3	98	124.00	188.00	118.00
3	99	52.00	189.00	70.50
3	100	52.00	190.00	70.50

GROUP NUMBER	MEAN RANK	SAMPLE SIZE	WEIGHTED DIFFERENCE
1	116.60	40	17808.40
2	127.48	50	51136.02
3	7107	100	59682.49

THE TIE CORRECTION FACTOR=   0.9999

H=   42.54

IF THE SAMPLE SIZES ARE > 5, THIS STATISTIC IS CHI-SQUARE DISTRIBUTED WITH 2 DEGREES OF FREEDOM.

RESULTS OF DUNN'S TEST

GROUP I	GROUP J	ABSOLUTE DIFFERENCE	CRITERION VALUE AT .05	.10
1	2	10.880	27.926	24.823
1	3	45.530	24.629	21.892
2	3	56.410	22.802	20.268

DO YOU WANT TO SOLVE ANOTHER PROBLEM?
IF YES, TYPE 1. IF NO, TYPE 0.

• 0

chapter ten

CONTEMPORARY SIMULATION APPLICATIONS

Digital Simulator for Allocating Firefighting Resources

John M. Carroll
Department of Computer Science
University of Western Ontario
London, Ontario, Canada N6A 5B9

ABSTRACT

The problem of allocating firefighting resources is complicated by the existence of numerous criteria for measuring the adequacy of service and by differences in the equipment and capabilities of service units. Our simulator consists of one program that generates simulated events from historical data and a second program that processes the simulated data and produces statistics that describe the availability and utilization of firefighting resources. We are able to improve the accuracy with which the simulator reproduces historical data by (1) factoring data concerning the occurrence of fire incidents into daily, monthly, and hourly components, (2) dividing a generalized exponential distribution of fire incident duration into response and time-dependent components, and (3) characterizing driving-time distributions as geographically dependent gamma functions. The simulator uses historical data available from fire departments. Example applications include evaluating alternative approaches to improving fire service in a particular area, investigating the possible impact of series of arson fires, and planning for urban growth.

INTRODUCTION

Our simulator is intended to help a municipality decide how many fire stations it needs and where they should be located. Each station typically has a pumper and an aerial-ladder truck. Some stations have chief's cars, rescue trucks, and tankers as well.

Fire department dispatchers decide which units to send to a fire, based on the distance from fire hall to fire, the type of fire, and the equipment available. In our simulation, the dispatchers' decisions are reflected in the data in a backup table. These data consist of the historical probabilities of station I (any station) having backed up or substituted for station J (any other station). These data take into account implicitly the geographical proximity and commonality of equipment. The goal of fire department planners is to minimize the number of events in which

1. Needed resources are already engaged. Obviously, such events can result in substantial property damage.
2. The first station called is engaged and a substitute must be called. This is called *primary interference*. It results in time being lost during the first critical minutes of a fire; damage is thereby increased greatly.
3. Other substitute resources must be used because of interference. This means another station was called because the station of choice was otherwise engaged.

The strategy of the simulation consists of three steps. First, we look at three years of actual history to identify the statistical parameters and distributions of quantities that describe fire events and the fire department's response to them.

Second, we run the simulator for ten years of simulated time with each proposed configuration of fire stations. This allows enough time for us to observe adverse effects of various combinations of events.

Finally, we select the configuration of fire stations that minimizes the number of undesirable events.

Note that we are *not* projecting experience ten years into the future; we are taking ten random samples of one year's experience. If we wanted to project experience into the future, we could use Gaussian linear regression but we should make several projections (say, ten) for each configuration to be tested and sum the results of each policy test.

The simulator consists of two computer programs. One, called EVENT, creates a file of simulated events that represent the occurrences of fires during any selected number of sample years.

The second program, SIMULATE, reads the events file, simulates the response of the fire department to each event, and summarizes the results.

The most creative part of the modeling process was the determination of the parameters and statistical distributions of the variables defining fire events and fire department responses. We had three years of historical data available to us. Reference [1] provides an excellent summary of prior work in this area and pertinent references.

Our validation strategy was to derive our parameters and distributions using data from 1976 and 1977, to sample from these distributions, and then to use chi-squared tests to compare our sample results with 1978 data.

The EVENT program requires four inputs for each historical event: date and time, geographical location (this determines the first station to be called), total number of stations called, and time spent at the scene.

The key element of the SIMULATE program is the backup table (derived from historical data). This table defines a Markov process establishing the order in which stations are called to substitute for the primary station if it is otherwise engaged and to back it up if more than one station is needed. Only after we have designated the primary station can we determine driving time to the fire. For this reason, we make driving time an input to SIMULATE and we derive it from historical data by subtracting the time of receipt of an alarm from the time of arrival at the scene.

Special Problems of Emergency Planning

The simulation and others like it that address the allocation of emergency resources (e.g., fire, police, ambulance) have a potential for abuse because they can reveal how a community might be vulnerable to attack. The performance of simulation studies can help warn municipal protective services, reduce existing vulnerabilities, and practice responses to possible attacks. At the same time, civic responsibility dictates that neither the results of such gaming nor the actual simulation programs should be made generally available outside the protective-service community; nor should any results be published until after indicated remedial action has been taken.

In a larger sense, this problem demonstrates that simulation and the computer science technology that supports it have a profound impact on society. This impact, in turn, suggests that simulation specialists should adhere to a code of professional ethics and should eventually establish a certification program similar to that established by other professions, recognizing, of course, that some simulation specialists are already members of licensed and regulated professions such as engineering.

BACKGROUND

The objectives of our work were (1) to develop a simulation program for use in planning future allocation of firefighting resources in the city of London, Ontario, Canada, (2) to develop a format for a fire department simulator that would be useful to other municipalities, and (3) to to generalize this format to handle other municipal emergency services.

In 1978 the population of London, Ontario, Canada, was 253,726. The city encompassed 68.64 square miles and 93,649 dwelling units. It employed 302 firefighters. There were 9 fire stations and 15 pieces of active apparatus.

Our principal source of data was a handwritten, bound ledger, the *Record of Fires, Alarms, and Operations*, for the years 1976, 1977, and 1978. The data fields we used were

Day and date (of incident)
Alarm received (time)
Arrival time
Company hours (hours worked)
Stations responding
Location code (nearest station).

These data are kept as a requirement of the Ministry of the Provincial Solicitor-General and would be available from any municipal fire department in Ontario.

EVENT PROGRAM

The simulation is driven by a chronological file of simulated fire incidents called the *events tape*. A program called EVENT creates this file. The term *fire-incident* includes false alarms, automobile collisions in which gasoline is spilled, drownings, electrical mishaps, etc., as well as actual fires.

Input

The file of historical data provides the input to EVENT. Figure 10-1 is a HIPO (hierarchical input/processing/ouput) diagram of the program. EVENT uses many statistics derived from the historical data: alarm means, prediction parameters, alarm frequency vectors, location frequency vector, multiple-alarm vector, and alarms-vs.-duration vector. We now describe each of these.

The *alarm means* consist of the total number of alarms from 1976 through 1978 (10,053) and the average number of alarms per year (3351), per day of the week (478.71), per month (279.25), per hour of the day (139.63), and per 15-minute interval (0.095 634).

The *prediction parameters* are the intercept (a) and the slope (b) of a linear Gaussian regression of annual alarms on time:

$$A(t) = a + bt$$

where $A(t)$ is the number of alarms in year t, a is equal to 2828.5, b is equal to 151.8, and t equals 1 for the year 1976.

We chose to break up the fire alarm history by month of year, day of week, and hour of day because this classification proved to be significant in a previous study of the London police force. [2] We found that fire alarms followed the same general trends as police calls. The day-of-week effect was not as pronounced as in

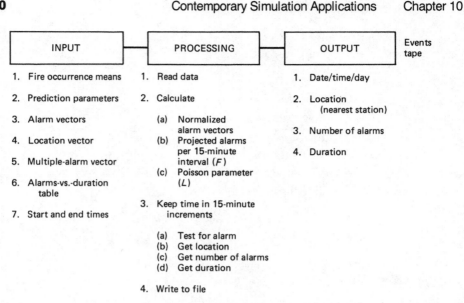

Figure 10-1   HIPO (hierarchical input/processing/output) diagram of program EVENT

the police study, but it too was a rough sawtooth peaking on Saturday. Tables 10-1 through 10-3 show the predictably higher probabilities of fire alarms occurring on weekends, in the spring and summer months, and in the early evening hours.

There are thus three *alarm frequency vectors:*

1. The *day-of-week vector* (Table 10-1) consists of the average annual number of alarms received on each day of the week.
2. The *months vector* (Table 10-2) consists of the average annual number of alarms received in each month of the year.
3. The *hour vector* (Table 10-3) consists of the average annual number of alarms received in each hour of the day.

TABLE 10-1   THE DAY-OF-WEEK VECTOR

Day	Average number of alarms received
Mondays	473
Tuesdays	437
Wednesdays	473
Thursdays	456
Fridays	497
Saturdays	554
Sundays	461

TABLE 10-2   THE MONTHS
VECTOR

Month	Average number of alarms received
January	247
February	237
March	276
April	320
May	305
June	289
July	336
August	288
September	271
October	276
November	247
December	258

TABLE 10-3   THE HOUR VECTOR

Hour	Average number of alarms received
01:00	136
02:00	135
03:00	96
04:00	70
05:00	56
06:00	41
07:00	51
08:00	69
09:00	80
10:00	106
11:00	115
12:00	142
13:00	143
14:00	163
15:00	156
16:00	181
17:00	199
18:00	197
19:00	192
20:00	214
21:00	238
22:00	214
23:00	196
24:00	161

TABLE 10-4   THE LOCATION
FREQUENCY VECTOR

Station number	Average annual number of alarms
1	611.56
2	461.43
3	333.76
4	261.71
5	361.57
6	184.98
7	401.45
8	259.37
9	475.17

The *location frequency vector* (see Table 10-4 and Figure 10-2) consists of the average annual number of alarms originating in the area served by each fire station. The bars in Figure 10-2 are located as if they were superimposed on a crude map of the city.

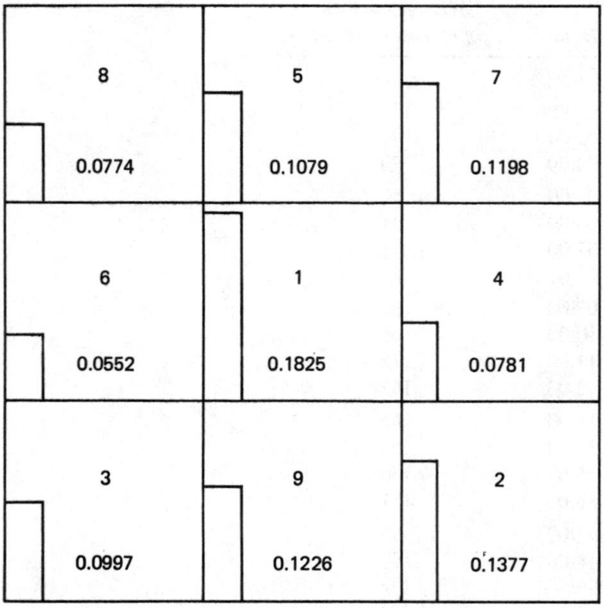

8	5	7
0.0774	0.1079	0.1198
6	1	4
0.0552	0.1825	0.0781
3	9	2
0.0997	0.1226	0.1377

Figure 10-2   Geographical distribution of fire incidents by area serviced by stations 1 through 9 (north is at the top of the page)

The *multiple-alarm vector* (see Table 10-5) consists of the average annual number of alarms in which various numbers of stations were called. Figure 10-3 shows the frequency distribution of fire incidents by number of stations called.

TABLE 10-5 THE MULTIPLE-ALARM VECTOR

Number of stations called	Average annual number of alarms
1	2077.91
2	536.11
3	561.84
4	160.10
5	12.37
6	0.33
7	1.0
8	0.33
9	0.0

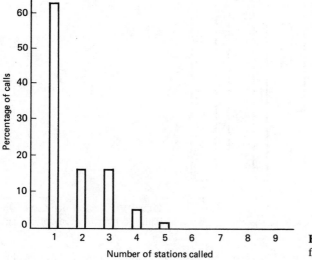

Figure 10-3 Frequency distribution of fire incidents by number of stations called

The *probability of duration longer than 300 minutes* was entered as a parameter equal to 0.99933 (based on the occurrence of seven big fires in three years). It is also possible to insert initializing parameters such as date and time.

The *alarms-vs.-duration vector* (see Table 10-6) consists of the mean duration in minutes for fire incidents requiring one, two, three, etc., stations. Figure 10-4 depicts the alarms-vs.-duration vector graphically.

Processing

The first step after reading in the data is to create *normalized alarm-frequency vectors* called $K1$ (day), $K2$ (month), and $K3$ (hour). This is done by dividing the vec-

TABLE 10-6   THE ALARM-VS.-DURATION VECTOR

Number of stations	Duration in minutes
1	21.19
2	31.94
3	33.24
4	35.03
5	39.09
6 or more	50.0

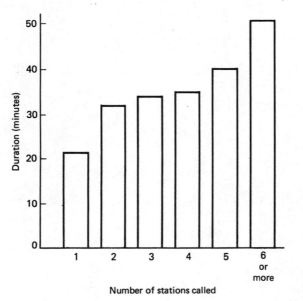

Number of stations called

Figure 10-4   Mean duration of fire incidents by number of stations called

tors in Tables 10-1, 10-2, and 10-3 by their average values. Figures 10-5 (K1), 10-6 (K2), and 10-7 (K3) show the three normalized alarm-frequency vectors. A projected number of fire incidents for each simulated year is calculated from the regression equation and the number is then normalized to a 15-minute period. We refer to the average number of incidents per period as F.

The program advances in uniform 15-minute increments. These intervals are used internally but are converted to show the simulated hour, month, and day of the week. We initialized the program to start at midnight on December 31, 1979.

To determine how many fire incidents occur in each period of simulated time, we obtain the appropriate values of K1, K2, and K3 and calculate

$$\lambda = F * K1 * K2 * K3$$

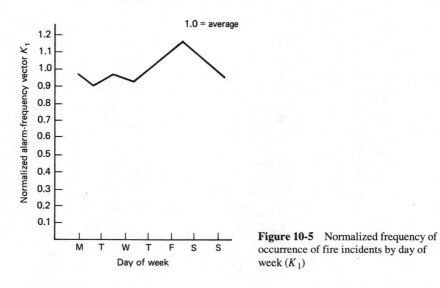

Figure 10-5 Normalized frequency of occurrence of fire incidents by day of week (K_1)

We use λ as the parameter in a Poisson subroutine

$$P_{(i)} = e^{-\lambda}\,\lambda^x/x\,!$$

In the simulated time period, $P_{(i)}$ is the probability of x fire incidents. If there is one or more fire incidents we make a record for each incident on the output file. If there

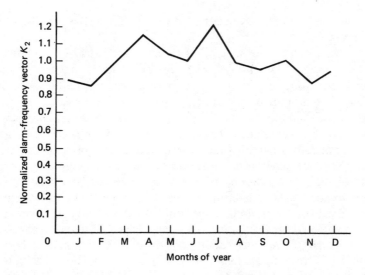

Figure 10-6 Normalized frequency of occurrence of fire incidents by month (K_2)

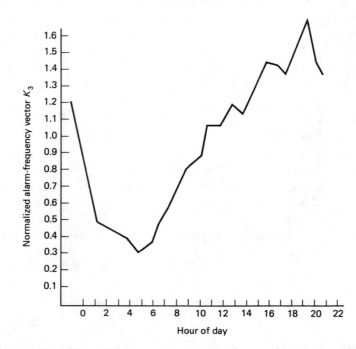

Figure 10-7   Normalized frequency of occurrence of fire incidents by hour-of-day (K_3)

is none, we simply advance the clock. Figures 10-8 and 10-9 contain logic flow charts of EVENT.

We validated our method of generating events by simulating ten years of data using average 1976 through 1978 historical data; that is, we did not project any increases. We then calculated the observed annual, daily, monthly, and hourly frequencies and applied chi-squared tests using the historical data as the values. The results are found in Table 10-7. The last three chi-squared entries in Table 10-7 are the average of ten values. These are all well within acceptable limits (i.e., within 5 percent of zero), and this suggested an acceptable goodness of fit.

Next we generate the location of the fire incident. The program converts the data of the input location vector into a cumulative frequency distribution, draws a random number, and determines where the fire will occur. We validated the location assignment by using a chi-squared test of the simulated data versus three years of actual data and obtained a value for χ^2 of 3.34 for eight degrees of freedom.

We get the numbers of stations called (i.e., alarms) by converting the data in Figure 10-3 into the same kind of cumulative frequency distribution that we used to find the location. We validated the multiple alarm subroutine by using a chi-squared test of simulated data versus three years of actual data and obtained a value for χ^2 of 2.73 for four degrees of freedom. (We had to bunch the expectations for 6, 7, and 8 alarm fires to get an expectation greater than 5.)

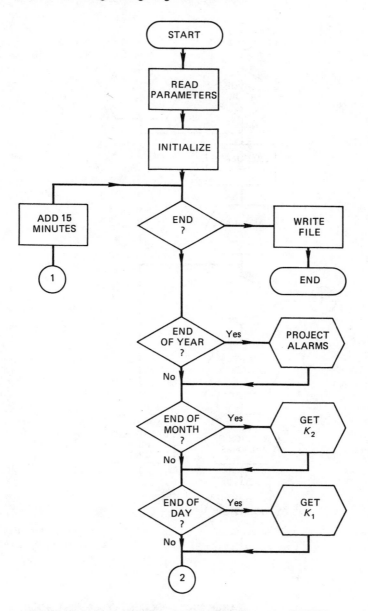

Figure 10-8 Part 1 of the logic flow chart of program EVENT

Duration of the fire incident for fires under 300 minutes appeared to be a function of the number of stations called. Extremely long incidents seemed to be independent of the apparatus requested. For example, a complicated rescue operation could take several hours and still require only a single rescue truck. Figures 10-10 and 10-11 show that duration appears to be distributed exponentially; actu-

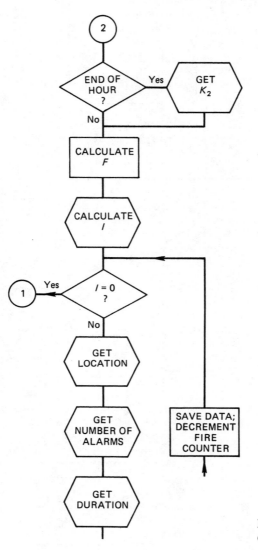

Figure 10-9 Part 2 of the logic flow chart of program EVENT

TABLE 10-7   PROJECTION OF HISTORICAL DATA

Type of data	Degrees of freedom	χ^2
Annual	9	10.12
Weekly	6	7.5
Monthly	11	17.66
Hourly	23	26.54

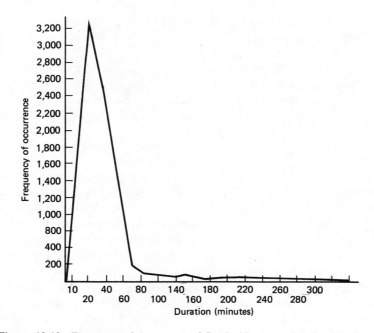

Figure 10-10 Frequency of occurrence of fire incidents vs. duration in minutes for incidents lasting less than 300 minutes

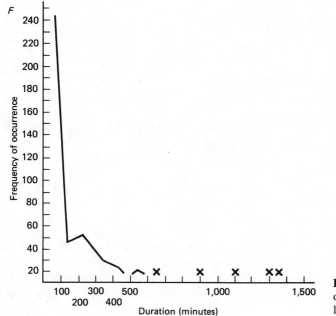

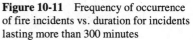

Figure 10-11 Frequency of occurrence of fire incidents vs. duration for incidents lasting more than 300 minutes

ally, it follows a generalized exponential distribution. There is an underlying exponential distribution for one-station fires, two-station fires, etc., and one for fires over 300 minutes in duration.

To obtain the duration of a fire we generate a random number. If it is less than 0.97901, we draw an observation from an exponential frequency distribution having a mean equal to the one corresponding to the number of stations called. (For example, if only one station is called, the mean M is 21.19 minutes.) Thus we have

$$D_{(f)} = \frac{1}{M} [\exp(-x/M)]$$

where $D_{(f)}$ is the duration of the fire in minutes. If the random number exceeds 0.99933, then we set $D_{(f)}$ to 300 plus the value of a random draw from a Poisson probability distribution with a parameter of 80. If the random number is between 0.97901 and 0.99933, we draw another random number. This double draw takes into account the fact that almost no fires last as long as between 200 and 300 minutes.

We validated the duration generator by comparing the means of incidents classified as 1, 2, 3, 4, 5, 6, and more station fires and those over 300 minutes in duration calculated from real and simulated data. Using a normal two-tailed test for differences of means we were unable to reject the hypothesis of equality of means in any of the cases (the maximum test value of z was 1.73, as compared with a critical value of 1.96).

Output

The events tape contains a record for each incident, consisting of the following data:

Date (year, month, day)
Time
Day of week
Location (i.e., first station called)
Duration in minutes

The cost was $9.03 (at $189 per CPU hour) to generate ten years of data on a CDC CYBER 73. The program is written in FORTRAN.

THE SIMULATE PROGRAM

Input

Figure 10-12 is the HIPO diagram for SIMULATE, the actual simulation program. The principal input is the events tape described above.

We also read in driving-time parameters (see Table 10-8). Figure 10-13 shows how these vary in different parts of the city. Times are shorter in the core area for

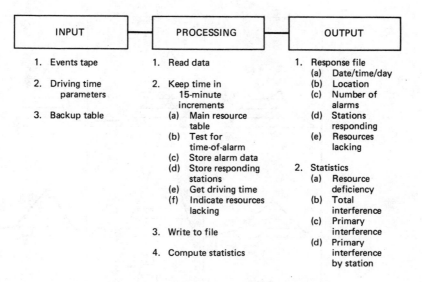

Figure 10-12 HIPO diagram for program SIMULATE

two reasons: (1) Distances are shorter downtown than in the suburbs, and (2) London has installed an Opticon system in the city core. In the Opticon system, a flash lamp on a fire engine activates photocell detectors that turn the traffic lights green.

The backup table (Table 10-9) contains historical information regarding which stations were called in multiple-alarm fires, or when the station of choice was engaged elsewhere. It is a 9-by-9 matrix. Each row corresponds to the station that requires either backup or a substitute. The columns contain the number of times the station in the column heading backed up the station in the row heading during the years 1976 through 1979.

TABLE 10-8 DRIVING-TIME PARAMETERS

Station number	Mean driving time	Standard deviation
1	3.9040	2.7600
2	5.1887	2.5943
3	3.7670	2.1748
4	4.2998	2.4825
5	3.3973	1.6987
6	5.1726	2.7017
7	4.5620	2.0402
8	5.3022	2.6512
9	4.7064	2.2758

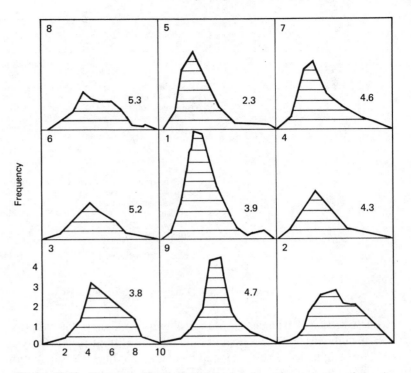

Figure 10-13 Frequency distribution of driving times for each area of the city, arranged geographically

TABLE 10-9   THE BACKUP TABLE (NUMBER OF RESPONSES)

Requesting station	Responding station								
	1	*2*	*3*	*4*	*5*	*6*	*7*	*8*	*9*
1	—	120	586	396	459	169	84	155	210
2	124	—	26	204	153	7	376	26	59
3	358	20	—	55	69	92	2	28	127
4	204	109	65	—	24	3	43	88	13
5	293	112	71	37	—	2	9	1	76
6	48	4	46	3	1	—	0	13	0
7	60	298	9	147	31	5	—	61	12
8	75	10	25	95	15	10	15	—	0
9	162	52	136	15	2	0	4	1	—

Processing

The simulation program is best understood by following the flow charts of Figures 10-14 and 10-15. The first step after initializing the parameters (not shown) is to read the events tape.

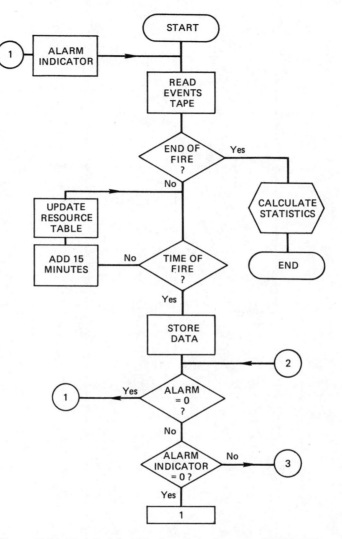

Figure 10-14   Part 1 of the logic flow chart of program SIMULATE

Next, we see if the time of a fire incident has arrived. If not, we increment the 15-minute clock and update the resource table. The resource table is a vector containing the times when each station will next be free. A zero entry means the station is free. Updating the table means decrementing each nonzero entry by 15 minutes.

If the time of a fire incident has arrived, we obtain the data describing it from the events tape. Then we test to see if all alarms have been answered. If so, we set the alarm indicator to zero and read the events tape again. The alarm indicator is a binary variable that is zero when the first alarm of a fire incident is being serviced and one otherwise. Its use is explained later.

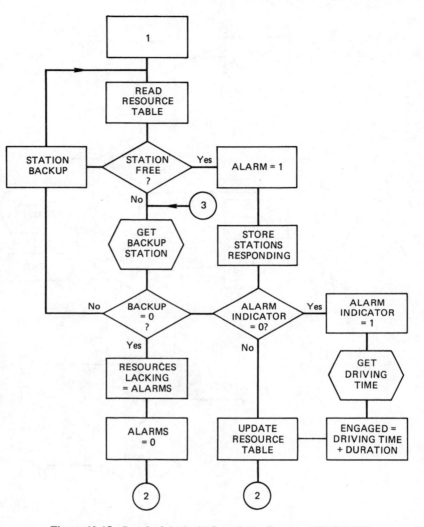

Figure 10-15 Part 2 of the logic flow chart of program SIMULATE

We next check the resource table to see if the station of choice is free. If not, we consult the backup matrix. This is a cumulative frequency distribution created from the backup table during initialization. We use a random number to determine which station will either substitute for or back up the station under consideration.

This procedure illustrates a major difference between police and fire simulators. In a fire operations simulator, the choice of a substitute or backup is governed by the kind of apparatus at the available stations as well as by distance. We simulate this additional consideration by making the choice stochastic instead of algorithmic. In police work, on the other hand, most vehicles provide identical service, so the backup can be chosen by an algorithm.

After selecting the substitute or backup, we again consult the resource table. If the substitute is engaged, we again go through the loop involving the backup matrix. If we cannot find a free substitute or backup, we mark the incident as one in which needed resources are lacking. The number of incidents of this kind is an important negative criterion in evaluating a proposed change in resource deployment policy. When such an incident occurs, we reset the alarms counter (which holds the number of alarms read from the events tape decremented by the number answered) and return to the main line of the program. If this is the first alarm for the fire incident under consideration, the alarm indicator will be zero. We set it to 1 to lock out this loop for subsequent alarms associated with the same fire incident.

We now get the driving time of the first station responding. We have hypothesized that driving time is a function of the geographical location of the first station answering a fire call and that it follows a gamma distribution.

$$D(d) = \beta^{\alpha} X^{(\alpha-1)} e^{\beta x} / (\alpha - 1)!$$

Driving time is $D(d)$, and the parameters α and β are found from the means and variances of the frequency distribution appropriate to the station under consideration.

$$\alpha = M^2 / \sigma^2$$

$$\beta = M / \sigma^2$$

We validated the driving-time generator by creating 1000 driving times for each station and comparing the means and variances of the simulated distribution with the means and variances of nine driving-time distributions. These were derived from three years of historical data. We established a hypothesis of equal means and equal variances, and tested these hypotheses by the normal test for means and the F-test for variances. We were unable to reject any of our nine hypotheses at the 5 percent level of significance.

Next, we add driving time to duration and update the resource table for the first station responding. We now test to see if all alarms have been serviced.

If more alarms remain to be serviced, we test the alarm indicator (which is now equal to 1) and go directly to the backup table to identify the backup station of choice. We traverse one backup loop until we find either a free station or an exit if there is none. When we find a free station, we decrement the alarm counter by 1 and test the alarm indicator. Since it is equal to 1, we update the resource table for the backup station with the same data we entered for the primary station and return to test the alarm counter again.

Output

When the end of the events tape is sensed, the program calculates statistics from the output file. The output file contains the following data for each fire incident recorded on the events tape:

Date/time/day
Location (first station called)
Number of alarms (i.e., stations called)
Identity of stations responding
Number of resources (i.e., stations) lacking

We used the following statistics in our experimental work:
For every station:

Station number
First calls received
First calls serviced
First calls missed
Backup calls received
Backup calls serviced
Backup calls missed
Minutes engaged
Minutes idle

For the entire department:

First calls received
First calls serviced
First calls missed
Backup calls received
Backup calls serviced
Backup calls missed
Resources lacking by first station called

SIMULATE cost $7.61 to run on the CYBER 73 using as input the events tape previously described. This program was written in FORTRAN.

APPLICATIONS

One of the complaints about fire protection in London is that service is not adequate in the East End. This section of the city is highly industrialized with a few large, older plants and many small operations such as metal fabricating shops. These structures are interspersed among numerous older frame dwellings. It is, in general, a low-rent residential area. Losses caused by structural fires are high. The area is served principally by Station 7.

As a first approach, we considered moving Station 2 to the East End where it would share the load with Station 7. To simulate this move we changed the first-call probabilities of Stations 1, 2, and 5 as follows:

Station	Change in first-call probability
1	+0.0164
2	−0.0277
5	+0.138

The second policy we investigated was building a new fire station in the East End (Station 10). We simulated this by giving the new station the driving-time distribution of Station 7 and by arranging for the new station to take half the calls of Station 7 on a probabilistic basis—both first calls and also calls to act as a backup or substitute. We ran the programs to simulate ten years of fire incidents and we evaluated the policies according to these four criteria:

1. Fire incidents for which at least some apparatus requested is unavailable.
2. Fire incidents for which the station of choice as primary, backup, or substitute is unavailable and a substitute had to be sent.
3. Fire incidents for which the first station is unavailable. (This is called primary interference. It is especially important because the time wasted in obtaining a substitute station, the additional driving time incurred, and the possible unsuitability of the apparatus of the substitute station all contribute to allowing a fire to get out of control in the crucial first minutes after the alarm occurs.)
4. The distribution of instances of primary interference among stations and their possible consequences in terms of loss of life or property.

We ran the ten-year simulation with the existing deployment of resources. For the two proposed policies: for Policy 1, Station 2 moved to the East end, and for Policy 2, Station 10 was built in the East End.

The results (see Table 10-10 and Figures 10-16 and 10-17) are based upon 40,670 simulated fire incidents. The cross-hatched areas in Figures 10-16 and 10-17 indicate where the contemplated change in the resource allocation policy would create an increase in primary interference; the double cross-hatching signifies a significant increase. The outlining indicates the area in which a decrease in primary interference would result from implementing the policy under consideration.

Moving or building a fire hall is a political and economic decision as well as an operational one. Information such as this simulation can provide is only one of many inputs used by decision makers. Nevertheless, scrutinizing Figures 10-16 and 10-17 can provide insight useful in the decision-making process.

The significant increase in primary interference in the core area seen in Figure 10-16 results when the central fire station must pick up the load currently carried by Station 2. Such an increase in fire-loss potential in the densely populated, highly

TABLE 10-10   RESULTS FROM THE POLICY EXPERIMENT

	Station	Base conditions	Policy 1	Policy 2
Criterion 1:				
Incidents with resources lacking		16	2	16
Criterion 2:				
Incidents in which substitute resources were used		7644	4673	6248
Criterion 3:				
Incidents in which primary interference occurred		1874	1882	1853
Criterion 4:				
Incidents of primary	1	614	971	514
interference broken down	2	273	222	279
by station	3	1172	15	209
	4	59	6	143
	5	100	159	179
	6	27	29	36
	7	257	251	62
	8	69	61	116
	9	363	168	253
	10	0	0	62

assessed core area is unacceptable. Furthermore, Policy 1 does not improve service in the area served by Station 7, the major goal of the proposed change.

On the other hand, implementing Policy 2 (see Figure 10-17) not only improves service in Station 7's former area but also improves service both in the core area and in the area serviced by Station 9. The latter region is a growth area.

The significant increase in primary interference in Sections 3 and 8 appears to arise from upsetting the backup pattern by making the backup stations vacate their

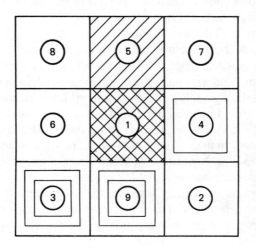

Figure 10-16   Simulated results of moving Station 2 due east. Cross-hatched region shows where primary interference increased (undesirable); double cross-hatching shows a significant (greater than 5 percent) increase. (Outlined area shows where primary interference declined; double outlines mean the decline was significant.)

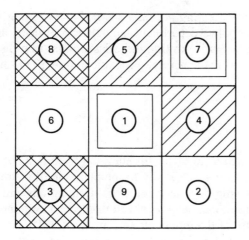

Figure 10-17 Simulated results of building a new station adjacent to the area served by Station 7 (shading has the same meaning as in Figure 10-16)

own areas and thus become unavailable for service there. Recent growth in the area served by Station 3 will create a need to improve fire protection there anyway. Although we cannot claim credit, we are pleased to report that the city has decided to build a tenth fire hall in the East End adjacent to the area presently served by Station 7 sometime in the early 1980s.

We used the simulator in three studies of a more speculative nature. These were interesting, but their results are less reliable than the foregoing results.

In the first study, we investigated the impact of a series of set fires like those that the city of Kitchener experienced during the summer of 1979. We adopted the hypothesis that the arsonist's objective is to cause maximal loss of life and property and disruption of public services. We introduced random entries on the events tape and observed the results, determining that, in general, the fire department would be able to cope with the challenge. On the other hand, if the arsonist concentrated his or her efforts in time and space, the potential for catastrophic interference would exist. We believe it would be irresponsible to report the probabilities in this paper.

Next, we investigated the consequences to resource planning if the growth of the city were to be concentrated in its southwest quadrant. In this exercise, the outcome was heavily dependent on how we changed the probabilities of fire incidents occurring to reflect the hypothesized changes in the distributions of population and structures. Before we can have any confidence in such a study, we will have to replace our equation that predicts fire incidents with a multiple regression model that predicts incidents by area of the city and that takes into account population density and distribution of structures as well as the historical occurrence of incidents.

Finally, we studied resource allocation for rescue situations. Only two fire stations (1 and 7) have rescue units. We found that we could expect 56 incidents a year in which rescue units would not be available when needed. On the surface, this appears to be a grim statistic. However, our work in this area is flawed because of our inability to categorize rescue situations adequately (i.e., we know *when* the rescue truck went out, but not usually *why*). In future work, we want to address the rescue situation directly by investigating not only the activities of the fire department but also parts played by the police department, ambulance services, and other agencies. Interesting case histories arose during the Mississauga train disaster of 1979, the Woodstock tornado of 1979, the natural gas explosion of 1974, and the blizzard of 1972.

CONCLUSIONS

We have developed and tested a simulator for studying operations of municipal fire departments and have applied it successfully to the planning of resource allocation. The model rests upon both empirical and theoretical foundations and has been validated against historical data. It can be generalized to depict a large number of urban environments and makes use of the input data that are generally available from most fire departments. These economical computer programs are written in FORTRAN and have been run successfully on both DEC and CDC computers. Development cost was minimal inasmuch as the project was undertaken as part of a senior laboratory course in computer science.

ACKNOWLEDGMENTS

The author thanks Chief Ray Morley and the officers and men of the London City Fire Department for their cooperation. Also, the following people participated in data reduction, systems analysis, and programming: D. L. Belletti, B. Dynin, K. R. Hill, M. S. Kwan, M. W. Lahn, G. S. McQuade, A. D. Marshall, R. A. P. Sinnaeve, S. L. Szeto, F. Y. Tam, and D. G. Wiseman. The support of the National Science and Engineering Research Council (Canada) under grant number A7132 is gratefully acknowledged.

REFERENCES

[1] ISHERWOOD, G. D., AND Y. WARD. "The Relationship between Response Time and Fire Property Losses." *Infor*, Vol. 17, no. 4, November 1979, pp. 373–389.
[2] CARROLL, J. M., AND P. G. LAURIN. "Using Simulation to Assign Police Patrol Zones." *Simulation*, Vol. 36, no. 1, January 1981, pp. 1–12.
[3] MILLER, I., AND J. E. FREUND, *Probability and Statistics for Engineers*, Englewood Cliffs, N.J.: Prentice-Hall, 1965.
[4] NAYLOR, T. H., J. E. BALINTFY, D. S. BURDICK, AND K. CHU. *Computer Simulation Techniques*. New York: John Wiley, 1966.
[5] SHANNON, R. E. *Systems Simulation*. Englewood Cliffs, N.J.: Prentice-Hall, 1975.

QUESTIONS

1. What statistical techniques were used to calibrate and validate this model? Discuss the purpose of each technique.
2. Why might the modeler have chosen a time unit of 15 minutes? Discuss.
3. Survey the studies in which this model has been used.

Coping with Business Risk Through Probabilistic Financial   Statements

Pamela K. Coats
Department of Finance
College of Business
The Florida State University
Tallahassee, Florida 32306

and

Delton L. Chesser
John R. Nolen & Associates
409 Continental Building
San Angelo, Texas 75903

ABSTRACT

Today's financial environment is a broad arena with many uncertain economic, political, and operational factors. Because of this widespread uncertainty, corporations of all sizes are turning to financial forecasting models. For the most part, however, these models are deterministic and do not consider the riskiness, or likelihood, of the projections fed to them. Our model includes risk explicitly through the introduction of Monte Carlo techniques into a traditional corporate model. The resulting model produces both standard financial reports and the associated probabilities of occurrence, confidence intervals, and standard deviations. The model, which has been validated by historical comparisons, allows the planner to test many scenarios and determine the likelihood of satisfactory financial performance.

©1982 The Society for Computer Simulation. Reprinted with permission, from *Simulation*, Vol. 38, no. 4, April 1982, pp. 111−121.

INTRODUCTION

All businesses face the problems of inflation, liquidity, rising trade barriers, sudden shifts in interest rates and foreign exchange, changing tax policies, and the ever-varying energy situation. To cope more effectively with these problems, firms have begun to use corporate planning models: The adoption of modeling has been widespread with more than 2000 corporations reported recently [15] to be using, developing, or intending to develop a corporate planning model.

Despite the wide acceptance of corporate planning, it often does not provide a reliable basis for assessing current challenges. The reasons for unreliable results are numerous; however, one of the principal causes is that over 90 percent [1, 13, 15] of currently used models are deterministic. Since deterministic models use only single estimates of key forecasting variables, they are best suited to relatively stable economic periods, such as the late 1950s and the early 1960s. The use of deterministic models in the current unstable period is myopic at best. In today's economic environment, managers must be able to evaluate a wide range of options in probabilistic terms. Results from deterministic models may be misleading and confusing because the actual values for decision variables such as sales growth, interest rates, and price/earnings ratios will inevitably differ from any single set of estimated values. A need exists for a practical approach that results in decisions having the greatest chance of achieving objectives.

A PROBABILISTIC PLANNING MODEL

This study describes the formulation of a probabilistic corporate planning model that projects a company's future financial condition by considering risk explicitly.

The model is applicable to a wide range of businesses. In a later example we have used public data from Sears, Roebuck and Company (Sears, for short) to demonstrate how any firm can use risk analysis to increase the probability of profitable results. We have focused on financial operations because a comprehensive model describing the firm's total environment generally has questionable merit. Comprehensive models are either too detailed and, consequently, applicable only to a specific firm, or so general that they are not applicable to any particular firm.

Our corporate model is not tied to a general economic model. Rather, it captures and incorporates effects from the overall economy through the user's judgmental assignment of values to its external variables. Econometric models have had poor track records recently, causing users to express strong dissatisfaction with their forecasting accuracy, their computer costs, and the subscription fees of econometric service bureaus. [5] Regarding 1980 forecasts, economist Stephen McNees notes that econometric modelers were not only consistently wrong, but they also consistently changed their forecasts in the wrong direction. [10] Indeed, University of Chicago economist Victor Zarnowitz contends that, in over twenty years of observation, he has been unable to discover any real difference between econometric and judgmental forecasts. [10]

REQUIREMENTS AND CONCEPTUAL FRAMEWORK OF THE MODEL

From an examination of current corporate models and a review of the state of the art, we determined that the major requirements of an effective corporate financial planning model are

1. Realistic quantification of uncertainty (assignment of probabilities)
2. Reflection of the unpredictability of future events (repeated random selection of key variables)
3. Recognition of the critical interdependencies among variables and events (simultaneous equations)
4. Utilization of the sales projection as the base for formulating the remaining equations
5. Employment of time as an explicit variable to make the model dynamic
6. Generation of clear, concise reports in usable formats
7. Flexibility to meet current needs as well as ability to adapt to future demands
8. Speed, accuracy, and reasonable computer needs.

Based on these requirements, we concluded that an effective corporate model must be designed with randomness, simulation, probabilities, and simultaneous equations in mind. The element of chance is introduced by using Monte Carlo simulation which employs random sampling to approximate unknown effects. Another reason for selecting Monte Carlo simulation is its adaptability to computer implementation.

Risk is recognized by acknowledging that decision making involves a choice among uncertain alternatives. Consequently, no particular value can be assigned with complete certainty to any decision variable. Instead, for each variable the decision maker is provided with a range of possible values and associated probabilities. Additionally, the planner must be aware of the effect of one variable on other variables. These interdependent effects among a model's variables are assessed by employing simultaneous equations. The result of jointly using random sampling, computer simulation, probabilities, and simultaneous equations is a reliable estimate of the likelihood of achieving a performance standard, such as a desired return on investment (ROI). This process for considering risk explicitly is called *risk analysis*.

MODEL FORMULATION

We formulated our probabilistic financial planning model by applying the above requirements and concepts to the model proposed by Warren and Shelton. [9] We used this model because its equations conform to accepted accounting standards for a balance sheet, income statement, and other widely understood financial analyses. However, our model is fundamentally different from the earlier model. The most important difference is that Warren and Shelton's model is deterministic and uses only a single best estimate to represent each input variable. Conversely, our model is probabilistic and uses a random selection of 100 values for each input variable, based on its range of possible values described by a set of probabilities. To accommodate this selection process, we had to expand each input and output variable to store 100 simulated values for each year.

The inclusion of risk analysis is valuable because of the reliability that it adds to the results. Reliability is gained by describing each forecasted variable in terms of (1) its expected values, (2) its standard deviation, (3) the probability that its actual value will be within the limits of the mean class, and (4) the probability that its actual value will be within one standard deviation of its expected mean value. Additionally, the model user may request the printing of a risk profile which shows the probability of any forecasted variable attaining a particular value. Using risk-adjusted information provides the firm with a greater chance of meeting its financial objectives.

The current model's driving force and, consequently, its most important input is the firm's sales projection. As shown in Table 10-11, all other variables depend either directly or indirectly on sales. The purpose of the model's 23 simultaneous equations is to produce clear, concise reports in usable formats which are sensible and easy to interpret. The model provides reliable projections for

1. The firm's sales, operating and net income, interest costs, and funds available for dividends (see Table 10-12)
2. The firm's pro forma financial condition (see Table 10-13)
3. The firm's asset requirements and capital needs to achieve the above projections (see Table 10-14)
4. The effects of the results on the firm's common stock earnings ratio, shares issued, earnings per share data, rate of return to investors, and share data, rate of return to investors, and capital structure (see Table 10-14)
5. The probability of achieving various rates of return on investments, the financial measure upon which most of the implications depend (see Table 10-15 and Figures 10-18 and 10-19).

TABLE 10-11   SIMULTANEOUS EQUATIONS USED IN THE CORPORATE MODEL [SUBSCRIPT $_0$ INDICATES PERIOD $(t - 1)$]

Section 1:   Generation of sales and earnings before interest and taxes for period t
   (1) $SALES = SALES_0 \times (1 + GSALES)$
   (2) $EBIT = REBIT \times SALES$

Section 2:   Generation of total assets required for period t
   (3) $CA = RCA \times SALES$
   (4) $FA = RFA \times SALES$
   (5) $A = CA \times FA$

TABLE 10-11 (cont.)

Section 3: Financing the desired level of assets for period t

(6) $CL = RCL \times SALES$

(7) $NF = (A - CL - PFDSK) - (L_0 - LR) - S_0 - R_0 - b\{(1 - T)$
     $[EBIT - i_0(L_0 - LR)] - PFDIV\}$

(8) $NF + b(1 - T)[(IE)(NL) + (UL)(NL)] = NL + NS$

(9) $L = L_0 - LR + NL$

(10) $S = S + NS$

(11) $R = R_0 + b\{(1 - T)[EBIT - i(L) - (UL)(NL)] - PFDIV\}$

(12) $i = i_0[(L_0 - LR)/L] + IE(NL/L)$

(13) $K = L/(S + R)$

Section 4: Generation of per share data for period t

(14) $EAFCD = (1 - T)[EBIT - (i)(L) - (UL)(NL)] - PFDIV$

(15) $CMDIV = (1 - b)EAFCD$

(16) $NUMCS = NUMCS_0 + NEWCS$

(17) $NEWCS = NS/[(1 - US)P]$

(18) $P = (m)(EPS)$

(19) $EPS = EAFCD/NUMCS$

(20) $DPS = CMDIV/NUMCS$

Section 5: Indicators of profitability for period t

(21) $EPSGR = (EPS - EPS_0)/EPS_0$

(22) $ROA = \{EAFCD + [(UL)(NL)(1 - T)]\}/A$

(23) $ROI = \{[(EPSGR + 1)(EAFCD/NUMCS)](NUMCS + NEWCS/P)\} - EAFCE$
      $+ \{(i)[(1 - T)(NL)]\}/[(b)(EAFCD) + NEWCS + NL]$

TABLE 10-12 SEARS, ROEBUCK AND COMPANY INCOME STATEMENT FOR YEARS ENDING JANUARY 31 (MILLIONS OF DOLLARS)

	1978	1979				1980			
	Initial value	Mean (limits of mean class)	Standard deviation	P1	P2	Mean (limits of mean class)	Standard deviation	P1	P2
Sales	17,224.00	19,267.91 (19,261.0– 19,284.5)	37.57	39	76	21,747.35 (21,732.3– 21,785.1)	84.46	39	76
Operating income		1,708.75 (1,700.9– 1,729.2)	45.38	39	76	1,928.73 (1,020.0– 1,954.3)	54.99	38	76
Interest expense		181.58 (178.1– 182.4)	6.65	40	77	191.14 (190.5– 192.2)	3.51	18	87
Income before taxes		1,527.18 (1,514.1– 1,546.8)	51.98	41	76	1,737.60 (1,718.9– 1,752.2)	52.41	47	76

TABLE 10-12 (cont.)

Taxes	622.18 (617.6– 635.0)	27.86	38	76	707.88 (700.9– 719.1)	28.92	42	76
Net income	905.00 (896.5– 911.8)	24.13	46	76	1,029.72 (1,018.0– 1,033.1)	23.51	47	76
Available for common stock dividends	905.00 (896.5– 911.8)	24.13	46	76	1,029.72 (1,018.0– 1,033.1)	23.51	47	76
Earnings retained	504.43 (499.9– 510.8)	17.39	42	76	573.93 (567.5– 578.6)	17.57	48	76
Common stock dividends	400.58 (396.6– 401.0)	6.75	44	75	455.81 (454.5– 458.4)	5.97	45	76
Debt repayments	1,866.40 (1,825.0– 1,961.2)	217.72	39	76	1,866.40 (1,825.0– 1,961.2)	217.72	39	76

Note: Values appear for the initial year of certain variables. These values are provided by management because they are required by the model's system.

$P1$: Probability that the actual value will be within the limits of the mean class.

$P2$: Probability that the actual value will be within one standard deviation of the projected mean value.

TABLE 10-13   SEARS, ROEBUCK AND COMPANY BALANCE SHEET FOR YEARS ENDING JANUARY 31 (MILLIONS OF DOLLARS)

	1978	1979				1980			
	Initial value	Mean (limits of mean class)	Standard deviation	Probabilities P1	P2	Mean (limits of mean class)	Standard deviation	Probabilities P1	P2
Assets									
Current assets		9,223.32 (9,212.7– 9,250.2)	60.4	39	76	9,866.6 (9,852.3– 9,905.8)	85.81	38	76
Fixed assets		5,851.39 (5,842.0– 5,875.4)	58.46	39	76	6,726.90 (6,718.5– 6,749.6)	49.73	38	76
Total assets		15,074.61 (15,054.6– 15,125.5)	113.50	39	76	16,593.35 (16,570.8– 16,655.4)	135.54	38	76
Liabilities and net worth									
Spontaneous liabilities		5,650.63 (5,612.3– 5,750.1)	220.53	39	76	6,378.22 (6,336.9– 6,500.0)	261.39	37	76

TABLE 10-13 (cont.)

Other liabilities	2,000.00	2,214.66 (2,214.4– 2,215.2)	1.59	39	76	2,400.56 (2,398.6– 2,400.7)	3.95	21	77
Common stock	675.00	705.00 (654.0– 731.1)	123.06	37	88	736.37 (671.6– 770.1)	157.00	39	76
Retained earnings	6,000.0	6,504.39 (6,499.9– 6,510.8)	17.39	42	76	7,078.32 (7,067.4– 7,089.5)	34.96	46	76
Total liabilities and net worth		15,074.61 (15,054.6– 15,125.5)	113.50	39	76	16,593.35 (16,570.8– 16,655.4)	135.54	38	76

P1: Probability that the actual value will be within the limits of the mean class.
P2: Probability that the actual value will be within one standard deviation of the projected mean value.

TABLE 10-14 SEARS, ROEBUCK AND COMPANY FINANCIAL ANALYSIS FOR YEARS ENDING JANUARY 31 (IN MILLIONS OF DOLLARS UNLESS OTHERWISE NOTED BY *)

	1978	1979				1980			
	Initial value	*Mean (limits of mean class)*	*Standard deviation*	*P1*	*P2*	*Mean (limits of mean class)*	*Standard deviation*	*P1*	*P2*
External funds needed		2,056.85 (2,038.3– 2,092.2)	86.11	38	76	2,030.18 (1,995.6– 2,104.6)	174.20	39	76
Value new debt issued		2,081.08 (2,036.3– 2,171.8)	216.36	39	76	2,052.31 (2,010.1– 2,144.9)	215.28	39	76
Value new common stock issued		30.01 (−21.0– 56.1)	123.6	39	76	31.36 (17.7– 39.0)	33.94	39	76
Interest rate on new debt*		0.07868 (0.07825– 0.07963)	0.00218	39	87	0.07868 (0.07826– 0.07963)	0.00218	39	87
Interest rate on total debt*	0.13	0.08199 (0.08058– 0.08251)	0.00295	42	89	0.07962 (0.07929– 0.08008)	0.00159	21	87
Number of new common shares issued		1.19 (−0.378– 2.200)	3.85	50	76	0.91 (0.511– 1.173)	0.98	51	76
Total number common shares outstanding	319.84	321.03 (319.462– 322.040)	3.85	50	76	321.94 (319.972– 323.212)	4.83	50	76

TABLE 10-14 (cont.)

Earnings per share of common stock*	2.62	2.8203 (2.776– 2.845)	0.1085	48	87	3.2003 (3.138– 3.215)	0.1203	46	82
Retention rate*		0.55725 (0.55643– 0.55914)	0.00434	39	87	0.55725 (0.55643– 0.55914)	0.00434	39	87
Debt per share of common stock		1.248 (1.225– 1.250)	0.036	48	87	1.416 (1.412– 1.438)	0.039	39	87
Market price of a share of common stock*		33.138 (32.805– 34.367)	2.510	36	77	37.601 (37.080– 38.837)	2.810	39	77
Price/earnings (P/E) ratio		11.7328 (11.650– 11.922)	0.4354	39	87	11.7328 (11.650– 11.922)	0.4354	39	82
Debt-equity ratio		0.3073 (0.30643– 0.30914)	0.0043	39	87	0.3073 (0.30643– 0.30914)	0.0043	39	87
Tax rate		0.4073 (0.40643– 0.40914)	0.0043	39	86	0.4073 (0.40643– 0.40914)	0.0043	39	87
Growth rate in sales		0.1187 (0.11825– 0.11963)	0.0022	39	87	0.1287 (0.12826– 0.12963)	0.0022	39	87
Operating income to sales		0.0887 (0.08826– 0.08963)	0.0022	39	87	0.0887 (0.08826– 0.08963)	0.0022	39	87
Current assets to sales		0.4787 (0.47825– 0.47963)	0.0022	39	87	0.4537 (0.45326– 0.45463)	0.0022	39	87
Fixed assets to sales		0.3037 (0.30326– 0.30463)	0.0022	39	87	0.3093 (0.30911– 0.30979)	0.0011	39	87
Spontaneous liabilities to sales		0.2932 (0.29118– 0.29797)	0.0109	39	87	0.2932 (0.29118– 0.29797)	0.0109	39	87
Return on total assets		0.0672 (0.06687– 0.06737)	0.0008	48	87	0.0689 (0.06846– 0.06913)	0.0009	46	87
Return on new investment		0.0645 (0.05805– 0.06616)	0.0126	47	87	0.0897 (0.08942– 0.09110)	0.0026	44	89

$P1$: Probability that the actual value will be within the limits of the mean class.

$P2$: Probability that the actual value will be within one standard deviation of the projected mean value.

TABLE 10-15 SEARS, ROEBUCK AND COMPANY SUMMARY OF RISK PROFILE STATISTICS FOR RETURN ON INVESTMENT (ROI) FOR 1979 AND 1980

Year	ROI output ranges (%)	Probability of occurrence (%)*	Cumulative probability (%)†
1979	9.0513 to 9.8629	3	3
	8.2397 to 9.0513	6	9
	7.4282 to 8.2397	4	13
	6.6166 to 7.4282	24	37
	5.8050 to 6.6166	47	84
	4.9934 to 5.8050	5	89
	4.1818 to 4.9934	6	95
	3.3702 to 4.1818	3	98
	2.5586 to 3.3702	1	99
	1.7470 to 2.5586	1	100
	Arithmetic mean value = 6.4533%		
	Expected standard deviation = 1.2626%		
1980	9.7789 to 9.9461	1	1
	9.6116 to 9.7789	1	2
	9.4443 to 9.6116	3	5
	9.2771 to 9.4443	6	11
	9.1098 to 9.2771	5	16
	8.9425 to 9.1098	44	60
	8.7752 to 8.9425	26	86
	8.6080 to 8.7752	4	90
	8.4407 to 8.6080	6	96
	8.2734 to 8.4407	4	100
	Arithmetic mean value = 8.9678%		
	Expected standard deviation = 0.2639%		

*Probability that the actual value will be within the designated class interval.

†Probability that the actual value will exceed the applicable minimum class value.

EXAMPLE APPLICATION

One Forecasted Scenario

We evaluated the effectiveness of our model by applying it to Sears' 1979 and 1980 operations. Our sources for data on Sears were *Moody's Industrial Manuals* and the *Sears, Roebuck and Company Annual Reports* (10-K) [8] published prior to 1979. Our forecasts in no way reflect the opinion or position of the management of Sears. A major reason for our choice of Sears as an example is its use in the article by War-

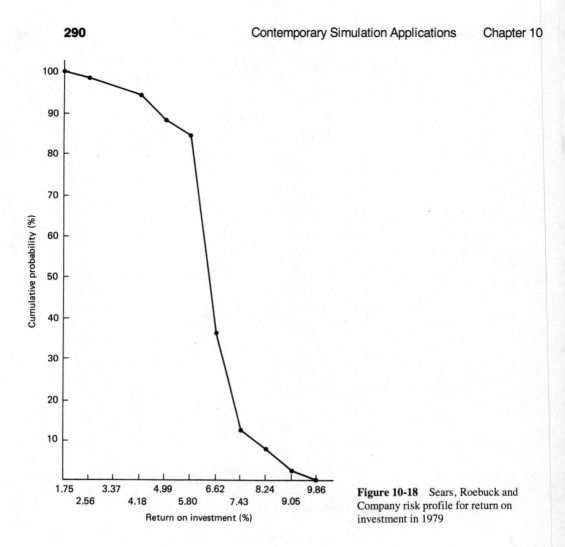

Figure 10-18 Sears, Roebuck and Company risk profile for return on investment in 1979

ren and Shelton. [9] Our results may therefore be compared directly to their results from a deterministic model.

In preparing the forecast, we analyzed the data with the objective of providing reliable projections for the five previously mentioned major decision areas in easily understood formats. Attaining that objective requires an eclectic approach that combines traditional financial analysis with risk analysis. Tables 10-12, 10-13, and 10-14 illustrate our approach. Even though Tables 10-12 and 10-13 basically follow the format of a typical income statement and balance sheet, giving critical projections of the firm's operations and pro forma financial position, these results differ fundamentally from those provided by traditional statements. The yearly values reported for each variable represent an arithmetic mean or expected value derived from 100 possible situations. The model considers risk by acknowledging that no

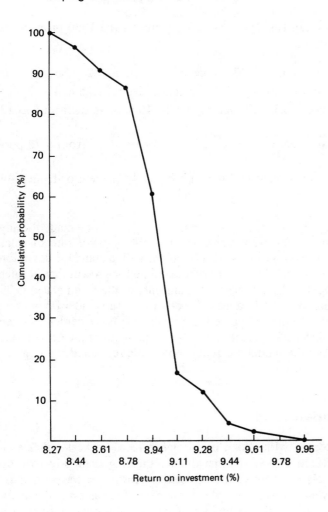

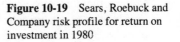

Figure 10-19 Sears, Roebuck and Company risk profile for return on investment in 1980

particular outcome for a specific event is completely certain. Recognition that risk affects real-world outcomes distinguishes the current model from most others.

Each forecasted variable is also described by its standard deviation which measures statistically the amount of dispersion about the expected value. The standard deviation contributes perspective to a planner's outlook by indicating how widely possible outcomes vary from the expected value. Perhaps even more enlightening information is provided by the probabilities that the actual value of each variable will be within either the range of the mean class or within one standard deviation of the projected mean value. Presentation of this data allows the planner to better assess the impact of uncertainty by identifying the variables whose expected values are less likely to occur and the impact that deviations will have on the forecasted results.

Table 10-14 summarizes the effects of Sears' operating results and current financial condition on its overall financial state. For example, the 1980 projections indicate that

1. Earnings per share will increase by 13.47 percent or from $2.82 to $3.20
2. Dividends per share will increase by 13.60 percent or from $1.25 to $1.42
3. Market price of common stock will increase by 13.46 percent or from $33.13 to $37.59
4. Rate of return on total assets will increase by 2.53 percent or from 6.72 percent to 6.89 percent
5. Rate of return on new investment will increase by 39.14 percent or from 6.44 percent to 8.96 percent.

As shown in Table 10-14, a more comprehensive analysis can be made for any desired variable. For example, an examination of the 1980 growth rate in sales shows that its expected value E(G Sales) is 12.87 percent with a standard deviation $\alpha_{G\ Sales}$ of 0.22 percent. The probability that the actual value for growth rate in sales will be between 12.83 percent and 12.96 percent (the limits of the mean class) is 39 percent. The probability that the actual value will be within one standard deviation (12.65 to 13.09 percent) of its expected value (12.87 percent) is 87 percent. Given these results, we conclude that 12.87 percent is a reliable estimate of the growth rate in 1980 sales. Since sales are the model's driving force, this restricted range gives credence to other projected results.

The Featured Performance Variable ROI

Management is often concerned with which performance measure best indicates a firm's ability to compete effectively. Given that financial management's overriding responsibility is to use assets efficiently, the selected performance measure must reflect both the productivity and profitability of capital. Since the easiest and usually the quickest way to improve profitability is by increasing capital productivity, return on investment (ROI) is generally selected as the critical performance measure.

A recent survey [6] of large American corporations (the so-called *Fortune* 1000) supports the idea that ROI is the accepted measure of performance. Of the 620 firms responding, 459 (74 percent) used investment centers to assess performance and 427 (93 percent) of the 459 used either ROI or the combination of ROI and residual income to evaluate the productivity of their investment centers. ROI is used as the performance measure because of the interrelated effects of inflation, capital scarcity, and increased capital costs. As capital costs increase, so must the return on capital; if a company cannot earn a return that exceeds its cost of funds, the company is not economically profitable and thus cannot compete effectively.

Reece and Cool [6] specify the advantages of adopting ROI as the primary performance measure:

1. Since ROI is a percentage-return measurement, it is consistent with how companies measure their costs of capital.
2. ROI allows comparisons of diverse results. Since the return is a ratio, it normalizes results for various operations.
3. ROI uses generally accepted definitions of profit, investment, and value in its calculations, so it is easy to interpret.
4. ROI is useful to people outside the firm, because it can be calculated by financial analysts evaluating the economic performance of a company and in comparing the performance of different companies. [7]

The current study uses two rate-of-return measures: rate of return on total assets (ROA) and rate of return of new investment (ROI). Since management has limited influence over long-term capital funds commited in previous periods, the more relevant measure is return on new investment, because the productivity of new capital has the greater effect on projected results. Thus when we refer to ROI, we mean the rate of return on new investment unless otherwise stated.

Since ROI is the featured performance variable, our model provides more detailed information on the probability of its achieving a particular value through the preparation of a risk profile as shown in Table 10-15 and Figures 10-18 and 10-19. ROI's effectiveness as a performance criterion is strengthened when it is used with decision rules to estimate the firm's likelihood of attaining its financial goals. Otherwise, such goals must be evaluated in terms of both expected values and the probability of achieving these expected values. For illustrative purposes, assume that the following decision rules specify Sears' major financial goal:

1. Expected ROI must have a minimum value of 6 percent. Note that we are calculating ROI on an after-tax basis, so that it can be compared directly with the firm's after-tax cost of capital.
2. There must be a minimum probability of 80 percent that ROI will be at least 6 percent.
3. There must be a minimum probability of 90 percent that ROI will exceed the firm's after-tax cost of capital (assumed to be 5 percent).

By examining Table 10-15 and comparing its reported results with the decision rules, we conclude that Sears' 1979 and 1980 projected financial performance is acceptable. The precise criteria established are largely dependent upon management's willingness to accept risk and the general economic outlook.

Examining Alternative Scenarios

Knowing the expected results is not sufficient. To compete effectively, managers must be able to meet contingencies: they need reliable answers to difficult "what if" questions that critically affect the firm's financial well-being. For example, what will the projected effects on the 1980 results be if, for each of the two fore-casted years (1) sales grow at only 10 percent, (2) the ratio of fixed assets to sales increases to 33 percent, (3) the ratio of current assets to sales increases to 55 per-cent, (4) spontaneous liabilities (low cost or interest-free debt) fall to 20 percent of sales, and (5) the rate of interest on long-term debt capital increases to 10 percent. Although 10 percent seems low by 1982 standards, it would have been a sound guess based upon the prevailing 1978 rates. The model can analyze, simultaneously or individually, these and other effects in a matter of minutes.

Table 10-16 highlights the model's projection of the simultaneous effects of the above events on 1980 operations. The most significant effects occur in the mean values and, consequently, the limits of the mean class. The mean value of each vari-able declines significantly. However, the firm's financial performance is still acceptable according to our decision rules, because ROI exceeds both the 5 percent and 6 percent requirements.

TABLE 10-16   COMPARISON OF CONTINGENT AND EXPECTED 1980 RESULTS

	Mean value [contingent (expected)]	*Limit of mean class (expected)*	*Standard deviation (expected)*	Probabilities	
				P_1 [contingent (expected)]	*P_2 [contingent (expected)]*
Earnings per share of common stock	$1.91 ($3.20)	$1.86–$1.93 ($3.14–$3.22)	$0.11 ($0.12)	50% (45%)	87% (82%)
Debt per share of common stock	$0.84 ($1.42)	$0.82–$0.85 ($1.41–$1.44)	$0.04 ($0.04)	44% (29%)	87% (87%)
Market price of a share of common stock	$22.00 ($38.00)	$22.00–$23.00 (37.00–$39.00)	$2.11 ($2.81)	36% (49%)	76% (79%)
Return on total assets	5.98% (6.89%)	5.94%–6.00% (6.86%–6.91%)	0.1% (0.1%)	44% (46%)	87% (87%)
Return on new investment	6.91% (8.96%)	6.88%–6.94% (8.94%–9.11%)	0.1% (0.3%)	37% (44%)	89% (89%)

P_1: Probability that the actual value will be within the limits of the mean class.

P_2: Probability that the actual value will be within one standard deviation of the projected mean value.

The ability to evaluate quickly the interactive and simultaneous effects of pos-sible changes such as these is essential in helping to minimize future problems and to take advantage of opportunities. Perhaps the greatest advantage of this type of modeling is the freedom it provides for experimenting with ideas and alternatives without committing actual resources. Such experimentation promotes understand-ing, and a corporation's understanding of its points of leverage and vulnerability plays a critical role in its survival.

Since effective decision making results from considering several scenarios, rather than from just analyzing a single course of action, many assumptions must be tested. However, the cost of testing and evaluating assumptions manually is prohibitive. Such analysis is feasible only with the aid of computer-based models that offer a practical means for evaluating proposed courses of action before the actual utilization of resources. Our model is designed specifically to permit this type of analysis.

VALIDATION

The model's performance was tested by evaluating its accuracy in replicating the financial position of Sears during the four two-year periods 1964—1965, 1969—1970, 1974—1975, and 1976—1977. Table 10-17 compares the model's simulated results and the firm's actual results for the major variables. An examination of Table 10-17 shows close agreement between the actual and simulated results. Such agreement lends credence to the model's effectiveness as a planning tool. We must recognize, however, that accurate forecasts for past periods do not guarantee the occurrence of accurate forecasts in future periods. Future accuracy is predicated upon a correct analysis of impending conditions.

TABLE 10-17 COMPARISON OF SIMULATED AND ACTUAL RESULTS FOR YEARS ENDING JANUARY 31 (IN MILLIONS OF DOLLARS UNLESS DESIGNATED BY *)

Variable	1964 Simulated (actual)	1965 Simulated (actual)	1969 Simulated (actual)	1970 Simulated (actual)	1974 Simulated (actual)	1975 Siumlated (actual)	1976 Simulated (actual)	1977 Simulated (actual)
Sales	$5,152.62 (5,152.00)	$5,783.81 (5,783.00)	$8,311.78 (8,311.74)	$8,976.17 (8,976.14)	$12,543.03 (12,543.00)	$13,292.03 (13,292.00)	$13,639.56 (13,639.89)	$14,949.69 (14,950.21)
Operating income	$ 483.83 (484.00)	$ 600.68 (601.00)	$ 891.68 (891.87)	$ 992.68 (992.87)	$ 1,366.04 (1,367.00)	$ 1,777.07 (1,777.00)	$ 1,057.13 (1,057.08)	$ 1,083.14 (1,083.25)
Net income	$ 261.12 (261.02)	$ 304.60 (304.09)	$ 419.44 (418.03)	$ 439.17 (440.95)	$ 679.44 (679.90)	$ 511.80 (511.40)	$ 915.37 (915.10)	$ 1,067.95 (1,076.07)
Total assets	$3,669.18 (3,668.60)	$4,271.92 (4,271.43)	$6,507.73 (6,507.58)	$7,078.82 (7,079.30)	$10,429.16 (10,428.00)	$11,338.83 (11,339.00)	$11,576.16 (11,576.00)	$12,711.00 (12,711.00)
Earnings per share	$ 3.46 (2.91)	$ 4.04 (2.55)	$ 2.73 (2.72)	$ 2.85 (2.85)	$ 4.32 (4.33)	$ 3.24 (3.25)	$ 3.20 (3.30)	$ 4.15 (4.35)
Dividends per share*	$ 1.76 (1.75)	$ 2.05 (2.00)	$ 1.29 (1.30)	$ 1.33 (1.35)	$ 1.75 (1.75)	$ 1.84 (1.85)	$ 1.79 (1.85)	$ 1.52 (1.60)
Market price of common stock*	$ 115.40 ($96—136)	$ 130.67 ($124—154)	$ 63.85 ($60—75)	$ 66.95 ($51.77)	$ 102.39 ($78—123)	$ 63.62 ($41—90)	$ 59.57 ($48—74)	$ 66.80 ($61—79)
Price/earnings ratio*	33.90 (33.90)	32.38 (33.37)	23.00 (23.00)	23.00 (23.00)	23.00 (23.00)	20.00 (20.00)	18.6 (18.6)	16.1 (16.1)
Return on total assets*	8.0% (8.0%)	8.0% (8.0%)	7.0% (7.0%)	7.0% (7.0%)	7.0% (7.0%)	6.0% (6.0%)	5.0% (5.0%)	5.0% (5.0%)
Return on additional investment*	3.0% (3.0%)	8.0% (7.0%)	3.0% (3.0%)	4.0% (4.0%)	4.0% (4.0%)	3.0% (3.0%)	3.0% (3.0%)	10.0% (10.0%)

A strength of our model is its utilization of fundamental accounting equations that are unlikely to change. Thus any unexpected outcomes are attributable to mis-judgment of numerical values, not to inaccurate functional relationships. The likeli-hood of this type of error occurring emphasizes again the importance of providing the decision maker with a probabilistic statement regarding the outcome of critical variables.

The model itself consists of approximately 1200 lines of FORTRAN code. It stores over 50 variables, each dimensioned as a 3×100 array. On the average, a two-year forecast requires 108K bytes of core and 1 minute, 10 seconds of execu-tion time on an IBM 370/148. The batch run cost is about $5. We are in the process of redesigning and recoding the model in BASIC to operate interactively on an Atari 800 microcomputer with 48K bytes of RAM. We would be pleased to make the orig-inal FORTRAN code available to readers upon request.

SUMMARY AND CONCLUSION

Since a firm's future is generally thought of in terms of its financial performance, corporate financial planning is experiencing rapid, widespread adoption. Such plan-ning usually involves the use of a model of the firm's financial structure. Financial models are an effective planning tool because their quantitative nature permits management to communicate more clearly the firm's goals and needs.

The reliability and usefulness of corporate financial models is related directly to their ability to consider explicitly the probable effects of risk on the firm's future operations. Models must also be oriented to the user's understanding and needs. With these requirements in mind, the current study involved the formulation and empirical testing of a risk analysis model using the example of several years' opera-tions for Sears, Roebuck and Company.

In conclusion, we should note that the American Institute of Certified Public Accountants (AICPA) and the Securities and Exchange Commission (SEC) are currently forming guidelines for the publication of financial forecasts. [2] Both authoritative bodies are pointing toward the use of risk analysis techniques to ensure the reliability of forecasts. In essence, the AICPA and SEC are requiring forward-looking reports to be objective, logical, supportable projections of a firm's most prob-able financial results, supplemented by ranges of probabilistic statements. Clearly, risk analysis provides a suitable medium for satisfying these key requirements.

REFERENCES

[1] COMBS, D. *On Choosing a Financial Modeling Package.* New York: CUFFS Planning & Models, September 9, 1977.

[2] MENTZEL, A. J. AND L. PROSCIA. "Financial Forecasts—State of the Art." *CPA Jour-nal,* August 1980, pp. 12–p18.

[3] MEYER. H. I. *Corporate Financial Planning Models*. New York: John Wiley, 1977.

[4] *Moody's Industrial Manual*(s). New York: Moody's Investor Service, 1963–1978.

[5] NAYLOR. T. "Third Generation Corporate Simulation Model," *Simulation with Discrete Models: A State-of-the-Art View*. New York: IEEE, 1980, pp. 131–141.

[6] REECE. J. S., AND W. R. COOL. "Measuring Investment Center Performance." *Harvard Business Review*, May–June 1978, pp. 28 ff.

[7] SEARBY. F. W. "Return to Return on Investment." *Harvard Business Review*, March–April 1975, pp. 113–119.

[8] *Sears, Roebuck and Company Annual Report*(s). Chicago: Sears, Roebuck and Company, 1963–1978.

[9] WARREN. J. M., AND J. P. SHELTON. "A Simultaneous Equation Approach to Financial Planning." *Journal of Finance*, December 1971, pp. 1123–1142.

[10] "Where the Big Econometric Models Go Wrong." *Business Week*, March 30, 1981, pp. 70–77.

SUGGESTED READING

Cases and Models Using IFPS. Austin, Tex: Execucom Systems Corp., December 1979.

"Corporate Planning: Piercing Future Fog in the Executive Suite." *Business Week*, April 28, 1975, pp. 46–54.

DRUCKER, P. *Management: Tasks, Responsibilities, Practices*. New York: Harper & Row, 1947, p. 437.

Financial Forecasting and Modeling System, SFL200178. Atlanta: Management Science America, 1978.

FRANCIS, J. C., AND D. R. ROWELL. "A Simultaneous Equation Model of the Firm for Financial Analysis and Planning." *Financial Management*, Spring 1978, pp. 29–44.

GENTRY, J. A., AND S. A. PYHRR. "Simulating an EPS Growth Model." *Financial Management*, Summer 1973, pp. 68–75.

GERSHEFSKI, G. W. "Building a Corporate Financial Model." *Harvard Business Review*, July–August 1973, pp. 61–72.

HERTZ, D. B. "Risk Analysis in Capital Investments." *Harvard Business Review*, January 1964, pp. 95–106.

MABERT, V. A., AND R. C. RADCLIFFE. "Forecasting—A Systematic Modeling Methodology." *Financial Management*, Autumn 1974, pp. 59–67.

MAYS, R. B. *Corporate Planning and Modeling with SIMPLAN*. Boston: Addison-Wesley, 1979.

NAYLOR, T., AND H. SCHAULAND. "Experience with Corporate Simulation Models—A Survey." *Long Range Planning*, April 1976, pp. 94–100.

NAYLOR, T., ed. *The Politics of Corporate Planning and Modeling*. Oxford, Ohio: Planning Executives Institute, 1978, p. 127.

———, ed. *Simulation Models in Corporate Planning*. New York: Praeger, 1979, p. 294.

———, AND J. M. VERNON. *Corporate Economics*. New York: McGraw-Hill, 1981.

SCHRIEBER, A. N., ed. *Corporate Simulation Models*. Seattle: Univ. Washington Press, 1970, p. 614.

SCOTT, D. F., JR., et al. "Implementation of a Cash Budget Simulator at Air Canada." *Financial Management*, Summer 1979, pp. 46–52.

WAGNER, G. R. "Strategic Thinking Supported by Risk Analysis." *Long Range Planning*, Spring 1980, pp. 61–68.

QUESTIONS

1. Discuss the advantages of probabilistic financial statements as contrasted to traditional deterministic financial statements.
2. What is the principal independent variable in this model? What are some of the dependent variables?
3. Compare this forecasting technique with regression analysis.

APPENDIX

DEFINITIONS USED IN THE CORPORATE FINANCIAL PLANNING MODEL
(SUBSCRIPT $_0$ INDICATES PERIOD $(t - 1)$

Variables: Values provided by management		*Variables: Values determined by model*	
$SALES_0$	Sales in previous period (dollars)	$SALES$	Sales (dollars)
$GSALES$	Growth rate in sales	CA	Current assets (dollars)
RCA	Current assets as a percentage of sales	FA	Fixed assets (dollars)
RFA	Fixed assets as a percentage of sales	A	Total assets (dollars)
RCL	Spontaneous liabilities (liabilities that increase automatically with sales)	CL	Spontaneous liabilities (dollars)
		NF	Needed funds (dollars)
$PFDSK$	Shares of preferred stock outstanding	$EBIT$	Earnings before interest and income taxes (dollars) or operating income
$PFDIV$	Preferred dividends (dollars)		
L_0	Debt in previous period (dollars)	NL	New debt (dollars)
LR	Debt repayment (dollars)	NS	New stock (dollars)
S_0	Common stock in previous period (dollars)	L	Other liabilities (dollars or liabilities that do not automatically increase with sales)
R_0	Retained earnings in previous period (dollars)		
b	Retention rate of earnings	S	Common stock (dollars)
T	Average tax rate	R	Retained earnings (dollars)
i_0	Average interest rate in previous period	i	Interest rate on debt
IE	Expected interest rate on new debt	$EAFCD$	Earnings available for common stock dividends (dollars)
$REBIT$	Operating income as a percentage of sales		
UD	Underwriting costs of debt (rate)	$CMDIV$	Common stock dividends (dollars)
US	Underwriting cost of equity (rate)	$NUMCS$	Number of common shares outstanding
K	Ratio of debt to equity	$NEWCS$	New common shares issued
$NUMCS_0$	Common shares outstanding in previous period	P	Price per share (dollars)
m	Price/earnings ratio	EPS	Earnings per share (dollars)
P_0	Price per share (dollars)	DPS	Dividends per share (dollars)
		$EPSGR$	Earnings per share growth rate
		ROA	Rate of return on total assets
		ROI	Rate of return on new investment

A Planning Model to Forecast Market Demand for Two Lines of Durable Products

Sharon H. Manning
Westinghouse R&D Center
Pittsburgh, PA 15235

ABSTRACT

This paper describes a planning model which forecasts annual domestic market demand by end use for a Westinghouse division manufacturing two lines of durable goods. This model takes an unusual approach in solving the problems that arise in analyzing any mature market: it combines traditional econometric techniques with a structured theory of the market. The structured approach was used where data such as saturation or replacement rates were available. The adequacy of the structured portions was determined by historical industry data and the division's marketing experience. Where appropriate inputs for the structured model could not be found, traditional econometric techniques were employed. This combination of approaches allows the model to reflect the overall market structure, market maturity, economic influences, and division insights. A computer program has been written to allow periodic updates and easy simulation of alternative economic scenarios for planning purposes.

INTRODUCTION

Modeling the market for a durable good presents a unique problem—its "market" is actually the sum of two or more distinct parts. We can usually assume the product (which might be a major appliance, a furnace, a water heater, etc.) stays with the building in which it is originally installed until it is worn out. This fact implies that the market can be divided into at least two segments: (1) installations in new buildings and (2) replacements for worn-out units. This presents a problem for someone modeling the entire market since growth in these two segments is driven by different forces. New installations are directly related to new construction, which in turn is tied to the economy. Replacements, however, are based primarily on the condition of the stock of existing units.

When a product is fairly new, replacements account for an insignificant part of total sales. Then, as the market matures (i.e., when "everybody" has one), sales of new units fall off, and replacement becomes a major part of the market. If a model is to track the changing relationship between these two segments, each one must be considered separately.

The preceding observations became the basis for developing a planning model to forecast domestic market demand for a Westinghouse division which manufactures two lines of durable goods. These two lines will be referred to as Lines A and B. Line A consists of over half a dozen individual product types, but was examined in the aggregate. Line B was examined by its three product types: x, y, and z.*

Previous attempts to model this market have been based on quarterly historical installations by type of unit. Applying traditional econometric methods to these data failed to produce a reasonable model. (This is not too surprising given the discussion above.) This failure led to the approach documented in this paper. Instead of a purely econometric model, a more structured model was designed, based on inputs such as saturation rates and product lives.

Historical data was gathered for Line A from the appropriate industry association (IA1) publications. It was available by product type and by end use (where installed) for 1958 to 1978. Data for Line B was gathered from its industry association (IA2) publications, available by product type only, for 1972 to 1978.

These data, plus the division's "feel" for the market, determined how well the structured model performed historically. For some submodels, inputs needed to retain the structured approach, such as a particular saturation rate, could not be obtained. In these cases, modeling reverted to traditional econometric methods, employing macroeconomic variables supplied by Data Resources, Inc. (DRI). The following two sections describe the models for Line A and for Line B. The final section tells how the model can be used by management in strategic planning.

MODEL FOR PRODUCT LINE A

The first step in developing a structured model is to build a theory of what drives the market. For new construction, housing (or commercial/industrial starts) and saturation rates can adequately explain demand. For replacement, demand can be estimated by how many units have been installed in the past, along with some measure of the life expectancy of those units. Ideally, the model would track sales of Line A units both by product type and by end use. Unfortunately, due to inconsistencies in the industry data, the market could only be analyzed for total Line A units of all types by end use. The primary end use categories are residential and commercial.

Residential

New residential.   The model for new residential units is multiplicative, combining housing starts and a corresponding saturation rate (see Figure 10-20). Although this module is structured, it still reflects economic influences through

---

*A product type might be distinguished from others of the same line by size, fuel source, quality, or some similar factor.

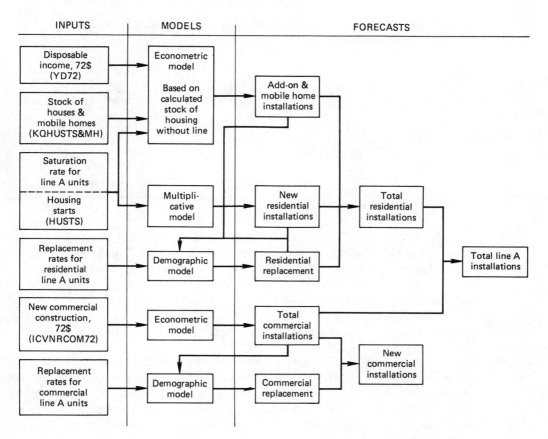

Figure 10-20   Schematic diagram of model for Line A

housing start expectations. The DRI housing start variable for single- and multifamily dwellings is called HUSTS. The relevant saturation rate was obtained initially from government census figures (1974 to 1977), but was extended back to 1953 and forward to 1990 using division judgment. This rate gives the percent of new single- and multifamily dwellings in which Line A units were installed.

   Replacement.   Inputs to the replacement module are the modeled total residential installations in prior years and the residential replacement rates (see Table 10-18). The replacement rates, which reflect average or expected product lives, were obtained from engineering estimates based on both equipment design and field experience. More rigorous approaches for estimating replacement rates have been used elsewhere. (See, for example, R. B. Fechtel et al., *Energy Capital in the U.S. Economy*, Part B, MTSC., Inc., November 1980, pp. A.7–A.11.) In this case, however, the cost of such a rigorous method was not justified.

TABLE 10-18   REPLACEMENT RATES FOR RESIDENTIAL LINE A UNITS (PERCENT)

	Age of unit in years								
	16	*15*	*14*	*13*	*12*	*11*	*10*	*9*	*8*
1950–1955	16	22	24	22	16				
1955–1960	16	22	24	22	16				
1960–1965		16	22	24	22	16			
1965–1970			16	22	24	22	16		
1970–1975				16	22	24	22	16	
1975–1980					16	22	24	22	16
1980–1985					16	22	24	22	16

For a given year, the replacement chart shows what percent of units of a certain age are due to be replaced. For example, in 1976, 16 percent of the 12-year-old units were replaced, 22 percent of the 11-year-old units were replaced, 24 percent of the 10-year-olds, etc. The rates are given for five-year periods to provide for a trend toward shorter product life. The model ages the installed units, computing how many are to be replaced in each year. Hence, this model assumes that those units that need to be replaced are replaced, and that they are replaced with a similar product.

Add-on mobile homes. Mobile homes might logically be included either with add-ons or with other new installations. (Typically, a Line A unit is installed when a mobile home is new, yet very few come with "factory installed" units.) Since we did not have a good data source from which to determine mobile home saturation rates, we included mobile homes with add-ons. The add-on mobile-home module is based on an estimate of the stock of all types of housing which do not have Line A units, i.e., the potential market. This stock variable (KWOA) is computed by calculating the stock of housing *with* Line A units (an accumulation of new residential installations), and subtracting that from the total stock of houses and mobile homes (DRI variable KQHUSTS&MH). Given the potential market, a corresponding saturation rate was needed to complete the model. When none could be found, an econometric model had to be substituted.

The best linear model tried employed the independent variables KWOA (stock of housing without Line A units) and YD72 lag 1 (disposable income in 1972 dollars, lagged one year). This model had respectable statistics $\bar{R}^2 = 0.8024$, Durbin-Watson $= 1.59$), but the forecast dropped below zero in the 1980s. Although division personnel felt that this market was maturing and would decrease gradually in the forecast period, there was no reason to think the market would disappear entirely. Hence, the linear model was replaced with its logarithmic counterpart. The $\bar{R}^2$ for the log model was much better (0.8701), but the Durbin-Watson statistic was poor (1.08). The behavior of the forecast, however, was reasonable enough to allow acceptance of the latter model.

Total residential.   Total residential installations are computed as the sum of new, replacement, and add-on mobile homes. The residential model and a forecast computed in early 1980 are plotted in Figure 10-21.

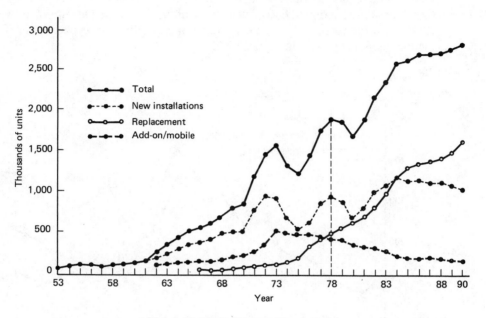

Figure 10-21   Residential Line A model

Commercial

New commercial.   Although the theoretical commercial module was similar to that for residential, data problems prevented me from retaining that structure. Specifically, gross inconsistencies were evident in IA1's split between new and replacement installations. Hence, an econometric model was estimated for total commercial installations. Then, new installations were computed as the difference between total and replacement.

Replacement.   Initial attempts to model commercial replacement were based on the assumption that the residential replacement rates could apply equally well to the total and commercial categories. Hence, commercial replacement was calculated as the difference between total Line A replacement and residential replacement. This method failed to give reasonable results. Instead, the commercial replacement rates were estimated by the division and applied to the modeled commercial installations. The rates (or, rather, the product lives) were then adjusted up and down until a back-forcast was developed which yielded a historical replacement mix consistent with division experience.

Total commercial.   As stated above, there were data problems involved in modeling new commercial, so total commercial was modeled instead. The best econometric model was based on the DRI variable ICVNRCOM72 (investment in new commercial construction in 1972 dollars). The results for this model paralleled those for add-on mobile. The $\bar{R}^2$ was fairly high (0.82), the Durbin-Watson was low (0.40), but the forecast tapered off, conforming to division expectations. The commercial model is plotted in Figure 10-22.

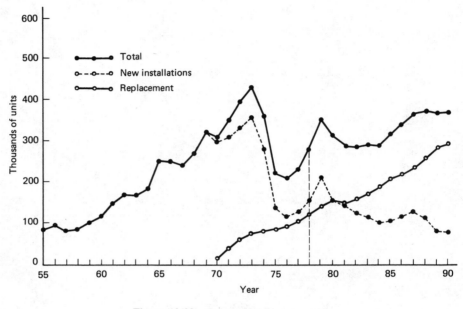

Figure 10-22   Commercial Line A model

Total Line A

Total Line A is calculated as the sum of total residential and total commercial installations. As implied above, the entire modeling process for the Line A portion was guided by a lack of confidence in the industry breakdowns by end-use segments. However, IA1's historical total Line A figures are based on actual manufacturers' shipments. Hence, the goodness of fit of the total Line A model can be compared to the actual historical shipments. Figure 10-23 shows that, overall, the model compares well with history. In addition, the general trend of the forecast is in line with division expectations.

MODEL FOR PRODUCT LINE B

The total Line B market, as it applied to this division, is composed of only a subset of its industry's market. Data for this market (IA2 figures) are split into product types, x, y, and z. Given the previous trouble with modeling the Line A portion by type, my

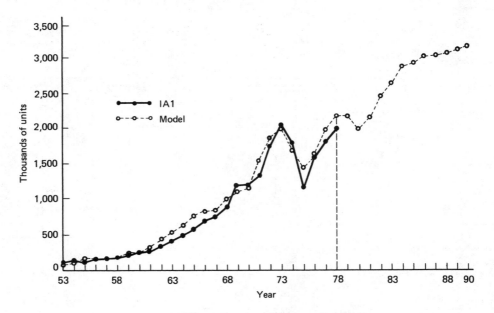

Figure 10-23 Line A model vs. industry historical installations

first try to model Line B ignored this segmentation. However, in doing the analysis, it was found that market behavior could not be adequately explained without the breakdowns. This is primarily due to a large difference in the life expectancies of the three unit types. In addition, retaining the breakdowns highlights the current dramatic drop in demand for product *y* resulting from recent economic events.

New Installations

As for the Line A model, the main stumbling block to a totally structured model for new Line B installations was a lack of appropriate inputs. The government census figures do provide a saturation rate for Line B. Unfortunately, it is for the total Line B market in the strictest sense, not simply this division's portion. Hence, the total new Line B had to be computed as modeled total minus modeled replacement (see Figure 10-24).

To get segmentation once total new installations were computed, the division provided historical and expected percentages of total for products *x*, *y*, and *z*. The predicted percentages were based on the division's expectations for the pattern of the switch away from product *y*. Applying these percentages to total new gave new Line B by product type. The model breakdowns and total new installations are plotted in Figure 10-25.

Replacement

As indicated at the beginning of this section, Line B replacement had to be computed by product type because of the wide dispersion in product lives. Since segmentation

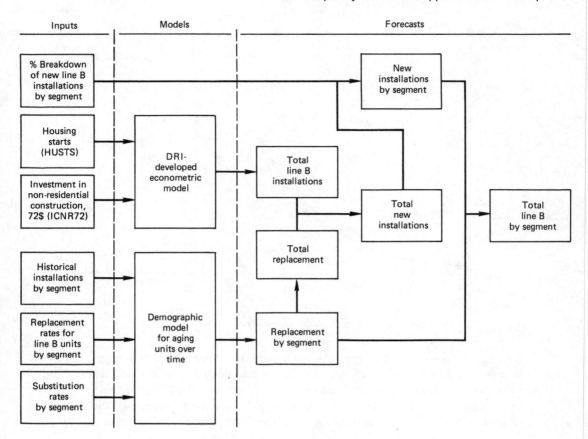

Figure 10-24   Schematic diagram of model for Line B

was incorporated into the model, an additional market phenomenon could also be included—substitution of one type unit for another. Once the number of units due to be replaced are calculated in the usual manner, the substitution rates can be applied to determine what types of replacement units are actually installed. For example, in 1980, the division expects that only 18 percent of the y units needing to be replaced will be replaced with another y unit. Seventy percent of them will be replaced with x units. (The remaining 12 percent represent unit types other than x, y, or z.)

Before 1977, the substitution phenomenon was insignificant. Recently, however, economic and technological changes have made substitution rates a critical factor in determining the replacement market. The replacement market is plotted in Figure 10-26.

Total Line B

Since no adequate model could be found for new Line B installations, an econometric model was needed for total Line B. Concurrent with this project, Cindy

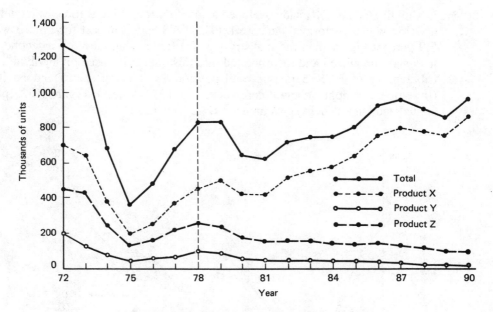

Figure 10-25   Model for new Line B installations

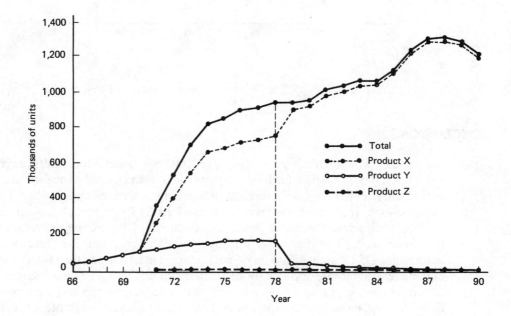

Figure 10-26   Model for Line B replacements

L. Cunningham of DRI had developed a quarterly total Line B forecast model for the division which performed very well. (The 1978 back-forecast for Line B was only 5.9 percent higher than the industry figure.) Rather than develop a second model, hers was annualized and incorporated into this one. It is based on a polynomial distributed lag of HUSTS (single- and multifamily housing starts) and on ICNR72 (investment in nonresidential construction in 1972 dollars). Figure 10-27 plots the annualized DRI model versus industry (IA2) actuals.

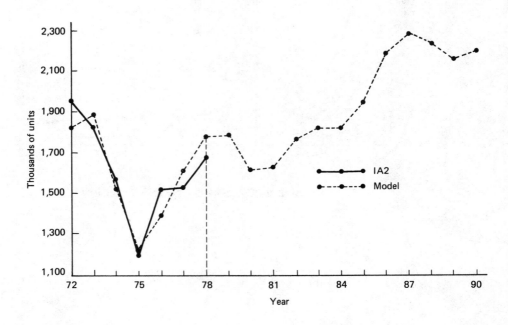

Figure 10-27   Line B model vs. industry historical installations

CONCLUSIONS

A computer program has been written to generate the model's forecast through 1990 and print it out in compact form for presentation to management. The forecast can be periodically revised as DRI's forecast of economic variables changes to reflect the latest outlook. In addition to providing a ''most likely'' forecast, the model can also be used to simulate alternative economic conditions. DRI provides several alternative forecasts for each base case forecast. By inserting alternative forecasts of the appropriate DRI variables into this model, management can see the effect of various extreme economic possibilities on the market. In addition, if other changes to the structural parts of the model become apparent, adjustments can easily be made. For example, replacement rates for future years may be changed if life expectancies shorten or lengthen.

By combining traditional econometric techniques and a structured approach, this model obtains considerable advantage over one based on a single approach. Clearly, pure econometrics could not describe the replacement phenomenon. Yet the economy does effect the number of new installations. Only a combination of methods can adequately reflect overall market structure, market maturity, economic influences, and business insight.

QUESTIONS

1. Explain why traditional econometric models failed to forecast demand accurately.
2. Describe the components of this model.
3. What benefits can be obtained from this modeling effort?

A High-Level Simulation Model of a Networked Computer System

William H. Hochstettler and Lawrence L. Rose
BATTELLE
Columbus Laboratories
505 King Avenue
Columbus, Ohio 43201

ABSTRACT

The objective of this research was to design and develop a baseline planning tool for the OCLC networked computer system. This planning tool was necessitated by the dynamic, fast-growing on-line bibliographic needs of OCLC's remote library users. In this paper we show simulation to be the best research approach and justify the IPSS language as appropriate to this application. The resultant model of OCLC's networked computer system is highlighted. This tool was developed in a timely manner to meet the immediate goals of OCLC planners and provides a standard baseline model for future extensions.

INTRODUCTION

OCLC, Incorporated, is an organization that provides computer-based library services to over 2200 member libraries in 50 states, the District of Columbia, Puerto Rico, and several countries. Computerized services currently offered include a Cata-

©1980 IEEE. Reprinted, with permission, from *Proceedings of the 1980 Winter Simulation Conference*, December 3–5, 1980, pp. 275–289.

loging Subsystem, Interlibrary Loan Subsystem, and Serials Control Subsystem. An on-line Acquisitions Subsystem will be available in late 1980.

Background

OCLC services are offered through a telecommunications network of dedicated leased and dial-access telephone lines to the OCLC computer center in Columbus, Ohio. Over 1900 member libraries use special terminals manufactured for OCLC to access the On-line System; other libraries use dial-access terminals. Member libraries can thus catalog books, serials and other library materials, order custom-printed catalog cards, create machine-readable data files, maintain location information on library materials, and facilitate interlibrary lending. Not a single computer, the OCLC On-line System is a unique configuration of heterogeneous computers. Due to the expanding nature of the OCLC user community and supporting data bases (the current bibliographic data base has over 6 million records with an average record length of 560 characters), the computer network has undergone many changes and rapid expansion since its inception in 1971.

Until late in 1978, OCLC used Xerox Sigma computers to do all on-line processing, as shown in Figure 10-28. As the number of users, terminals, bibliographic services, and data base records increased, OCLC expanded its computer system until it consisted of 12 communications processors feeding the four Xerox Sigma 9 computers that shared access to common memory and the secondary storage media containing the data base. Since further expansion of the data base and processing power by adding more Sigma computers was limited by the multiaccess capabilities of the secondary storage media, a major change in the hardware configuration was needed to facilitate expansion.

The Network Configuration

OCLC chose in 1978 to implement a network approach, as shown in Figure 10-29. The data base function was removed to a separate computer system, the Data Base Processor. Another separate computer system, the Network Supervisor, was added to manage the message transmission between the Communication Processors and the Application Processor (the Sigma 9 computers). Since this network approach was adopted two years ago, growth in both secondary storage capacity and computer processing power (so far) has been readily accomplished.

A brief overview of this network is necessary to present the problem scenario properly. The terminals connected to the OCLC On-line System are the primary input source of messages to the system. These terminals transmit data to the OCLC On-line System by an OCLC special ASCII multipoint protocol. The terminals are based on an 8080 microprocessor having 2K bytes of display memory.

The Communications Processor serves as a line concentrator to over 140 2400-baud synchronous multidrop lines. Each processor in the Communication

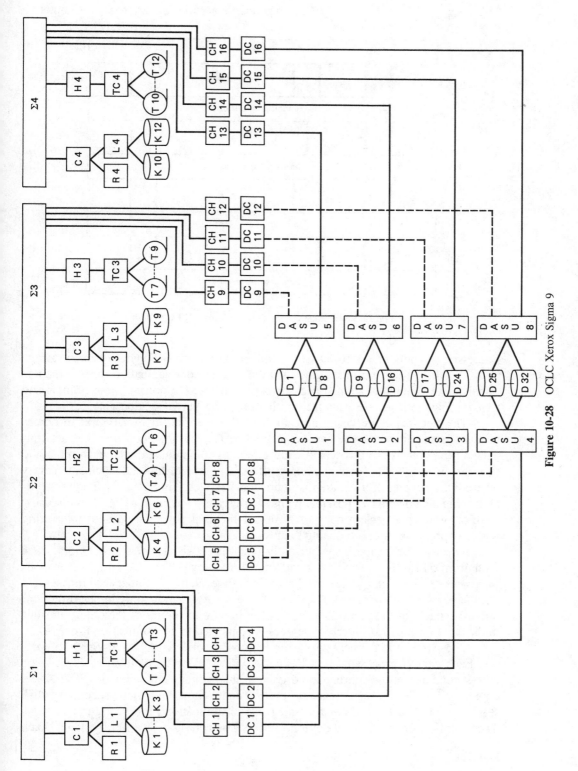

Figure 10-28 OCLC Xerox Sigma 9

311

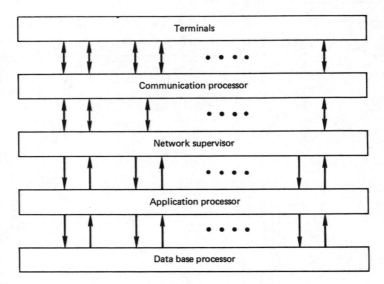

Figure 10-29   Logical relationship of the OCLC system

Processor component is a 16-bit word minicomputer with 64K bytes of main storage with 11 2400 bit/sec multidrop lines and dial-access communication lines to the terminals. The Communication Processor is responsible for polling the terminals and transmitting messages between the terminals and the Network Supervisor.

The Network Supervisor and Data Base Processor are both networks of homogeneous minicomputers, the Tandem T16. The Tandem system processors are connected by a high-speed bus which allows them to communicate with each other and to transmit data. The Tandem system offers a mirrored data management facility whereby two copies of a file are maintained at all times. This not only increases reliability but also offers two possible paths to access any record in a file. It is also possible to produce a backup copy of a file for archival purposes without interfering with any processes currently using the file.

The Network Supervisor currently consists of six Tandem processors, each with from 512K bytes to 704K bytes of main storage. There are 12 synchronous interfaces to the Communication Processor. Five Network Supervisor processors each have two custom interface devices. These usually operate only in one direction, but are capable of bidirectional transmission in case of failure of the other interface in the pair. Each pair of interface devices is connected to an Application Processor.

The Application Processor currently consists of five Xerox Sigma 9 computers. Each central processing unit has 128K words (32 bits) of main storage, some disk secondary storage, three tape drives, a line printer, and a console. The Xerox computers share two card readers, two card punches, and more secondary disk storage used for off-line data processing. Each Application Processor currently runs three application tasks—cataloging, serials control, and interlibrary loan. Each

Application Processor is connected through two custom interface devices to the Data Base Processor.

The Data Base Processor is a 14-processor Tandem network. Each processor has at least 864K bytes and up to a maximum of 1.1M bytes of main storage. The entire system has two consoles (one for back), a line printer, and 28 logical 260-megabyte mirrored disk drives for data base storage.

As depicted in Figure 10-29, the OCLC On-line System functions as follows. Each processor in the Communication Processors has up to 250 terminals connected through leased lines. Eight asynchronous dial access ports also are available which may vary the load from processor to processor, depending on which terminals are actively being used. A message entered on a terminal has a static path to a Communication Processor, which in turn has a static path to one of six tasks executing on the Network Supervisor to receive inbound messages. The message is transmitted over a static path to a particular Sigma 9 Application Processor. If the message is a request for a data base operation (some messages such as LOG-ON or LOG-OFF do not require Data Base Processor involvement), it is transmitted through a static path to one of six Data Base Communications Processes. The Data Base Communication Process then transmits the request to one of several (currently 13) identical tasks to perform the actual data base access. The response to the inbound message is transmitted to the terminal through the same path just described.

Despite multiple paths possible through the entire On-line System, messages are routed by static paths defined by the OCLC system operators. The only exception is the input/output tasks of the Data Base Processor, which run on several central processor units. Thus, the tasks in the various components of the On-line System are statically connected for message transmission and are modeled in a similar manner.

PROBLEM STATEMENT

The OCLC support system—hardware, software, and data base—has grown dramatically in its first decade. With the recent trend toward on-line cataloging and interlibrary loan and the constant growth of published materials, the OCLC work load continues to increase, with no plateau in sight. The hardware and software support systems are becoming more and more complex and sophisticated by necessity. Planning for expansion to incorporate more users, increased traffic from each user, and more library services is a must. Yet the impact of migrating to new hardware configurations or utilizing different data base organizations or reassigning software processes to the CPUs in the OCLC network becomes increasingly more difficult to predict or assess.

Within this dynamic environment OCLC must remain competitive with other on-line library facilities. Competitive factors include (1) cost per user transaction (for instance, the cost of a catalog card or authority reference)—relates to overall hardware cost and CPU time; (2) data base coverage (one cannot charge for a ''not found'' answer)—relates to storage size and organization; (3) variety of services available to the client—relates to robustness of software processes; and (4) response time—performance characteristics of the system must be maintained. Response time can be singled out as critical: once a user is accustomed to a 5-second response, a 20-second response becomes unacceptable. Thus, OCLC's growth in data base size, services, and customers must not sacrifice response time.

A clear need exists for a planning tool for management decision making at OCLC: a tool which, in a cost effective and timely manner, can be utilized to assess the impact of suggested system changes to the behavior of the OCLC On-line System at a global level.

SIMULATION AS A PLANNING TOOL

To date, the major tool for evaluating system changes at OCLC has been experimentation with the actual system, either off-line after normal production hours or online during production hours. Experimentation lowers the reliability of the system and requires time and effort to implement proposed changes. If a change is determined unsatisfactory, the effort is a loss. Therefore, a more reliable tool for planning and evaluation at all stages of system use was required.

Analytic models are often utilized to characterize queueing systems (Gross and Harris, 1974), but with varying degrees of success. While an analytic model (if derivable for OCLC's system) would be effective in assessing growth and change in user traffic and usage rates, it could not readily cope with modifications to consider data base organization changes, additional software processes to support new on-line uses, hardware reconfigurations or upgrades, etc. The fact that the OCLC system is unique and has not reached steady state further complicates matters (Graybeal and Pooch, 1980).

Simulation models and languages, both continuous and discrete, thus received prime consideration by the authors. Planning is one of the primary objectives of simulation, as noted by author R. E. Shannon (1975):

> Simulation is the process of designing a model for a real system and conducting experiments with this model for the purpose either of understanding the behavior of the system or of evaluating various strategies (within the limits imposed by a criteria) for the operation of the system.

As an alternative to actual system experimentation, simulation offers several distinct advantages:

1. *Cost*—Construction and use of a simulation model is considerably less costly than actual experimentation, in terms of acquisition costs, operational costs, and manhours.

2. *Flexibility*—A properly designed and parameterized simulation model enables the decision maker to explore several alternative solutions; actual experimentation offers limited flexibility.

3. *Timeliness*—A simulation model enables timely consideration of alternatives, and timely data is information of use to the decision maker.

4. *Ease of use*—A good simulator provides an unambiguous definition of the process being modeled. The more closely the user understands the basic concept of the model, the more easily he can use it and have confidence in its results.

Over the past 15 years, simulation has become better understood and utilized as a planning tool. Successful related implementations, not only at OCLC (Wong, 1979), but at large agencies such as the U.S. Army (Brownsmith, Carson, and Hochstettler, 1979; Rose, 1979) and the U.S. Social Security Administration (Maisel and Gnognoli, 1972; Rose and Carr, 1979) have paved the way and provided a firm basis for further implementation of simulation models.

CHOOSING AN APPROPRIATE LANGUAGE

Computer languages provide capable and flexible support for simulation activities that are characterized by properties such as complexity, magnitude, time variance, and repetition. These languages are generally classified into three categories: general purpose, continuous simulation, and discrete simulation.

General-purpose languages, such as FORTRAN, PL/1, and ALGOL have been extensively used in simulation. However, they offer no particular environment or "world views" for the modeler—one must build the model from scratch. As simulation languages have increased in number, reliability, and sophistication, they have become more attractive to the modeler than have the general-purpose languages. For this research activity, timeliness and flexibility were requisites in the modeling process; hence an approach that would reduce the programming requirements was preferred.

Continuous simulation languages, such as CSL, DYNAMO, and CSMP offer the modeler an environment particular to the solution of user-defined difference or differential equations. The specification of a computer network, its users, its software, and its work load as a set of difference or differential equations is exceedingly difficult and can require simplifying assumptions not desired in a high level of detail. These languages are not particularly suitable for computer systems simulation.

Discrete Simulation Languages

Discrete simulation languages have been under development for over 15 years, and in the last 5 years they have proved extremely valuable to the modeler. They provide constructs such as a time clock, next events queue, data structures, statistics, dynamic transactions, etc., automatically to the modeler. Thus, one can, when utilizing one of these languages, concentrate on the definition of the model as opposed to the implementation of the model on a computer in some alien language. One defines a discrete simulation model by detailing all of the events/acts pertinent to the system and the actions which trigger them. The model is defined at a macro or micro level depending on the detail of the design or the requisite precision of the output. Four types of discrete simulation languages can be distinguished: (1) activity oriented, (2) event oriented, (3) process oriented, and (4) transaction oriented.

Activity-oriented languages. Activity-oriented languages represent time-dependent activities as instantaneous occurrences in simulated time. Thus, one does not schedule occurrences; rather one specifies under what conditions they can happen. The modeler program is composed of two major sections: a test section to determine what activities can now occur and an action section to update state and time conditions. The language implicitly controls the invocation of the test and action sections. CSL (Buxton, 1966) is an example of an activity-oriented language.

Event-oriented languages. Event-oriented languages represent an event as an instantaneous occurrence in simulated time, automatically scheduled to occur when it is known by the model definition that the proper conditions exist for its occurrence. All events and their interactions are defined independently; an executive program can automatically sequence all scheduled events for the modeler. GASP (Pritsker, 1975), a FORTRAN extension, and SIMSCRIPT (Kiviat, Villanueva, and Markowitz, 1975) are the two main event-oriented simulation languages used today.

Process-oriented languages. Process-oriented languages are a hybrid derived from the concise notation of activity-oriented languages and the efficiencies of event-oriented languages. A process is a set of events; it is dynamic and can exist over time. Processes can be interrupted, have subprocesses, and can be reactivated. The executive program controlling these processes is necessarily more complex than that required for event/activity-oriented languages. SIMULA (Dahl and Nygaard, 1966) is an ALGOL extension that exemplifies the process orientation.

Transaction-oriented languages. Transaction flow-oriented languages use the block structure of flow charts to describe the simulation, with transactions flowing through activity and time altering blocks. Each block specifies specific actions, with restrictions on parallel execution, etc. These languages are extremely easy to use but are much less flexible than the other types. GPSS (Gordon, 1975) is the best known language of this type, but recently SLAM (Pritsker and Pegden, 1979) was introduced as a transaction-oriented extension to GASP.

Desired Language Features

Given this wide range of discrete-event simulation languages, how does one determine that subset of languages that is suitable for ones needs? Mihram (1972) notes five desirable features for the chosen subset as follows:

1. Ability to characterize initial model status
2. Flexibility in model definition

3. Report and statistical generation capability
4. Design flexibility for validation and analysis
5. World view consistent with modeling views of process

A sixth notable feature is added by Pritsker (1979):

6. Self-documentability of the code

From our standpoint of computer systems simulators, the "world view" is paramount; yet none of these features can be ignored.

A quick examination of the four language types revealed that, for this problem domain, the process or event-oriented languages were preferable to the transaction or activity-oriented languages such as GPSS and CSL. The process orientation of SIMULA is very attractive due to our view of a computer system as one of interacting, interruptible processes—including manual, software, and hardware. System priorities create hierarchies of control, and those, too, are clearly modeled by processes and subprocesses. The event orientation of SIMSCRIPT or GASP provides code modularity and varying levels of detail. Unfortunately, the events form a flat rather than a hierarchical control structure as required. These three languages satisfy Mihram's aforementioned initial four criteria reasonably well, but the world view of each language is extremely general. And the self-documentability of the code is questionable. These deficiencies are due to the fact that none of these languages is computer-systems oriented. There are no language statements that directly relate to hardware, software, data, jobs, etc.; thus, the modeler faces a stern test to construct a valid model of a computer system given only these general simulation tools.

Special-Purpose Simulation Languages

Recent research has led to the development of several simulation languages specific to computer systems simulation. Three of these languages are SIMTRAN (Wong, 1975), ECSS (Kosy, 1975), and IPSS (DeLutis, 1978). It is notable that SIMTRAN and ECSS are extensions to event-oriented GASP and SIMSCRIPT, respectively, and that IPSS is an ALGOL-like process-oriented language. Thus, these three languages are in our preferred class of process/event-oriented tools. Each provides initialization characterization, flexibility in definition, report generation, and design flexibility, albeit in differing forms.

Besides satisfying the criteria of their predecessors, these languages provide a world view in light of computer systems and code that relates to computer science terminology. Thus, we find that these specialized languages satisfy the last two criteria (which the general-purpose simulation languages failed to do). These languages enable one to describe a computer-based information system in a meaningful and straightforward manner with considerably less effort than would otherwise be possible.

SIMTRAN is notable for its hardware characterization, ECSS for its software and operating system characterization, and IPSS for its data base characterization.

Each has components to model all aspects of an information system, but their capabilities and inherent levels of detail differ significantly.

The IPSS language was chosen for this project for several compelling reasons. First, SIMTRAN is a proprietary product of General Electric and is available only on their user network. Second, ECSS requires the SIMSCRIPT compiler which was unavailable on the OCLC Sigma system configuration. Last, OCLC has had some prior successful experience using IPSS to model OCLC applications and encouraged its further use.

THE IPSS LANGUAGE

The Information Processing System Simulator (IPSS) is a discrete event simulation language with special features to support the modeling of information systems. While the IPSS is capable of concisely modeling complex information systems, it is general enough in nature to model naturally occurring systems. Special attention to the hardware components and the data management facilities of a computer system and data base structures makes IPSS especially appropriate for information system simulation.

The Methodological View

The specification of a model in IPSS is a three-step-transformation process as depicted in Figure 10-30. The first step is the transformation of knowledge about the system into four logical components (services, processors, information stores, and user work load). The components may be specified precisely or imprecisely. A combination of levels of detail enables the focus of the model to be on the modeler's areas of special interest. This component specification does not use a simulation language, but is a methodological division of the system into a logical component structure for modeling. It provides a meaningful interface between the user and the modeler.

Services component.   The specification of the system tasks is accomplished in the services component. These services may be manual tasks or software modules of a computer information system. These are the processes that are invoked by the job stream (work load component).

Hardware resource component.   The hardware resource component identifies those computer, human, and other processors available to do the system's work. It describes the physical means for information transmission and storage.

Information storage component.   The information storage component is the specification of the logical relationships between data and the physical storage system. This logical-physical linking is one of IPSS's strengths.

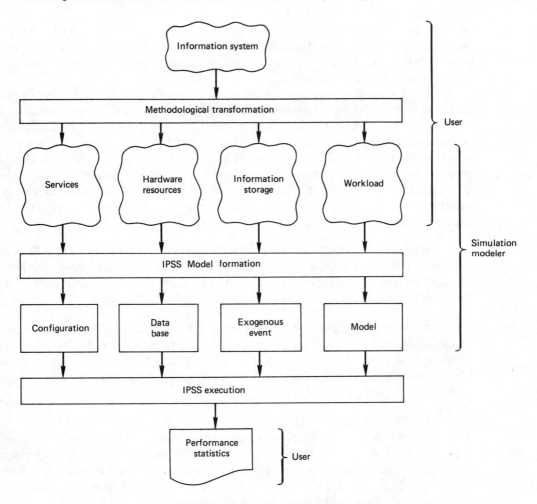

Figure 10-30   Modeling methodology

IPSS Model Formation

The second transformation process takes the four logical component descriptions and produces four components written in the IPSS language. This process requires a simulation modeler with IPSS technical expertise. The model specified in the IPSS language consists of up to four separate components: (1) configuration, (2) exogenous event, (3) data base, and (4) model.

Configuration Component.   The Configuration Component defines the hardware elements and software services of the information system. The IPSS has several special facilities that allow the modeler to gather statistics on the usage of the facilities. Particular hardware devices may be named, such as central processor, data channel, and secondary storage devices (see the appendix). Special queues or control points also may be defined.

The Configuration Component defines the exogenous and endogenous services that constitute the model. An exogenous service is invoked by the occurrence of an external event (such as the arrival of a message), while an endogenous event is invoked as a consequence of the exogenous event (such as the processing of the message just arrived). These two types of services, as defined in the IPSS language, may contain special IPSS statements that allow the queueing, seizing, and releasing of named facilities, as well as the creation and destruction of files. FORTRAN code intermixed with the IPSS statements provides further processing, control, or input/output.

Exogenous Event Component.   The Exogenous Event Component defines the information system's service requests. Here the modeler describes the distribution of the occurrences of the events that will drive the model. This service component is later mapped to the exogenous services defined in the Configuration Component.

Data Base Component.   The Data Base Component defines the various files of the information system (see the appendix). Extensions to the IPSS to allow the definition of data base schemas has been proposed but not implemented (Brownsmith, 1979). These definitions are later mapped onto the physical devices defined in the Configuration Component.

Model Component.   The final section of an IPSS model, the Model Component, allows the modeler to associate exogenous events with exogenous services. Files from the Data Base Component are assigned devices within the Configuration Component. The ability to equate these entities from the different components allows the modeler to substitute different components as the model changes and experimentation proceeds. Finally, the modeler specifies the options for the model's execution, including halting criteria, trace options, and statistical generation and reporting.

IPSS Execution

The final transformation depicted in Figure 10-30 is the execution of the model by the IPSS Execution Facility to produce performance statistics. The statistics can then be evaluated by the system modeler for validation and by the simulation modeler for verification.

IPSS Model Parameters and Output

The user of the IPSS model provides a work load for the system specified by the interarrival time of exogenous events. Changes to the hardware components or services may be made to investigate a proposed soluion to a problem. The IPSS simulation facility produces up to eight classes of statistics at the conclusion of the simulation; some statistics are provided automatically while others are user triggered by statistical collection statements. The classes of statistics produced are

1. General statistics
2. Request stream statistics
3. I/0 activity statistics
4. Queueing statistics
5. Utilization statistics
6. Wait statistics
7. Service statistics
8. Task/activity statistics

THE BASELINE MODEL

Despite the existence of detailed models of the OCLC System (Wong, 1979; DeLutis, Wong, Rush, 1979), a new approach was needed to produce a model general enough to be changed easily by the system modeler and yet still capable of being verified and validated. Previous models had included the storage structure of the OCLC data base in great detail. However, the new OCLC model represents the major On-line System components with the capability of adding further detail uniformly or to a particular component of the system, depending on the interest and modeling needs of the modeler. This high-level model was designed for construction and validation, particularly since statistics on the On-line System's performance are already compiled and available at a level of detail commensurate with the model.

Consistent with the methodology of the previous section, the major systems of the OCLC On-line System were identified for modeling. The logical components were initially identified by hardware function. Referring to Figure 10-29, five classes of hardware are illustrated (Terminals, Communication Processors, Network Supervisors, Application Processors, and the Data Base Processors). An initial model consisted of a service for each of the subsystems.

Several of the logical components have major software programs that have responsibility for a single task. For example, the Network Supervisor consists of a DYNCIO task that synchronizes message transmission to and from the Communication Processors. The CNIO task handles message transmission to and from the Application processors. Thus, the Network Supervisor is modeled by two services.

The OCLC On-line System is message driven from the terminals. The work load is defined by the number and frequency of messages generated. As the arrows indicate in Figure 10-29, messages progress through the system components from the terminal to the Data Base Processor and back to the terminal.

Now that the methodological transformation is complete, the model can be translated into IPSS. The model is constructed by top-down refinement. First, a service that describes the system operation in terms of logical pathways through the components is defined. Then, each system component is defined by one or more additional services. Later, each component service may be expanded or split into more subcomponent services adding detail at each level. Using this technique with the IPSS, one can describe the system by a straightforward mapping from the actual system modules to services in the model.

The OCLC model has one major exogenous service which traces the transmission of the message through the system by queueing, seizing, invoking, and releasing the various software tasks that are defined as endogenous services in the model. Thus, each OCLC On-line System component—Communication Processor, Network Supervisor, Application Processor, and Data Base Processor is modeled as one or more endogenous services.

Hardware Definition

Models in the IPSS are organized into components describing the environment or facilities of the information system and services provided. Defining the computer hardware is straightforward. Shown below is a portion of the Configuration Component which defines the central processing units of the OCLC network. CP, NS. Host, and DBP are the names of the central processing units for the Communication Processor, Network Supervisor, Application Processor, and Data Base Processor, respectively.

```
CENTRAL  PROCESSOR:    ID=(CP, 15)
CENTRAL  PROCESSOR:    ID=(NS, 6)
CENTRAL  PROCESSOR:    ID=(HOST, 5)
CENTRAL  PROCESSOR:    ID=(DBP, 14)
```

Note that more than one central processing unit is defined in each case. As the system expands by adding more central processing units, the number in the definition is increased. All facilities can be defined in the same manner.

Information Services Definition

The definition of the services in the IPSS follows the natural organization of the information system being modeled. As previously stated, the services are divided into two classes—exogenous and endogenous. The exogenous services handle the events that trigger the system, and the endogenous services describe these operations that result from the system being activated.

Figure 10-31 illustrates the Data Link Control (DLC) task which is the software module in the Communication Processor that transmits messages between the terminals and the Network Supervisor.

The endogenous service DLC is seized and released so that the IPSS will provide utilization statistics at the end of the simulation. The processor number of the

```
ENDO SERVICE:     ID=DLC, NAME=DLC$, PARAMETER LIST= (DLCNOX) ,
          SAVE AREA SIZE=9
          INTEGER DLCNO, DLCNOX
          EQUIVALENCE ($SAVE (4) ,DLCNO)
          INTEGER AVAIL/0/, INUSE/1/
END:    DECLARATIONS:
C       SAVE THE DLC NUMBER FOR THIS INVOCATION
        DLCNO=DLCNOX
END:    INITIALIZATION;
C       COLLECT UTILIZATION STATS ON THIS ENDO SERVICE.
        SEIZE:              FACILITY=DLCTCP (DLCNO)
C       QUEUE TO GET CP CPU BEFORE TRANSMITTING TO NS.
        QUEUE:              FACILITY=CP (DLCNO) ;
        WAIT FACILITY:      FACILITY=CP (DLCNO) ;      STATUS=AVAIL;
        DEPART QUEUE:       FACILITY=CP (DLCNO) ;
        SET STATUS:         FACILITY=CP (DLCNO) ;      STATUS=INUSE;
        SEIZE:              FACILITY=CP (DLCNO) ;
C       TRANSMIT MESSAGE TO NS.   (ASSUME .02 MILLISECOND)
        PROCESS:                  TIME=.2E+1, CENTRAL PROCESSOR=CP (DLCNO) ;
C       FREE DLC CPU AND RETURN.
        RELEASE:            FACILITY=CP (DLCNO) ;
        SET STATUS:         FACILITY=CP (DLCNO) ;      STATUS=AVAIL;
        RELEASE:            FACILITY=DLCTCP (DLCNO) ;
END:    ENDO SERVICE:
```

Figure 10-31 DLC endogenous service

Communication Processor that this DLC task runs on has been determined previously by the invoking service and is passed as a parameter. The DLC task advances the clock a fixed time quantum to simulate the processing time expended on the message transmitted. However, a distribution of times could be sampled or statements could be substituted which actually model the message processing the DLC performs. Another possibility would be to replace the PROCESS statement with an invocation of a lower-level endogenous service. Thus, a natural model of the actual system operation has been synthesized by the division of the DLC task into an input/output task and message processing task.

Work Load Definition

The exogenous event stream which defines the arrival of the events that trigger the exogenous services defined in the configuration section consists of one major event—the generation of a message at a terminal. In this model the arrival time is based on peak time usage when a message enters the system (on the average every 0.03 seconds). Figure 10-32 is the IPSS listing of this exogenous event procedure, which is translated into a FORTRAN subroutine, as are all IPSS procedures.

```
EXOGENOUS EVENT:          ID=EXMSG,                    TIME=PROC(EXMSGS);
PROCEDURE:                         NAME=EXMSGS,               TYPE=SUBROUTINE;
      COMMON /MSGCNT/ MSGCNT, MIGTRC
      INTEGER MSGCNT
      LOGICAL MSGTRC
      REAL *8 ITIME
C     START AT 9.00AM (9*60*60*10000)    --  TIME IS IN TENTHS OF MS.
      DATA ITIME/32400.0D+4/
      SYSCOM(1) = 0
C *** A MESSAGE ARRIVES ON THE AVERAGE EVERY .03 SECONDS
C     STORE INTERARRIVAL TIME OF MESSAGES IN TENTHS OF MILLISECONDS
      ITIME = 0.3D+4

      CALL $SREAL(ITIME,SYSCOM,34)
      RETURN
END:  PROCEDURE
END:  EXOGENOUS EVENT STREAM:
```

Figure 10-32 IPSS Exogenous event routine

Extension is straightforward. The constant average time could be replaced by a sampling of a distribution, or the clock could be examined and a different distribution used for nonpeak system loading. Further, another exogenous event could be supplied that would change the distribution table this exogenous service uses. Thus, a separation between the occurrence of the event (exogenous event routine) and the action from the occurrence (processing of the exogenous service) is established in the IPSS language. This separation promotes natural modeling and allows the modeler to experiment by simply changing a small exogenous event stream procedure independent of the actual processes the exogenous event triggers.

MODEL UTILITY

Since the OCLC On-line System is a dedicated real-time computer system, the primary performance measure is response time. OCLC corporate policy mandates a minimum response time. This high-level model produces this vital statistic in addition to the intermediate statistics such as queueing times and facility utilizations. Despite the lack of great detail, the model is valuable for experimenting with system changes that can be succinctly expressed. The most obvious example is a change in computer hardware, for which the model would immediately provide statistics for evaluation. These statistics are provided without endangering the safe operation of the On-line System and can be done before purchasing new equipment, thereby allowing cost-benefit analysis based on performance modeling.

Changes in the OCLC On-line System are generally the result of changes in the work load, types of services, or types and number of processors. This baseline model can be readily modified to support these types of changes. The top-down design of the model necessitates changes in only these services affected without disturbing the rest of the model.

The user work load is defined in the model by the service representing the terminals. The baseline model contained a constant interarrival time; however, another service can be written to model the user actually keying the message. The service can also be altered to reflect changes in system usage. Users may enter fewer extended searches because the system penalizes such requests with long response times.

The OCLC services change as more features to the On-line System are added. In 1979, the Interlibrary loan subsystem became operational. The addition of this new service is reflected in changes to the service modeling, the application processors, and an addition to the Data Base Processor Service to reflect the Interlibrary loan data base.

Changes in processors are typical of system upgrades. Periodically, more processors and data base storage units are added as the system usage and data base grows. This is done by merely increasing the number of units in the hardware description as described in Section 6. This is done without disturbing the definition of the services which use these hardware components.

A more complex change may involve the addition of both new hardware and software. A new subsystem may be implemented on a different brand of host computers. (Currently, all subsystems are executing on all host computers.) The new computer is added to the hardware description and a service is rewritten to simulate the new subsystem. Minor modifications are made to the services simulating the Network Supervisor tasks to recognize the new message types requesting the new subsystem processing and to route them to the new host computer definition for simulated execution.

It is also possible to extend the model of the application tasks as new application subsystems are in development and planning. The IPSS model would produce a system impact statement on these new applications before their installation.

CONCLUSIONS

This research has demonstrated the feasibility of implementing a model of a complex information system running on an advanced networked computer system such as OCLC's, in the IPSS language. Due to the special features of the IPSS, the model can be developed faster and with less modeler effort than by using a general-purpose simulation language. The model written in IPSS is self-documenting and provides the modeler with greater understanding of the system being modeled. This research further points to the possibility of expansion of the model to support not only system performance and evaluation due to the growth of services (new library applications), changes in user habits, and growth in user population, but also modeling the user driving this on-line system in more detail than just the arrival of a message every so many seconds.

Several features of the IPSS were found most useful in constructing the OCLC model. An IPSS endogenous service can be treated as a facility and thus, it can be queued, seized, invoked, and released. This was particularly helpful in modeling the many identical tasks running on the multiple central processing units of the On-line System. For example, the computers in the Applications Processor each have a copy of the same tasks to do the application processing. The IPSS further allows the definition of Task Control Points. These are used as the facility name that is queued, seized, and released, thereby allowing one definition of a service to be modeled as being available as copies on many central processing units.

The goal of this OCLC On-line System model was to provide a prototype model for further development based on the analytical needs of system planners. The model's scope included all major system components (both hardware and software) at a "black box" level of detail to provide a broad and general basis for future modeling endeavors.

Due to the level of the model and the general availability of the information on the system, the model was quickly developed and verified within a three-month period. The design and implementation of this compact model was enhanced significantly due to the appropriateness of IPSS to the modeling of information systems.

The production of the model demonstrates the validity of high-level simulation of a network computer system to system analysts charged with the creation and maintenance of the On-line System. The model is appropriate for tuning, planning, system analysis, and design. Further, the modeling facility is capable of reacting swiftly to changes both of an actual or experimental nature.

Although the current version of this OCLC model does not make use of the data-base features of the IPSS, it will be possible at a later time to extend effectively the Data Base Processor endogenous services defined in the IPSS language to model the OCLC data bases. Further extensions may include the actual modeling of the various operating systems running on the separate components—Network Supervisor, Application Processor, and Data Base Processor. This promises to be a powerful planning tool and should aid the OCLC system engineers in maintaining a viable system to support OCLC's needs, both today and in the future.

REFERENCES

BROWNSMITH, J. D. *A Methodology for the Performance Evaluation of Data Base Systems: An Extension of the IPSS Methodology.* Ph.D. Dissertation, Columbus: The Ohio State University, 1979, 242 pp.

————, J. S. CARSON, AND W. H. HOCHSTETTLER. *An IPSS-Based Model-Building Methodology for Ranking and Evaluating Computer Hardware/Software Systems*, Final Report to AIRMICS. Atlanta: Georgia Institute of Technology, 1979, 162 pp.

BUXTON, J. N. "Writing Simulations in CSL." *Computer Journal*, Vol. 9, no. 2, 1966, pp. 137–143.

DAHL, O. AND K. NYGAARD. "SIMULA: An ALGOL-Based Simulation Language." *CACM*, Vol. 9 no. 9, 1966, pp. 671—678.

DELUTIS, T. G. *The Information Processing System Simulator (IPSS): Language Syntax and Semantics*, OSU-CISRC-TR-78–6, Columbus: The Ohio State University, 1978, 477 pp.

————, J. E. RUSH, AND P. M. K. WONG. "The Modeling of a Large On-Line, Real-Time Information System." *Proceedings of the 12th Annual Simulation Symposium*, Tampa, Florida, 1979, pp. 350–370.

FERRARI, D. *Computer Systems Performance Evaluation*. Englewood Cliffs, N.J.: Prentice-Hall, 1978, 554 pp.

GORDON, G. *The Application of GPSS V to Discrete Systems Simulation*. Englewood Cliffs, N.J.: Prentice-Hall, 1975, 389 pp.

GRAYBEAL, W. J., AND U. W. POOCH. *Simulation Principles and Methods*. Cambridge, Mass.: Winthrop Computer Systems Series, 1980, 249 pp.

GROSS, D., AND C. M. HARRIS. *Fundamentals of Queueing Theory*. New York: John Wiley, 1974, 556 pp.

KIVIAT, P. J., R. VILLANUEVA, AND H. M. MARKOWITZ. *SIMSCRIPT II.5 Programming Language*, E. C. Russell, ed. Santa Monica: Rand Corp., 384 pp.

KOSY, D. W. *The ECSS II Language for Simulating Computer Systems*, R-1895-GSA. Santa Monica, Calif.: Rand, 1975, 472 pp.

MAISEL, H., AND G. GNUGNOLI. *Simulation of Discrete Stochastic Systems*. Chicago: Science Research Associates, 1972, 465 pp.

MIHRAM, G. A. *Simulation: Statistical Foundations and Methodology*. New York: Academic Press, 1972, 526 pp.

PRITSKER, A. A. B. *The GASP IV Simulation Language*. New York: Wiley-Interscience, 1975, 451 pp.

————, AND C. C. PEGDEN. *Introduction to Simulation and SLAM*. New York: Halsted Press, 1979, 588 pp.

ROSE, L. L. *Computer Systems/Database Simulation*, Final Report to AIRMICS. Atlanta: Georgia Institute of Technology, 1978, 50 pp.

————, AND G. W. CARR. *Modeling the SSA Process*, OSU-CISRC-TR-78-7. The Ohio State University, 1979, 51 pp.

————, "IPSS: A Language and Methodology for Information Processing Systems Simulation." *Proceedings of the Simulation-Management Workshop*, AIRMICS, pp. 142–174.

SHANNON, R. E., *Systems Simulation: The Art and Science*, Englewood Cliffs, N.J.: Prentice-Hall, 1975, 387 pp.

WONG, G. "Computer System Simulation with GASP IV." *Proceedings of the 1975 Winter Simulation Conference*, 1975, pp. 205–209.

WONG, P. M. K. *A Methodology for the Definition of Data Base Workloads: An Extension to the IPSS Methodology*, Ph.D. Dissertation, Columbus: The Ohio State University, 1975, 269 pp.

APPENDIX

IPSS STATEMENTS

IPSS definitional statements	*IPSS procedural statements*
Configuration Component	Configuration Component
Access Mechanism	Allocate Data Set Extent
Buffer Pool	Create Data Set
Central Processor	Destroy Transactions
Control Unit	Find in Queue
Data Channel	Get Record Address
Data Set	Post Semaphore
Device	Process Time
Endogenous Service	Read Physical Record
Exogenous Service	Request Service
Procedure	Rewind Tape
Queue	Seek DASD Cylinder
Semaphore	Set DASD Track Sector
Volume	Set Facility Status
	Set System Parameter
Exogenous Event Component	Start Queue Statistics
Exogenous Event	Start Usage Statistics
Procedure	Stop Queue Statistics
	Stop Simulation
Data Base Component	Stop Usage Statistics
Area	
Segment	Model Definition Component
Organization Method	Equate Facility
Extent	Simulate
Record type	
Device	
Procedure	
Volume	
Model Definition Component	
Begin . . . End Model	

QUESTIONS

1. What are the major competitive factors affecting OCLC's success or failure?

2. Describe the criteria OCLC used to select a language for this model.

3. What are some components and outputs of this model in the IPSS language?

Assessing Defense Procurement Policies

Daniel L. Blakley
McLaren College of Business Administration
University of San Francisco
San Francisco, California

and

Kalman J. Cohen, Arie Y. Lewin, and Richard C. Morey
Fuqua School of Business
Duke University
Durham, North Carolina

ABSTRACT

Defense systems are typically over budget and behind schedule, and they often fail to meet performance specifications despite incentive schemes. Our study shows that these schemes are ineffective primarily because they assume (1) that a contractor's sole criterion is maximizing profit on a contract and (2) that a single decision maker determines criteria that do not change in the course of system development. In fact, defense contractors have many different criteria, and their relative importance often changes with time.

We have developed a dynamic computer simulation model that accounts for all the major motivations of the defense contractor, some of which are unrelated to any particular contract. It considers how the contractor interacts, at both the corporate and the project-manager levels, with decision makers at the U.S. Department of Defense (DOD).

The model is a relatively new type called a decision process model. A special feature of the model is that the user can modify the corporate-level and project-level goals of the contractor to better meet DOD goals as stated in the contract. The model can help planners assess the impact of alternative DOD incentive schemes and procurement policies on the performance of defense contractors.

The model is currently in the third stage of validation (statistical and sensitivity analysis). It has already yielded a few tentative results, most of which are strongly supported by field interviews with DOD and defense contractor personnel.

INTRODUCTION

Our research was motivated by the persistent problems experienced by the U.S. Department of Defense (DOD) in the procurement of advanced weapons systems— specifically, cost overruns, slippages in delivery schedules, and failure to meet

©1982 The Society for Computer Simulation. Reprinted, with permission, from *Simulation*, Vol. 38, no. 3, March 1982, pp. 75–83.

performance specifications. It is abundantly clear that standard contractual (profit-based) incentives have little effect on project outcomes. Our work has the twin objectives of investigating why DOD procurement policies have been ineffective and demonstrating how a simulation model may be used to assess alternative strategies.

Our model of defense contractor performance uses a new methodology, called decision process simulation, which has not been extensively reported in the literature. [2, 3, 14] The model simulates the decision making behavior of a hypothetical defense contract firm and its relationship with the DOD at the project and corporate organizational levels. Rather than using traditional profit-maximizing theories, we have developed a comprehensive model of contractor behavior based on important determinants of performance identified in our study of defense contracting organizations. We have incorporated these determinants into simulated decision processes within the contracting organization.

The decision process model (DPM) includes such basic elements as DOD project goals, DOD incentive mechanisms, contractor goals, and the response mechanisms of contractor organizations (and, of course, the availability of skilled workers and other operating constraints). Each of these elements, which collectively define the performance of the firm, are structured in a loosely coupled system of submodels and are parameterized to facilitate the analysis of different incentive schemes and behavioral variables.

In what follows, we discuss

1. The rationale for the decision process modeling approach and the motivation and goals of the defense contracting firm
2. The specification and decomposition of the contractor's operating environment into specific submodels
3. Model solution and validation
4. Tentative policy implications and ways in which the simulation may be used to improve DOD procurement policy

BACKGROUND

In awarding incentive contracts for the development and procurement of new weapon systems, the DOD has assumed that the primary aim of defense contracting firms is to maximize profits. [4] Research, however, suggests that the goals of defense contractors include survival, growth, market share, and reputation as well as profit. [1, 10, 11, 12] Furthermore, observations show that the relative importance of those goals depends significantly on the positions and responsibilities of the decision makers within the contracting organization. [6, 7]

Generally speaking, contractor management perceives survival as dependent upon achieving a level of performance on projects that enhances the reputation of the company and the ability to obtain future business. [5] Management believes that retaining high-quality technical and administrative personnel, even when business activity declines, is also critical to securing future large-scale contracts. [8, 13] A growing market share helps improve internal capabilities, develop barriers to entry by potential competitors, and spread fixed costs over a large base. In short, observations show that contractor management (at various levels) will voluntarily sacrifice short-run profits on a given contract in favor of

1. Improving opportunities for follow-on projects
2. Promoting the spinoff of technology to commercial business
3. Acquiring and retaining high-quality personnel in scarce disciplines
4. Engaging in research that promotes future product development.

We view the typical defense contracting firm as having a DOD business as well as an organizationally distinct commercial business (see Figure 10-33). In the following we assume that the defense business, which is managed by a single project manager (PM), consists of a number of projects awarded by the DOD. Likewise, the commercial business consists of a number of product lines, some of which benefit from technology developed under DOD contracts. [6] In addition, the contractor may transfer indirect costs of the commercial sector to DOD projects as a result of the establishment of a larger overhead base. [8]

While the commercial sector of the firm may attempt to maximize profit, the defense sector's goals are survival, growth, and prestige, subject to maintaining a necessary level of cash flow. While a manager in the commercial environment can affect revenues by controlling such variables as price, product mix, production lev-

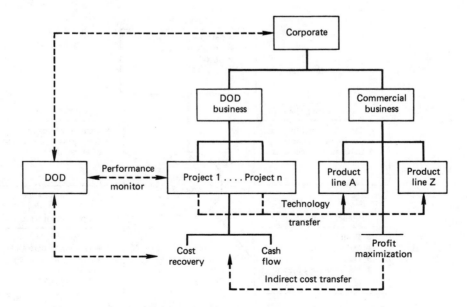

Figure 10-33 Defense contracting organization

els, and inventory, the general manager of DOD projects has little control over economic variables since they are largely determined by the contracts. The revenue from defense contracts is based on cost recovery—specifically, on the recovery of direct and indirect costs billed against existing projects. A fundamental problem in past research has been the failure to consider the differences between doing business in the traditional market environment and in the defense contracting environment. [11, 12]

SIMULATION ENVIRONMENT

Our decision process model consists of eight loosely coupled submodels which describe three levels of organization: project managers, corporate management, and the DOD. Each level of organization has its own goals, expectations, and decision processes which are simulated by the appropriate submodels. Operating conditions and project requirements are also important dimensions of the simulation environment. The model (see Figure 10-34) organizes these goals and constraints into an operating plan. Simon [13] and Cyert and March [3] have advocated and developed the method of analyzing organizational behavior by simulating the operating environment and decision processes of relevant decision makers.

The model simulates the actions of the project manager in a defense contracting firm and the interdependent actions of corporate and DOD management over varying time horizons. In the model, the project manager makes decisions on a monthly basis, while corporate managers make decisions quarterly based on the

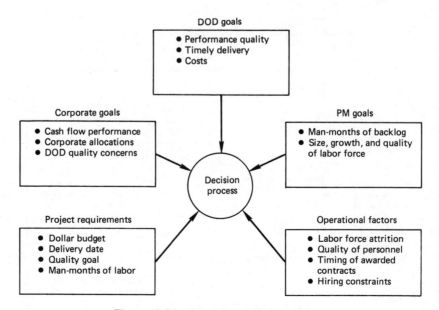

Figure 10-34   Decision-making environment

project manager's performance in the previous three months. The DOD is also interested in the project manager's actions, but only as they affect project performance. The DOD can monitor the performance of specific projects on a monthly (but delayed) basis.

DOD contracts consist of diverse projects with several distinguishing attributes. Each project is either major or a (significantly smaller) spinoff with a dollar budget, delivery date, and an amount of standard quality man-months (SQMMs) of work to be performed. In addition, major projects require the achievement of a specific standard of quality before the DOD will accept delivery of the product.

The project manager tries to hold the backlog of awarded projects within upper and lower bounds. Besides hiring and firing workers, the manager can affect the backlog by assigning workers to indirect activities. By adding to the indirect work force, the project manager can increase the number of proposals submitted to the DOD. Likewise, by assigning better qualified people to writing proposals, the project manager can win more contracts. The model assumes that the quantity and quality of submitted proposals, in addition to performance on previous contracts, determines the subsequent awarding of new projects by the DOD. [14] Thus, by having more or better qualified people write proposals, the project manager can increase the number of projects awarded and, other things being equal, the man-months of backlog.

A secondary, but related, concern of the project manager is the quality, size, and growth of technical staff. In the model, the work force consists of high- and low-quality personnel. The project manager knows which is which and uses this knowledge in assigning, hiring, and firing workers. The number of high-quality workers is limited, although the project manager can hire low-quality personnel as necessary to meet labor requirements. Higher attrition rates, high salary rates, and higher productivity characterize high-quality workers. In the model, the project manager pursues backlog and staffing goals, subject to various operational constraints and pressures, by assigning, hiring, and firing personnel on a monthly basis. High-quality workers are assumed to be more productive and therefore fewer are required to complete a given amount of work.

The corporate level, on the other hand, is concerned with cash flow, significant cost overruns, and the financial position of the firm. If the quarterly performance on defense contracts is unsatisfactory, corporate managers apply pressure to the project manager to increase cash flow. The project manager can increase cash flow by increasing the amount of billable time charged against existing contracts (i.e., increasing the number of workers on direct activities) and by decreasing current cash expenses (i.e., firing workers). Corporate management is also concerned with expenditures for research and development and administrative expenses (overhead) charged to the project manager's organization.

Finally, the DOD monitors the performance of the project manager on the individual projects under his or her control. Specifically, the DOD is concerned with

1. Total costs as determined by cumulative billings against each contract
2. Performance quality as reflected in the average quality of personnel used in

satisfying the standard quality man-months (SQMM) or work which must be performed on each project

3. The time of delivery relative to the contracted delivery date.

The exhaustion of the SQMM balance does not necessarily coincide with the exhaustion of the dollar budget, and, depending on the type of contract, excess costs may not be fully billable to DOD. When the DOD identifies quality or scheduling problems, it applies pressure to the project manager. If this pressure does not result in improved performance, the DOD may convey its concern to corporate management.

In summary, a realistic model of a defense project must include levels of management and their associated goals and expectations as well as operating factors and constraints. Our model not only includes these interactions, but also allows for changes in the goals of the project manager and corporate management in response to prolonged success or failure. (DOD goals are assumed to be static and are determined with the awarding of a contract.) Figure 10-35 illustrates how our model describes the dynamic interaction of the project manager, corporate management, and the DOD.

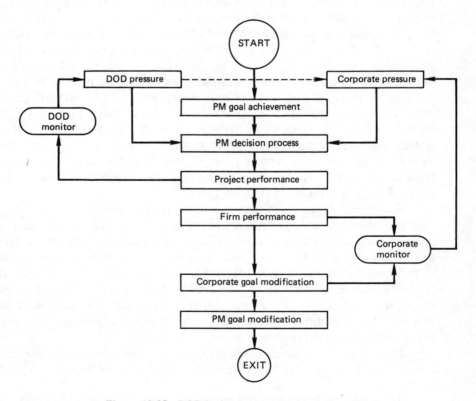

Figure 10-35   DOD project manager/corporate interaction

MODEL DECOMPOSITION

Table 10-19 provides a listing and ranking of the pressures and goals facing the project manager in the current model. The paramount concerns of the project manager are corporate cash flow pressure, DOD schedule pressure, and deficient quality on a completed project. The project manager is assumed to address these before considering other problems. The manager does not ignore other pressures and goals, but rather gives them secondary consideration.

TABLE 10-19 DETERMINANTS OF PROJECT MANAGER BEHAVIOR

Level 1:	Corporate cash flow pressure
	DOD schedule pressure
	Deficient quality on completed project
Level 2:	Corporate quality pressure
Level 3:	Backlog goal
Level 4:	DOD quality pressure
Level 5:	Volume (staff) goal

The next most important influence on the project manager is DOD complaints about quality as relayed from the corporate level, i.e., Corporate Quality Pressure. The DOD (as mentioned earlier) voices its concern about quality to corporate management only if previous pressure on the project manager has been ineffective. As shown in Table 10-19, this may occur when the project manager tries to build up backlog rather than place quality personnel on current projects. Finally, the lowest priority goals of the project manager are to maintain manpower levels and project volume.

Project Manager Manpower Assignment Submodel

The Project Manager Manpower Assignment Submodel assigns each high- and low-quality worker to indirect proposal activities or to a specific project (direct charge). As discussed previously, the project manager allocates personnel on the basis of DOD and corporate pressure, the achievement of personal goals, and the requirements of the backlog of incomplete projects.

Project Update Submodel

As the project manager allocates personnel, this submodel updates several running measures for each active project. Specifically, the submodel decreases the remaining dollars in each project's budget and the required SQMM balance and adjusts the quality index of work performed accordingly. Finally, the submodel calculates the time remaining before each project is due.

Personnel Submodel

The Personnel Submodel implements the project manager's chosen alterations in the size of the work force. This submodel includes labor-force attrition, which is assumed to occur simultaneously with the project manager's hiring and firing decisions. Thus the project manager must wait until the next period to compensate for attrition in the current period. The output of the Personnel Submodel is a description of the end-of-the-period labor force in terms of size and quality composition.

Backlog Determination Submodel

Using the output of the previous submodels, the Backlog Determination Submodel calculates the number of man-months of backlog (given current staffing levels) at the end of the current period. This involves determining whether a project has been awarded, and, if so, its dollar amount, delivery date, and SQMM size.

Cash Flow Determination Submodel

The Cash Flow Determination Submodel calculates the direct and indirect costs of the project manager's operation and the cumulative billings to the DOD allowed for the direct work performed on individual projects. Billings are allowed against a particular project only if its budget has not been depleted. Once the target cost on a given project has been reached, the contractually specified cost-sharing formula allocates subsequent cost overruns between the contractor and the DOD. If the ceiling cost is exceeded, the contractor must pay all further expenses incurred in satisfaction of SQMM requirements and quality standards.

Corporate Goal Adjustment Submodel

The Corporate Goal Adjustment Submodel determines the amount of overhead (corporate administrative and independent research and development expenses) that is charged to the project manager's operation. Also the submodel decides whether to apply cash flow pressure to the project manager.

Project Manager Goal Adjustment Submodel

The Project Manager Goal Adjustment Submodel modifies the project manager's personal backlog (measured in SQMM's) and volume goals based on past and current performance and pressure from corporate management and the DOD.

DOD Submodel

The DOD Submodel monitors progress made on each project and determines whether schedule or quality pressure should be applied to the project manager or to corporate management.

MODEL SOLUTION

The model consists of approximateley 900 lines coded in the SIMPLAN modeling language. [9] We chose SIMPLAN because of its multidimensional data base structure, text storage and editing capabilities, and statistical analysis features. The model currently runs at the Triangle University Computation Center on a system consisting of an Amdahl 470/V7 and an IBM 370/165. It requires 450K of core and the CPU time required for model solution (usually between 3 min 14 sec and 6 min) varies with the time horizon of the individual simulation.

All submodels have several routines which collectively describe the work flow during a run. Except for several routines in the Corporate Goal Adjustment Submodel (which are solved quarterly), each submodel is solved in each time period (month) of the simulation. The appendix contains a complete listing of the various submodels and routines.

As shown in Figure 10-36, the simulation begins with solution of the Project Manager Manpower Assignment Submodel. This involves the project manager's review of the operating environment (see Figure 10-35), taking note of pressure and

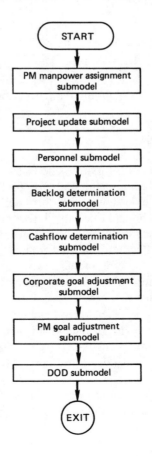

START

PM manpower assignment submodel

Project update submodel

Personnel submodel

Backlog determination submodel

Cashflow determination submodel

Corporate goal adjustment submodel

PM goal adjustment submodel

DOD submodel

EXIT

Figure 10-36 Submodel solution sequence

the achievement of goals. The manager divides the labor force between indirect and direct activities and assigns workers to specific tasks (either proposal writing or an incomplete project). The manager also decides on hiring and firing. After this sub-model is solved, the Project Update Submodel updates each project in the manager's backlog.

The Personnel Submodel then determines the size and quality composition of the labor force at the end of the period, taking into account the project manager's hiring and firing decisions and labor-force attrition. The Backlog Determination Submodel uses the output of the Personnel Submodel to calculate the actual man-months of backlog in the project manager's inventory at the end of the period. This also requires determining if a new project has been awarded in the current period.

After solution of the Cash Flow Determination Submodel, the Corporate and Project Manager Goal Adjustment Submodels are solved. Finally, the DOD Sub-model which monitors individual project performance, is solved.

For purposes of policy experimentation and general analysis, the model gen-erates reports at the end of a run (usually covering five years). The reports describe performance from the perspective of the DOD, project manager, and corporate lev-els; the output includes such data as

1. The average quality, duration, and costs of all complete projects
2. The size and quality of the work force and the average number of man-months of backlog
3. The monthly, annual, and cumulative cash flow.

VALIDATION AND APPLICATION

The model's structure and parameters must be tested and validated before it can be used to assess alternative procurement policies. The first phase of model validation involved conducting field interviews with the DOD and with contractor project managers. These interviews helped to establish the initial model specification. Field interviews after running the model allowed us to check its behavior against the judgement and experience of actual DOD and contractor personnel.

The second phase of validation included testing internal consistency by exam-ining various scenarios generated by modifying the model's attributes. By testing the effect of different simulated operating conditions, we were able to evaluate the overall plausibility of our model. Generally speaking, the behavior of our hypotheti-cal firm should reflect the behavior of real world firms in similar situations. Several scenarios have been thoroughly investigated (e.g., relaxing DOD monitoring capac-ity and increasing the availability of high-quality workers); the results are reason-able and consistent.

The third phase (currently underway) involves statistical exploration of the underlying characteristic of the simulation model. Sensitivity analysis, through the application of regression and hypothesis testing techniques, will yield information concerning the impact of different parameters and initial conditions on the model's performance. It will also provide greater insight into the workings of the model and provide a procedural description of its performance in various situations.

This research will demonstrate the use of the model as a laboratory for designing and comparing alternative DOD policies (e.g., contractual arrangements, monitoring procedures, pressure application) with the objective of improving procurement efficiency. The model could, for example, be used to

1. Explore ranges and sharing formulas for the target and ceiling cost specified in the contract
2. Assess the consequences of using preferred bidding lists (i.e., contractor rankings) and other methods of awarding contracts
3. Assess the effects of withholding progress payments
4. Study the efficiency of modifying the way the DOD interacts with contractor organizations (i.e., multilevel pressure application).

By conducting experiments with the model and identifying the operating factors and interactions which generate a particular outcome, we can gain valuable insight into the behavior of defense contractors. The decision process model becomes a useful management tool for identifying causes and solutions of the inefficiencies presently inherent in weapons acquisition.

Because of the complexity of the decision process model, various methods, including time-series plots and listings of key performance variables, are used to analyze performance output. It is beyond the scope of this paper to describe in detail the results of our various simulation experiments. However, a few qualitative remarks are in order. So far, the model's results have suggested the following policy implications:

1. Contractors respond to the DOD's current incentive schemes (i.e., making the contractor responsible for 10 percent to 20 percent of cost overruns) by increasing their internal capabilities (e.g., size and quality of work force) rather than by minimizing short-run production and development costs.
2. Although DOD monitoring of contractor performance does not necessarily produce satisfactory results from projects, its absence will result in more severe schedule slippages and cost overruns.
3. Discounting inflation, huge cost overruns in contracts for the development of large-scale weapons systems are most likely the result of poor initial cost forecasting or the deliberate underestimation of costs.
4. Significant fluctuations in demand for new projects by the DOD appear to be a primary instigator of poor project performance—both in expansionary periods as well as in business downturns.

CONCLUDING REMARKS

Although we have done a great deal of background work in model specification, we must still continue validation. A straightforward method of validating the model would involve observing in detail the decision-making process and operating environment of a specific defense contracting firm, fine-tuning the model, and comparing the actual and simulated results of specific policies. Before this can be done, however, we must obtain additional expert opinions from DOD and contractor personnel concerning the reasonableness of the model's specification and performance. Field work of this type is being conducted currently.

The underlying premise of this research has been that incentive contracts based on the traditional (profit maximizing) theory of the firm cannot deal with many real-world determinants of contractor performance. The simulation model formulated in this paper is an attempt to consider these factors explicitly and to analyze their effect on performance in a systematic fashion. Until broad-based research of this type is extended and refined, we can expect the impotence of DOD procurement policy and the resultant cost-overrun spiral to continue.

ACKNOWLEDGMENT

This research was supported by the USAF Business Research Center, Arie Y. Lewin, principal investigator.

REFERENCES

[1] Booz, Allen and Hamilton, Inc. *A Study of the Effectiveness of NASA Incentives Contracts.* Washington, D.C.: NASA Report no. NASW-1227, August 1966.

[2] Bonini, C. P. *Simulation of Information and Decision Systems in the Firm.* Englewood Cliffs, N.J.: Prentice-Hall, 1963.

[3] Cyert, R. M. and J. G. March. *A Behavioral Theory of the Firm.* Englewood Cliffs, N.J.: Prentice-Hall, 1963.

[4] Department of Defense . *Incentive Contracting Guide.* Washington, D.C.: U.S. Government Printing Office, October 1969.

[5] Fox, J. R. *Arming America: How the U.S. Buys Weapons.* Cambridge, Mass.: Harvard University Press, 1974.

[6] Hunt, R. G. *Extra-Contractual Influences in Government Contracting.* Buffalo, N.Y.: NASA Grant NGR33-015-061, State University of New York, 1971.

[7] Logistics Management Institute. *An Examination of the Foundation of Incentive Contracting.* Washington, D.C.: LMI Task 66-7, 1968.

[8] Lynch, P. J., and J. M. Pace. *An Analytical View of Advanced Incentivized Overhead Agreements in the Defense Industry,* MS thesis, Air Force Institute of Technology Grand Forks AFB, North Dakota, 1977.

[9] Mayo, R. B. *Corporate Planning and Modeling with SIMPLAN*. Reading, Mass.: Addison-Wesley, 1979,

[10] Oppedahl, P. E. *Understanding Contractor Motivation and Contract Incentives*, Study Project Report 77-1. Ft. Belvoir, Va.: Defense Systems Management College, 1977.

[11] Peck, M. J., and F. M. Scherer. *The Weapons Acquisition Process: An Economic Analysis*. Cambridge, Mass.: Harvard University Press, 1962.

[12] Scherer, F. M. *The Weapons Acquisition Process: Economic Incentives*. Cambridge,Mass.: Harvard University Press, 1964.

[13] Simon, H. A. "A Behavioral Model of Rational Choice." *Quarterly Journal of Economics*, Vol. 69, February 1955, pp. 99–118.

[14] ———. "From Substantive to Procedural Rationality." In S. J. Latsis, ed., *Method and Appraisal in Economics*. Cambridge, Mass.: 1976.

[15] Williamson, O. E. "The Economics of Defense Contracting: Incentives and Performance." In R. N. McKean, ed., *Issues in Defense Economics*. New York: Columbia University Press, 1967, pp. 217–256.

QUESTIONS

1. What criteria other than profit maximization are applied by defense contractors?
2. What levels of decision making and what time horizons are considered by the model? How do the goals of these levels differ?
3. Summarize the policy implications of this research.
4. What impediments do you foresee in the validation of the model? Discuss.

APPENDIX

PRINCIPAL SUBMODELS AND ROUTINES IN THE DPM

Project Manager Manpower Assignment Submodel

1. Goal Achievement and Pressure Check Routine
2. Direct/Indirect Manpower Allocation Routine
3. High/Low-Quality Manpower Allocation Routine
4. Project Specific Manpower Assignment Routine

Project Update Submodel

1. Schedule Update Routine
2. SQMM Balance Update Routine
3. Dollar Balance Update Routine
4. Quality of Performance Update Routine

Personnel Submodel

1. Hiring/Firing Routine
2. Labor-Force Attrition Routine
3. Quality of Labor-Force Update Routine

PRINCIPAL SUBMODELS AND ROUTINES IN THE DPM (*cont.*)

Backlog Determination Submodel

1. Capture Rate Determination Routine
2. New Proposal Generation Routine
3. New Contracts Awarded Routine
4. New Project Determination and Award Routine
5. New Project Attribute Assignment Routine
6. Accumulation of Existing Projects Routine
7. Backlog Calculation Routine

Cash Flow Determination Submodel

1. Direct Cost Determination Routine
2. Indirect Cost Determination Routine
3. DOD Billing Calculation Routine
4. Cash Flow Calculation Routine

Corporate Goal Adjustment Submodel

1. Project Manager Allocation Routine
2. Corporate Cash Flow Pressure Routine

Project Manager Goal Adjustment Submodel

1. Backlog Goal Modification Routine
2. Volume Goal Modification Routine

DOD Submodel

1. Schedule Pressure Routine
2. Project Manager Quality Pressure Routine
3. Corporate Quality Pressure Routine

Simulation Modeling Improves Operations, Planning, and Productivity of Fast Food Restaurants

William Swart and Luca Donno
Burger King Corporation
7360 North Kendall Drive
Miami, Florida 33152

ABSTRACT

This paper describes how a major fast food restaurant system uses simulation to dramatically improve efficiency, productivity, and sales in its more than 3000 restaurants worldwide. With a capacity to project and solve business problems, Burger

©1981 The Institute of Management Sciences. Reprinted, with permission, from *Interfaces*, Vol. 11, no. 6, December 1981, pp. 35–47.

King Corporation has been able to upgrade and streamline restaurant operations, contributing significantly to the continued growth of what is now the second largest restaurant system in the world. Among the substantial changes in the last five years, the introduction of drive-thru service and new menu items has transformed a once simple operation into a sophisticated production process. Consequently, management turned increasingly to Operations Research for answers to operational questions ranging from the most efficient restaurant design to the optimum number of employees needed to serve customers as sales vary. The impact of simulation models has produced millions of dollars in savings, or profits, in a number of operational, design, and procurement areas.

James McLamore opened the first Burger King restaurant in Miami in 1954 with a simple concept; he served a few variations of the basic hamburger and did not need a traditional kitchen. Because small businessmen could operate such a restaurant even without previous food experience, McLamore began to franchise the units.

Growth was deliberate and controlled. In less than 15 years, McLamore went from running a restaurant grossing under $100 a day to heading a company with $66 million in annual sales. The chain grew to 274 units by 1967, when it was acquired by The Pillsbury Company. Today the restaurant chain has systems sales worldwide of more than $2 billion. Average unit sales went from $254,000 in 1967 to $700,000 plus in 1980. Annual system sales have increased an average of 36 percent since Burger King's acquisition by Pillsbury.

There are now 3000 Burger King restaurants across the United States and in many other nations, and new units open at the rate of 300 annually. Eighty percent of the units are owned and operated by independent businessmen. They must adhere to a strict set of company policies but are otherwise free to run their business as they see fit.

The Corporation develops new products, systems, and procedures, but must in turn persuade franchisees to adopt them on a cost-to-benefit basis. Any change or new procedure developed for the system must demonstrate its validity to franchisees on the basis of increased sales and profits, and provide more than an adequate return on investment.

Burger King Corporation now employs more than 1000 at its Miami headquarters, and in excess of 130,000 are employed in company and franchised restaurants around the world. The company's growth has been attained not just by superior management and infusion of expansion capital but also, to a large degree, through a series of ongoing modifications to the original concept.

In the company's early stages, management focus was on expansion and development of the Burger King system. The original restaurant concept consisting of a single sandwich preparation board with a single, cafeteria-type line to serve customers was maintained. By definition this service system limited sales because of a limited delivery capacity.

The first major restaurant concept change was to go to a multichannel service, or "hospitality system." This system allowed for greatly increased sales delivery. To match this, additional production capacity was required. This was obtained by expanding the building backwards and placing the sandwich prep board perpendicular to the counter, allowing more sandwich prep employees to work simultaneously. These changes allowed reduced service time which created higher sales capacity to better accommodate peak-hour business. An example of one of the restaurant configurations is shown in Figure 10-37. Numbers indicate the order in which employees are assigned to handle customer load increases.

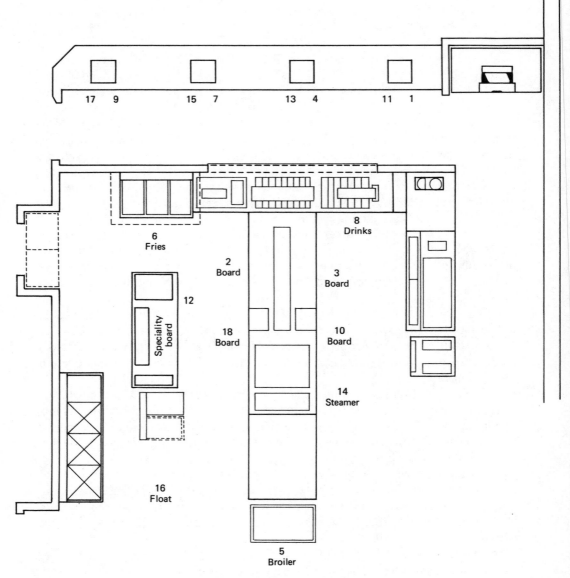

Figure 10-37   Positioning charts—"T" layout

As the take-out business increased substantially, Burger King was the first fast food hamburger system to introduce the drive-thru service lane. The impact on sales and operations was immediate and dramatic. Today, virtually 50 percent of the food sold in Burger King restaurants is through drive-thru service lanes. The drive-thru concept transformed the original single service system into two separate systems and required rearrangement of equipment and product so the counter crew and drive-thru service areas both had access to them.

Other variables impacted on the continuing growth of the Burger King system in 1973, to gain a marketing edge, the "Have It Your Way" concept was introduced.* The success of this concept was dramatic, but it did change the kitchen operations from mass production of similar items to a job-shop environment where product was prepared to individual customer specifications. The combined effect of increased average per store sales and the added complexity of kitchen operation required that, in many cases, additional production capacity be added to a very confined kitchen. In 1978 the Burger King system decided to significantly expand its product line through the introduction of Specialty Sandwiches, which now account for 20 percent of sales. The production of these sandwiches required different equipment and procedures which were implemented through an additional sandwich prep line. As a result of this diversification, sales again rose but the all important speed-of-service time was threatened, especially in peak periods, and original kitchen designs were no longer adequate.

Burger King Corporation projections call for average unit sales to reach $1 million by 1983, with restaurant operating profit to rise another 4 percent. It was clear, considering the increasing complexity of the production system and sophistication of the customer, that these goals could only be realized if the efficiency and productivity of the food delivery system was substantially increased for every restaurant in the Burger King System.

THE ROLE OF OPERATIONS RESEARCH

With a dramatically changing fast food hamburger restaurant business, it became clear that corporate productivity had to be a top priority. For this reason the Industrial Engineering and Operations Research functions were synergistically incorporated into one department at Burger King. That department is hereafter referred to simply as "Operations Research"; its primary mission is to enhance productivity throughout the system. The department established itself in a very short period of time through a series of high-impact studies, the first of which was the development of a computer model for the purchase of meat.

In late 1977 and early 1978, meat prices began to fluctuate widely. Because beef is the primary component of the food Burger King serves, cost pressure on margins was severe. Operations Research therefore developed a computer model to

Burger King Whopper Specialty Sandwiches and Have It Your Way are copyrighted trademarks of Burger King Corporation (1981).

determine what kind of meat to buy from which supplier so as to have the correct hamburger formulation at minimum cost. The result was a savings of almost 3/4 cents per pound. In a system that buys over 3 million pounds of hamburger meat a week, that 3/4 cents represents savings in excess of $22,000 each week.

The computer model was subsequently expanded not only to determine least-cost formulations at each individual packing plant, but also the least-cost distribution throughout the system. The new model optimized shipping and distribution costs based on meat availability nationwide and on anticipated demand for processed meat within the entire system. This expanded computer model saves Burger King Corporation at least $2 million annually.

But the impact of productivity improvement effort was most clearly demonstrated on the analysis of the drive-thru system. At first glance, the drive-thru concept is simple. The driver orders at the outside menu board. He then joins a waiting line (or stack) until it is his turn to pick up and pay for his order at the pick-up window. Staffing the drive-thru was originally determined by the restaurant manager. In most units, the drive-thru team usually consisted of one or two cashiers who would take the order, run to the sandwich chutes and drinks stations, assemble the order, bag it, and hand it to the customer.

Burger King established a standard transaction time of 30 seconds for the drive-thru window, but most units had service times in excess of that. During peak periods it simply was no longer possible for drivers wishing to use the drive-thru to even join the end of the car line. Sales were clearly being lost due to this problem. A system initially devised to provide customers convenience had become an inconvenience. Analysis at a number of units showed drive-thru transaction times were averaging 45 seconds. With a 45-second transaction time, the restaurant could handle a maximum of 80 cars an hour. With an average check of $2.44 per order, drive-thru sales were limited to a maximum of $195.00 per hour.

If the transaction time could be shortened to 30 seconds, cars served per hour could increase by 50 percent, and maximum sales would rise by almost $100 to $292 an hour. That represents an annual capacity benefit (or sales increase) of over $35,000 per restaurant. Working with franchisees, Operations Research devised a plan to improve speed of service at the drive-thru. The heart of the new system is the separation of drive-thru work into a series of distinct tasks.

One employee does nothing but take orders. The order taker gives the order to a runner/bagger who assembles the order and places it on an assembly shelf. The third member of the drive-thru team, the cashier, simply makes change and hands the order to a customer. The system allows for additional staffing when demand exceeds the ability of the three-person crew to maintain speed of service standards.

The Operations Research Department also recognized that customers waited an average of 11 seconds at the order station before being acknowledged. The rubber bell hose was therefore moved ahead of the order station so that the ordertaker was alerted to the customer's arrival prior to the car reaching the order station.

Today, all Burger King restaurants with a drive-thru have adopted the efficiency package. These restaurants have increased their annual sales capacity by over

$35,000. If each restaurant in the system gained only 50 percent of this, or $18,000 per unit in annual sales increases, the Burger King system would enjoy additional sales of $52 million annually.

Although most of the studies mentioned did involve the use of either simulation, optimization, or statistical models, these models were developed to solve specific problems. However, it soon became apparent that the increasing demands placed on the OR department by management would require the development of a comprehensive general-purpose restaurant model which could be used to address a wide variety of issues.

Top management's demands on the department increased because of the demonstrated success of the approach and the recognition that the current Burger King System is vastly more complex than it was when they were actively involved in day-to-day restaurant operations. The addition of drive-thrus, the increase in menu items, and the changes in building and equipment all contributed to making management's experience base, obtained in the original system, not suitable for decision making in today's environment. The only alternative to making decisions from an empirical base is through an analytical approach such as that provided by Operations Research.

While management was finding it increasingly difficult to set a firm stategy for the future, the key cost variables that impact on the ability of the restaurant to run profitably began to rise. These cost escalations were most severe in food and labor, which are the single largest cost components in operating a Burger King restaurant, accounting for nearly 40 percent of every sales dollar.

Faced with this situation, Burger King management in 1979 committed itself to develop a Productivity Improvement Program for both company and franchised restaurants. This program would clearly have great impact on the Burger King system. The success of the program would, to a large degree, determine the future success of the company itself. A continuing improvement in productivity would be required for each operator to maintain a satisfactory return on investment. In addition, restaurant productivity improvements are needed to maintain return on investment levels which, in turn, can be used to demonstrate to potential franchisees that a franchise is a sound investment. New franchised restaurants account for 15 percent of the system's growth annually.

Management's decision to entrust the development of this program to Operations Research provided the motivation and rationale to incorporate all knowledge gained previously and integrate it through the development of a comprehensive general purpose restaurant model.

MODEL DEVELOPMENT

In order to develop a general-purpose restaurant model, it was decided to view the restaurant as an operating system composed of three interrelated subsystems: The Customer System, The Production System, and The Delivery System.

In a typical store, a customer order is generated in the customer system (in-store and/or drive-thru). This order is transmitted to the kitchen via a CRT device. The four production areas in the kitchen (Drinks, Fryers, Main Sandwich Preparation Line, and Specialty Sandwich Preparation Line) respond if the order contains a product made in that area. The production action can be to replenish inventory of standard product or to prepare a custom sandwich (special order) for a waiting customer. Simultaneously, in-process inventories of fries, preassembled hamburgers, drinks, fish, chicken, lettuce, mayonnaise, etc., have to be checked to determine whether a replenishment action must take place.

Concurrent with the production activity, the delivery system is active in processing the customer by making change and assembling the order from inventories, whenever possible. The amount of time a customer waits, referred to as speed of service, is held to a minimum by using inventories whenever possible and by maximizing the lead time the kitchen has to produce special orders.

Although the above description has many analogies with typical manufacturing facilities, there are also some distinguishing characteristics: (1) the Burger King system is composed of 3000 relatively small manufacturing plants (stores) which together form a relatively large manufacturing organization; (2) sales volumes can vary by as much as 1000 percent within a 30-minute interval; (3) shelf life of most products is 10 minutes. This prevents the accumulation of inventory during slow periods to handle peak period demands and also creates the need for an extremely flexible production and manning strategy; (4) production, inventory, distribution, and a substantial percentage of consumption takes place under one roof.

Jim McLamore's original concept in founding Burger King was to serve quality food—quickly and courteously. Speed of service remains the keystone of the fast food hamburger restaurant. As shown in Figure 10-38, the peak service hour is noon luncheon business. For a given facility at a given location, the luncheon business usually can be increased in proportion to the increase in speed of service. Typically, a hamburger restaurant gains four times as much business a day as it gains for the lunch hour. So, if a Burger King restaurant is to raise sales but cannot, due to competitive pressures, increase prices substantially, it must increase the number of consumers served. And, to do that, the restaurant must increase its speed of service.

The equation is simple. The faster the service, the more people that can be served; therefore, the higher the restaurant's potential sales volume. For this reason the impact of suggested changes on speed of service is the principal criterion for operational decision making not only at Burger King but at most fast food companies.

The initial modeling efforts yielded a comprehensive linear programming model. Although valuable, it did not permit the analysis of the dynamic aspects of the operation, nor the consideration of the multiple interrelated queueing situations that arise in the customer, delivery, and production areas. Consequently, a simulation approach was selected.

Any general-purpose restaurant model for the Burger King System must have the capability of representing any restaurant type currently in the system as well as any potential design for a new restaurant. In addition to the different configurations,

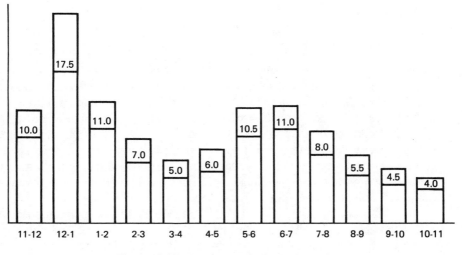

Figure 10-38 Percentage of sales by hour of day

the manufacturing system is subject to change during the course of any given day. Some production areas are closed down as demand lags and others that remain in operation are manned differently than at other sales hours.

The simulation model was built, therefore, using a modular approach. Each logical entity in the restaurant was defined and programmed in as self-contained a manner as possible. The modular model has maximum flexibility and efficiency in adapting to change and readily accommodates the various new products service concepts that are currently being tested or developed. Data collection for the program began with prototype studies in company-owned R & D restaurants, and measurements were then taken in more than 40 restaurants in 10 different regions of the country.

As the initial modeling efforts neared completion, it became essential to convince corporate management that the model actually predicted the behavior of real restaurants. Top management and selected franchisees were invited to witness and cooperate in a live calibration of the model. In this calibration, the manufacturing and service system of a typical unit was physically assembled in a warehouse. The "restaurant" was staffed with actual crew members from nearby units. Over the course of two days, the "restaurant" was subjected to a variety of customer arrival patterns matrixed against specific manning levels. The results were videotaped so that actual "restaurant" operations could be compared to the computer projections. In this manner, the reliability and accuracy of the model was established.

By allowing corporate management and franchisees to be actively involved in the calibration test, credibility was established beyond any doubt. Today virtually no operational decision at Burger King is made without having been subject to model analysis. The model currently averages 300 runs per month. (A typical application consists of 2500 GPSS blocks and runs approximately 60 seconds of CPU time on an IBM 370/3032.) Management's confidence in the model has been rewarded as its accuracy has been confirmed repeatedly in actual restaurant operation.

APPLICATIONS

Existing Restaurants

The prime motivation for the Productivity Improvement Program was to allow restaurants to handle current sales and future sales increases efficiently and profitably. The restaurant simulation models provided the analytical capability that allowed for the processing and evaluation of suggestions and alternatives from a Joint Franchise/Corporate Productivity Task Force. This resulted in the development of the Productivity Planning for Profit Kit (PPP), a two-phase process developed expressly for restaurant operators. Phase One allows them to recognize if productivity problems are limiting sales potential, specifically define those productivity problems, and indicate how these problems can be resolved.

After completion of Phase One, the PPP kit allows restaurant operators to project the achievable sales growth that could be reached in an individual restaurant. The kit then determines what changes must be effected in order to reach these goals. Finally, the kit allows the restaurant operator to project the return on investment and payback period of each of these productivity improvements.

At the conclusion of the PPP process, the operator has a five-year plan for the productivity upgrade of his restaurant. The PPP program is an ongoing process which is evaluated annually by the restaurant operator in conjunction with his franchise district manager. It begins with speed of service evaluations of both front counter and drive-thru. After work stations causing delays, if any, are identified, delay analyses examine each problem area to identify the specific bottleneck.

To cure these bottlenecks, the restaurant operator simply refers to one of over 100 Improvement Resource Cards. Each card indicates the maximum sales potential that can accrue from alleviating the productivity bottleneck. The restaurant operator, using the PPP kit, can project any combination of real and inflationary growth. Then, entering the Improvement Decision Path at current sales levels, the specific productivity improvements are clearly indicated.

The PPP kit allows the operator to predict his return on investment for each productivity upgrade, and by repeating the process, a complete five-year sales and productivity plan is assembled. The successful use of the kit spurred the use of the model to evaluate ongoing operations management decisions. Each change in facilities, procedures, equipment, and labor has to be justified in terms of its impact on the restaurant as a whole. This represented a fundamental change in the operating philosophy of the company. In the past year, over 20 changes have been evaluated with the model. Descriptions of several major changes follow:

- In an ongoing analysis of the drive-thru, Operations Research recognized that in many restaurants the stack size, or distance between the order station and the pick-up window, was too short and accommodated only two or three vehicles. The simulation models were used to determine what stack size gave the optimal lead time so that when a car reached the window its likelihood of waiting was minimized.

The longer stack, by drastically reducing waiting time, allows an additional 12 to 13 customers to be served in an hour. This adds $30 per hour in sales during the peak lunch hour, for an annual benefit of over $10,000 per restaurant. This change has been implemented throughout the system. Taking a conservative estimate of 1500 restaurants in which this change has been implemented, this would provide an additional $15 million in annual sales capacity.

- The second drive-thru window (placed in *series* to the first as opposed to a two-lane drive-thru), has been proposed as a means of expanding the sales capacity of the drive-thru during peak hours. The simulation model evaluating this modification projected a sales benefit of 15% during peak lunch hours. An experimental second window unit was installed in company-owned R & D restaurants. The observed sales benefit in actual operation was 14%. Based on average restaurant sales during the lunch hour and average drive-thru percentages, the second window adds $36.40 in sales each day, or more than $13,000 per year per restaurant. This revision has been introduced into the system. Estimating that 10% of the 300 new restaurants built each year will be able to implement the second drive-thru window, that represents an additional $390,000 in annual sales capacity.

- The model has proven especially valuable in examining the operational impact of the introduction of new products. For example, small specialty sandwiches were considered for introduction to supplement sales of the larger sandwiches. The computer simulation showed the introduction of these sandwiches would generate an average delay of 8 seconds per customer. On a yearly basis, this would represent a $13,000 loss in sales capacity for an average restaurant. On this basis, it was recommended that these sandwiches not be introduced. This resulted in the avoidance of a $39 million loss in capacity to the entire system.

New Restaurants

While the restaurant simulation model is based on experience in existing restaurants, it was designed to aid in the development of new restaurant configurations. As more and more fast food chains compete for fewer and fewer quality locations, it became clear that the size of the restaurant would have to be tailored to the trading area.

Positioning the correct size building in a specific location can substantially increase potential returns. Using the simulation model, the restaurant universe was segmented into three distinct segments: (1) the BK-500, (2) the BK-700, the new workhorse of the system, and (3) the much larger BK-900.

Traditional evaluations such as sales potential and kitchen capacity were not neglected, but increased emphasis was placed on designing the kitchen for human engineering to reach maximum efficiency for any given capacity with minimum labor. The computer model allowed speed of service to be plotted against sales, establishing a distinct operating range for the restaurants as currently configured.

For example, the BK-500, which is designed for trading areas of 5000 to 10,000 people, breaks even at just over $500,000. An average BK-500 easily handles sales of $675,000 a year with an hourly sales capacity of $550 per hour. As sales reach three quarters of a million, the BK-500 can no longer maintain our speed of service standards. At $820,000, sales are limited and real growth is no longer possible. The model, therefore, defines an operating range, from $500,000 to $800,000 plus. With this data, it is also possible to establish a distinct decision point at which the choice is made to build a BK-500 or another unit (Figure 10-39).

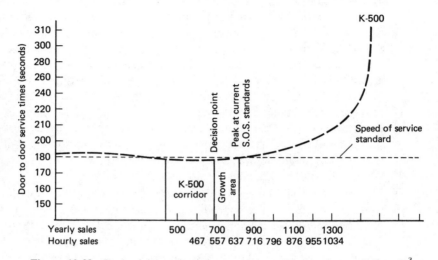

Figure 10-39 Productivity performance comparison: Yearly sales are in $ × 10³ and hourly sales are in $. Calculations for the decision point are made for 5% real growth and four years' operating life before need to upgrade kitchen to handle volume.

This decision must allow for sufficient growth before it becomes necessary to remodel and expand the restaurant to maintain speed of service.

In the case of the BK-500, this decision point allows sales volumes to achieve a real growth of 5 percent annually for four years before the $820,000 limit is reached. Projecting backward gives a decision point of $675,000.

Because the restaurant can be profitable at low volumes, the BK-500 is targeted to offer Burger King new opportunities in markets of a size which cannot sustain current restaurant designs.

Completion of similar analyses for the BK-700 and BK-900 allowed operators more clearly to match sales and service capacities to the needs of specific locations. More importantly, both the computer-designed BK-500 and the the BK-700 offer substantial labor savings. Whatever the sales volume, low, normal, or peak, both designs offer savings in excess of 1.5 to 2.0 percent in labor over today's designs. Completion of such analyses enables restaurant operators to select the most efficient restaurant configuration for each site and trading area.

Applications for Both New and Existing Restaurants

As stated earlier, labor is a major cost component in restaurant operation—one that is increasing rapidly due to federally mandated boosts in the minimum wage. From 1976 to 1981, the minimum wage increased 46 percent—from $2.30 to $3.35 per hour.

Although a restaurant operator cannot control the price of labor, he can control the amount of labor used to operate his restaurant. In establishing staffing levels, the operator has, in the past, been guided by the uniform Burger King Labor Standard, applied to all units.

The Operations Research Department realized that major bottom-line benefits could be achieved by tailoring the amount of labor restaurants required at various sales volumes to the restaurant configuration, product mix, and drive-thru percentage.

Through the use of the model it was possible for the first time to accurately project not only how many crew members were needed, but also the most effective positioning and division of labor within the restaurant.

The results of the simulation modeling are a series of staffing and crew positioning charts. Because staffing and work responsibilities are tailored to individual units, labor savings over the old standard are substantial. For instance, for a standard T restaurant with drive-thru, the savings occur at both peak and low volume hours (Figure 10-40), and are in excess of 1.5 percent of sales. For a typical restaurant with annual sales of three quarters of a million dollars, this represents an additional profit of $11,250.00.

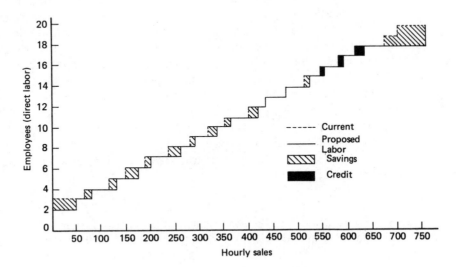

Figure 10-40   Labor savings for the "T" layout

IMPLEMENTATION AND BENEFITS

Lou Neeb, chairman and chief executive officer of Burger King Corporation, in a letter to the Chairman of the CPMS Competition, stated that

> The role of Burger King Corporation is not simply to operate restaurants, but to provide individual businessmen with an enterprise that offers an above-average chance of success. Providing this success requires that our restaurants achieve a more than adequate return on investment.
>
> The Operations Research activities the company has undertaken in the past few years are the primary means we have at our disposal to directly improve the profitability of each individual restaurant. Though the results of these efforts can total millions and millions of dollars for franchisees and the company, the real measure of success of these efforts is the continued fiscal vitality of the Burger King System itself. As our fast food competitors close literally hundreds of restaurants throughout the nation. Burger King restaurants continue to operate profitably . . . in large part due to the savings generated by improved productivity.

The primary means of achieving the productivity gains referred to for the 3000 existing restaurants in the Burger King System is the PPP kit. The corporation can only mandate productivity upgrades in the 400-plus company restaurants it owns. Participating among the franchise community is encouraged, however, by making the formulation of a productivity plan a mandatory item in each restaurant's annual franchise review. As part of this review, a franchisee must submit a minimum of a two-year productivity plan for each restaurant that has been in existence for more than two years. In this way, the corporation assures that all restaurants, both company and franchise, are participating in an ongoing productivity improvement program.

Because the productivity upgrades required by each restaurant depend on the unit's existing facilities, sales volume, and product mix, it is almost impossible to document a hard dollar benefit from the program. There is, however, every indication that the benefits to franchisees and the corporation are substantial.

In an earlier operational improvement program, called Grand Slam, franchisees invested an average of $40,000 per restaurant over the course of two years. Average restaurant sales as a result increased by approximately $200,000. Systemwide among the 2200 units then in operation, this represented a sales increase of $440 million. These sales increases returned royalty payments alone to Burger King Corporation of $15,400,000.

At the same time, company restaurant operating profits increased by more than 75 percent. If franchisees achieved the same percentage profit increases, the program generated almost $100 million, in additional profits systemwide.

Similar benefits are without a doubt being achieved through the implementation of the PPP kit. In the first year of the PPP process, most restaurants upgraded their microwave capacity and installed an auxiliary drink station, converted current drinks to automatic dispensing and front access, provided access to the shake machine from the front corner, and adopted the drive-thru efficiency package. The sales capacity benefits are in excess of $170,000—in line with those achieved during Grand Slam. Hence similar profit and sales increases can be anticipated.

In fact, there is every indication that the Productivity Improvement Program has substantially bolstered Burger King sales throughout the current economic downturn. Until the last quarter Burger King has consistently outperformed the industry, while both major competitors have posted negative gains in each of the last two years. Between August of 1979 and May of 1980 Burger King, according to Bernstein Research, posted real growth of 2.2 percent while McDonald's declined by 2.8 percent in real terms.

In the last two years, Burger King restaurants have gained market share at the rate of almost 1.0 percent a year—and each percentage point of market share represents more than $200 million. More importantly, on an individual restaurant basis, in the last three years, discounting the effects of the introduction of breakfast into the McDonald's menu, Burger King has halved the gap in individual restaurant sales. At this rate, Burger King restaurant sales will match those of the industry leader on a per-restaurant basis within the next five years.

All the new restaurants opened by Burger King today use kitchens designed according to the simulation model and therefore enjoy substantial labor savings from the moment they open. These 1.5 percent savings in labor are passed directly to the bottom line as incremental profit.

The use of the new simulation model in the design of the kitchens for the 300 new units opened each year provides an incremental profit of $3,375,000 annually. Over the 25-year life of the franchise agreement, this equals to a net present value of $84,375,000 (assuming that the real growth plus inflationary growth is equal to the corporate cost of money).

More importantly, Jerry Winter, Burger King Corporation senior vice president, Operations Systems, notes that

> Using the model to tailor the restaurants to particular markets opens up many trading areas which the company could not previously target for expansion. The development of the BK-500 alone opens up over 2,000 potential new trading areas for development. Without the development of this smaller 50-seat restaurant and its highly productive kitchen, our expansion efforts would be significantly curtailed.

Perhaps the single greatest impact of the simulation models is the establishment of the new labor standards. Initial projections were that these standards would produce a 1.5 percent of sales savings in labor costs. Actual field experience suggest that the savings may be substantially higher—in excess of 2 percent. However, even using the lower percentage, this represents a systemwide savings—or additional profit—of $32,625,000 annually. These savings are today reality as the system adopts the labor standards worldwide.

CONCLUSION

More and more individual franchisees, as the Corporation did earlier, are realizing that Operations Research and the restaurant simulation models provide the competitive advantage that leads to success.

Wally Crawford, who operates three franchised restaurants in Des Moines, Iowa, with his brother Dave, says

> In the past, there were two ways of being successful in the fast food industry. One, you were lucky enough to open in just the right location and there was nothing you could do but succeed. If you weren't lucky enough to select exactly the right location, you had to work extra hard . . . constantly trying new ideas to build the restaurant into a success. But there was no way to judge these ideas except by our own experience; experience and knowledge that as a newer franchisee you simply didn't possess. Now, with the restaurant models for the first time we have a tool that can evaluate new concepts as well as or better than even the most experienced operators. Operations Research increases the viability of each individual restaurant and in doing so assures that we will be around not just next year but 10 and 20 years from now.

Wally Crawford's thoughts are echoed by Mike Simmonds, who operates a group of highly successful Burger King restaurants in Omaha, Nebraska:

> To maintain the success of a restaurant even units with sales in excess of $1,250,000, you must constantly modify the restaurant to allow you to capture rather than cap your sales potential.
>
> Seat of the pants judgment is hardly the best means of managing a $1 million plus investment . . . even if one has worked in fast food restaurants most of one's life. Yet, until the introduction of the Productivity Improvement program . . . sophisticated guesswork was the only tool we had available.
>
> Operations Research today provides the means of evaluating the change before it is made . . . and predicting your return on investment.
>
> No other tool yet developed in the fast food industry . . . contributes so much to assuring a franchisee's peace of mind as he contemplates additional investments in existing or new restaurants.

Jerry Ruenheck, president of Burger King US, which is the largest of the Corporation's three decentralized operating divisions, says

> Our business has undergone dramatic changes. The importance of the simulation modeling program to all areas of operational and productivity planning is almost incalculable. The analytical knowledge and sophistication that simulation modeling gives Burger King, therefore, provides annual savings, or profits, in the millions of dollars for the system of company-owned and franchised restaurants.

ACKNOWLEDGMENTS

The authors wish to thank Dr. Douglas Hutchinson from the University of Tennessee at Knoxville for his contributions during the early stages of the model's development. In addition, special thanks are due to Jerry Winter, senior vice president for Operations Systems at Burger King Corporation, and Mike Guido, formerly with Burger King Corporation, for their early support of the modeling efforts.

1. What operations research techniques has Burger King used? For what purposes?
2. Describe the submodels in the PPP kit.
3. Discuss the results of Burger King's simulation efforts.

A Database-Supported Discrete Parts Manufacturing Simulation

William P. Rundgren M.S.
Manager of Data Processing,
Carver Pump Company
Muscatine, Iowa

Charles R. Standridge, Ph.D.
Assistant Professor
University of Iowa
Iowa City, Iowa

ABSTRACT

This paper discusses a prototype decision support system for discrete parts manufacturing. Basic decision support system concepts are presented. The context of this analysis effort within the entire technical operations of the firm is shown. The use of typical, real production data stored in a database for both traditional reporting purposes and as data input to a simulation model is discussed. Thus, the model can process both currently known future orders and generate currently unknown future orders in analyzing future production requirements. Furthermore, model outputs which measure the ability of the production system to meet future requirements are stored in the database. Thus, they may be analyzed and reported independently of the running of the model.

INTRODUCTION

This paper discusses a nucleus of decision support for discrete parts manufacturing. This nucleus consists of a simulation model supported by a relational database.

The body of the paper reviews essential features of the components of a decision support system and establishes the decision support context for the simulation model, details the logical data model and discusses the inner workings of the simulation model itself. It also discusses two distinct uses of the simulation model. On the one hand, a purely deterministic mode can be established. On the other, a mixed probabilistic/deterministic mode can be established which involves some features unique to database-supported simulation.

THE COMPONENTS OF DECISION SUPPORT

Database Management Systems

Dozens of database systems are available today. Gio Wiederhold surveys nearly one hundred. Every major hardware vendor and numerous software specialists have such products to market [Datapro Research Corporation]. They range from simple file management utilities to complete database management systems which include data dictionaries, data definition languages, data manipulation languages, query facilities, and report generators.

One relational database system which is designed especially for use in simulation modeling is the Simulation Data Language (SDL^{TM}) (Standridge).* In addition, SDL can be used as a tool for decision support. It was used in the course of this research not only as the respository of manufacturing data for conventional applications and as the destination for outputs from simulation runs, but also as a vehicle for generating random variates to be inputs to the simulation runs.

SDL is a FORTRAN-based data management system. It consists of three basic components. A data definition component allows for the creation of relations appropriate for holding the required data. (A relation can be thought of as a matrix of rows and columns.) It also allows for operations to be performed on the relations as the users require. A physical data management component allows the database implementor to account for his unique hardware requirements. The third component is a unique high-level batch programming language to create and manage the database called OIL.

One SDL command that is particularly useful in the context of this research samples from histograms created from either historical data or simulation results. This means that the simulation model can be free from specifications of distribution types for random sampling to assign values needed within the model. Instead, the distribution types are dynamically changing as users update the database. This was preferable to us since enough data existed to give confidence to the sampling.

*SDL is a trademark of Pritsker & Associates, Inc.

Management Science Models

Approaches to the problem of modeling production planning have included mixed integer programming, dynamic programming, and simulation (Baker, 1974; Johnson and Montgomery, 1974). The first two methods can provide correct solutions to the problem of optimizing some criteria of production but only in the smallest of cases. Problems encountered in the industrial world are of a scale which makes finding exact solutions formidable, if not altogether impossible, by MIP or DP methods. The method of digital simulation offers distinct advantages over other methods for aiding in production planning. It allows for the modeling of real-world problems while accommodating the dynamically changing data on which the production decisions rely. Disadvantages of the method are primarily cost related, as the model run requires significant computer time. However, with the cost of computer hardware continually decreasing, this disadvantage is likely to disappear altogether. The method offers great promise for incorporation into the production planning function for a wide range of manufacturing firms.

As the techniques of simulation are accepted into the firm's planning function, it will be natural to provide data from the firm's database to augment, if not replace, the data such as setup time or unit production time which conventionally is generated within the model by random sampling from assumed distributions. It will be seen that such data used by the firm for day-to-day report processing is in many cases the same data by which the simulation model should be driven.

In this research, a prototype decision support system for a discrete parts manufacturing operation was to be built. This prototype consisted of a GASP IV (Pritsker) simulation model of the operation and an SDL database containing typical information about the operation. This information included machine setup times, unit production times, and sequences of operations to be performed on particular products. The model was designed to make use of these data. Thus, when revisions in engineering and production plans are updated, then these data outputs of the model could automatically reflect the revisions. Finally, the outputs of the model were stored in the database.

Decision Support Systems

Neither the DBMS nor the MS model alone provide modern management with the type of tool it needs to operate. But together these components can constitute the nucleus of a decision support system.

Decision support systems constitute a fairly small but growing group of information management tools with unique characteristics. Keen and Morton (1974), Alter (1980), Sprague (1980), and others have helped to specify just what these systems are and how they can assist in complex decision making. DSS are distinct from the management information systems (MIS) of the 1970s. They emphasize the role of computers in assisting managers and technicians rather than replacing them. Their emphasis is more on flexibility than consistency and on support of semistructured tasks rather than clearly structured tasks. Whereas the MIS necessarily dealt with reporting information about the past, the DSS is oriented toward providing information for making decisions affecting the present and the future.

Structurally one might envision a general decision support system to resemble Figure 10-41. Most generally, it consists of a data management component storing information regarding the real-world system together with one or more application tools (statistical, financial or operations research for example) used to extrapolate from and report to the database. Obviously, a mature DDS would provide several such programs and their respective interfaces with the database. In this research a single simulation model was linked to the database demonstrating the usefulness of such techniques for discrete parts manufacturing.

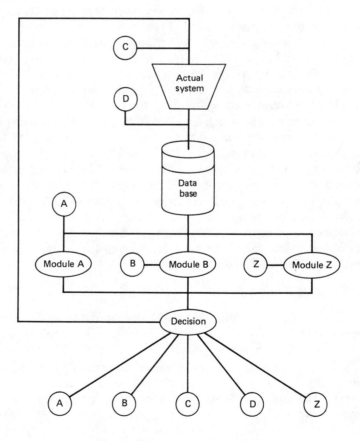

Figure 10-41   A generalized DSS

Current DSS are generally proprietary systems built to support unique tasks. While they differ considerably in their designs, they share what are coming to be seen as the essential features of a DSS:

They are management oriented rather than clerical.
They support—not replace—decision making.
They function in semistructured environments.

The system described in this research shares these features.

LAYOUT OF THE MODEL AND OF THE DATABASE

Context of the Model

It is useful to characterize the environment in which the model to be described was developed. The production facility is that of a medium-sized manufacturer of centrifugal pumps. It consists of about 60 conventional and numerically controlled machine tools processing discrete batches of parts. These batches move from machine to machine in predefined sequences which depend on the specific part, quantity, and material called for. Estimates of times at each machine have been made prior to releasing the batch for processing. Orders to make batches of distinct parts are prepared by production planners on the basis of a master schedule in advance of the date fabrication is to begin. These orders are then filed by week of expected release. Each week the appropriate orders are released for production. With the database in place, the orders are to be stored online instead of in a manual file, but the release method will remain the same.

Each week's group of orders consists of a different number. The number to be released in the immediate future generally is greater than that to be released in the distant future. On the appropriate date, that week's accumulated orders are released for production and join other orders in process. In addition, the attributes of typical orders were stored in the database. Samples from histograms of these attributes were used to build currently unknown, future orders. Thus the model uses actual production data when available and samples from histograms of that data when required to generate future, unknown order releases with attributes similar to those observed historically. Such a situation would be encountered when the actual number of orders to be released in a given week was less than the historical average.

The role of this model in the production planning function is to utilize the production data to predict shop status in some future period using deterministic data when available and probabilistic data as required. A unique feature of this approach is that the modeler need not hypothesize distributions from which the probabilistic data comes. In a business where history is accepted as an adequate predictor of future activities (this is true in the stable pump industry), this approach is to be pre-ferred. The model is indifferent to the type of distribution implied by the actual data, yet samples generated will accurately reflect the distribution type. In this sense the model and database work together to modify the model itself.

With support of manufacturing decisions as the overall objective, one can create a structure chart (Figure 10-42) which places the model in the context of deci-sion support. The node entitled ''Simulate Operations'' is to become the simulation model of the operations. The structure of the model is dictated by its intended use in conjunction with these other aspects of the DSS.

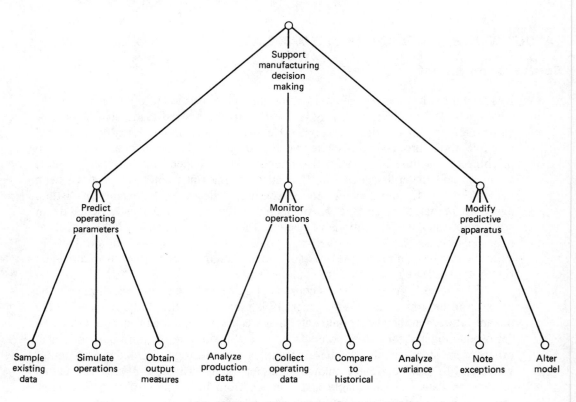

Figure 10-42   The simulation model in DSS context

The Logical Database

The global data model for the pump company is shown in Figure 10-43. Each node is a relation or group of related data items. Of particular concern to this research was data regarding product manufacture. This is the conventional routing data found in virtually all manufacturing systems. It is only part of the global data model, but it is of critical importance operationally and was therefore chosen to be used for this research. It can be seen that the use of any other corporate data would be similar to the use of production data. Only the size of the database and the global data model would change.

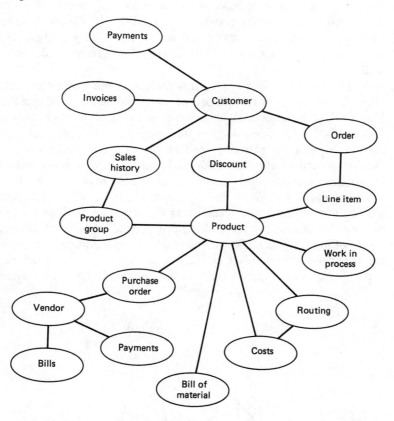

Figure 10-43   Global data model for pump manufacture

Along with this relation holding input data, a relation to hold model output data is required. These two relations form the core of the database. From them, subsets of rows and columns are created to form the basis for statistical analysis and histogram creation. It can be seen that data in these relations are available for a variety of purposes in addition to driving the simulation model. The production planner may access the relation holding data for a particular week and answer questions regarding that week's data. Last-minute changes due to new orders or canceled orders can be made without altering the model. This would not be the case in a conventional model. One would be forced to reexamine the data on which the chosen distributions were based and perhaps find better fits in alternative distributions. The database oriented model is thus more suited for use in the semistructured decision support environment where the decision maker himself effectively changes the structure of the model by introducing new inputs.

The Simulation Model

With the basic assumptions outlined above, a model of shop operation was developed. The language chosen was GASP IV (Pritsker). Compatability with the FORTRAN-based SDL was an important factor in the choice. Other languages which support invocation of FORTRAN subprograms would undoubtedly work as well.

The model itself consists of a main segment and several subroutines. Their interrelations are shown in Figure 10-44. It should be noted that the main module of Figure 10-44 is functionally the same as the module "Simulation operations" of Figure 10-42. The main body of the model (GSPMAIN) reads parameters describing the length of the planning horizon, the average time between order releases, the number of actual orders to be released, the historical average order size and a vector of machine identification numbers which itself may change from run to run. The next segment (ARVSHOP) then opens the database of production data and reads the attributes of orders to be released on the given week. Those attributes include the part identification number, its description, material of construction, product code, a vector of machine identification numbers indicating its path through the shop, a vector of expected times to set up the job on each respective machine, and a vector of expected times to produce a unit of specified product on each respective machine.

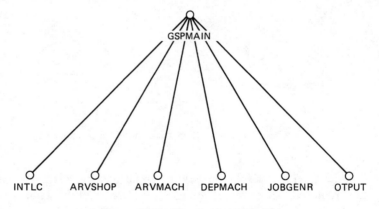

Figure 10-44   The simulation model

If the number of actual orders released at this time is equal to or less than the historical average, the initial arrival of a group of orders is complete. They are then assigned to the machine of first operation and processing begins. However, if the number of actual orders released is less than the historical average, the model generates artificial orders to be processed along with the actual orders. They act as place holders for orders expected but not yet prepared on the date of the model run. The attributes of these artificial orders are generated by sampling from histograms of data in the production database. These orders also are assigned to machines of first operation and processing begins.

The arrival to the machines specified by the production data is handled in a separate subroutine (ARVMACH). This subroutine checks the status of the machine required. If it is idle, the order is immediately set up and production begins. Since the expected time of completion is specified, an event is scheduled to represent completion of the order at the designated machine. If the machine is already busy, the released order is put into a queue until the machine is free.

Departure from a machine is a separate subroutine as well (DEPMACH). If no jobs are queued, the machine is set idle. If jobs are queued, one is dispatched and run. The rule for removal from the queue is shortest processing time first. This dispatching rule can be changed at the discretion of the modeler, of course, but is probably the most commonly used and is appropriate here. If the departure is from the machine of last operation, the job has finished its processing and is removed from the system. If not, it moves to the next scheduled machine.

The subroutine which generates artificial orders (JOBGENR) is a unique feature of this database-oriented model. Here an SDL function is invoked to sample from histograms of pertinent production data (sequence of operations, machine set up and machine operation times). Values are returned which are characteristic of the actual orders planners expect to release and reflect the distributions from which the data come.

If the time horizon of the simulation is less than the average time between order releases, the model generates only one release of orders, processes the parts, and terminates, writing appropriate output measures to the database. Otherwise the next batch of orders is released and the existing job mix is expanded. Processing then continues until the time horizon of the simulation is reached.

TWO SIMULATION MODES

To develop a usable database for inquiry is in itself a project of more technical than academic interest. But to lay out the database in such a way that the same empirical data used for inquiry is available for prediction and decision making via the simulation model is a project of academic as well as technical interest. By choosing the appropriate parameters for the model run, one is able to simulate the operation of the shop in either of two modes. These simulation uses of the empirical data are what distinguishes the database-supported model from the conventional model.

Purely Deterministic Mode

If the user so desires, he may set the parameters of the model to run in a purely deterministic mode. This would be done specifically by setting the number of actual orders to be released equal to the average order size. Except for the time spent in queue at each machine, this data would be available from conventional analysis of production data. The output from the simulation includes time in queues as well as time in the system, machine utilization and number of open orders. There are occasions when this information is all the planner requires.

Mixed Deterministic/Probabilistic Mode

On other occasions a planner may require information which can only come from running the model in mixed mode. He may have to know the implication of introducing a new product line which resembles existing products to some degree and is to be introduced at a future date. He would then set the model parameters to generate representative orders similar but not identical to orders for existing products. A simulation is in order since the increased load may or may not be possible in the required time frame. In this case the same number of actual orders are to be released but the average order size is greater so the model generates representative attributes for typical orders and releases them together with the actual orders on the required order release date.

During each model run, system performance information (specifically machine queue times) is collected and written to the database. Typically queue times are sought in manufacturing system modeling as a measure of the ability of the system to meet its production requirements. They were used as representative model outputs for this research as well.

SUMMARY

The research described here was done to show the value of linking DBMS with an MS model for use in manufacturing decision support. The characteristics of the broader system of which they are a part.

The database management system was seen to be of critical importance. Features of SDL make it a valuable DBMS for the purposes outlined here. It not only provides use in the conventional manner of data definition, inquiry, and reporting but also provides important functions for histogram creation and for subsequent sampling from those histograms.

The simulation model was also seen to be of critical importance. In addition to providing a vehicle for semistructured problem solving, it also demonstrates a unique mode of selfmodification. By working in conjunction with the database, it circumvents the conventional specification of distribution types and allows distribution types to be handled dynamically.

REFERENCES

[1] ALTER, STEVEN. *Decision Support Systems: Current Practice and Continuing Challenges*. Reading, Mass.: Addison-Wesley, 1980.

[2] BAKER, KENNETH R. *Introduction to Sequencing and Scheduling*. New York: John Wiley, 1974.

[3] COMPUTER AIDED MANUFACTURING-INTERNATIONAL. *Functional Specifications for an Advanced Factory Manaagement System*. Arlington, Tex., 1979.

[4] DATAPRO RESEARCH CORPORATION. *Datapro Directory of Software*. Delran, N. J., 1980.

[5] JOHNSON, LYNWOOD A., AND DOUGLAS C. MONTGOMERY. *Operations Research in Production Planning, Scheduling and Inventory Control*. New York: John Wiley, 1974.

[6] KEEN, PETER G. W., AND MICHAEL S. SCOTT MORTON. *Decision Support Systems: An Organizational Perspective*, Reading, Mass.: Addison Wesley, 1978.

[7] PRITSKER, A. ALAN B. *The GASP IV Simulation Language*. New York: John Wiley, 1974.

[8] ———, AND C. ELLIOTT SIGAL. *Management Decision Making: A Network Approach*. Englewood Cliffs, N.J.: Prentice-Hall, 1982.

[9] RUNDGREN, WILLIAM P. "Database Supported Manufacturing Modeling and Simulation," Unpublished Masters Thesis, The University of Iowa, Iowa City, 1981.

[10] SPRAGUE, RALPH H. "A Framework for the Development of Decision Support Systems." *Management Information Systems Quarterly*, Vol. 4, no. 4, 1980.

[11] STANDRIDGE, CHARLES R. *The Simulation Data Language (SDL) Language Reference Manual*. West Lafayette, Ind.: Pritsker and Associates, 1980(a).

[12] ———. "The Simulation Data Language: Fundamental Concepts For Its Use in Simulation Studies." *Simulation*, September 1981(b).

[13] ———. "The Simulation Data Language: Applications and Examples of Its Use in Simulation Studies." *Simulation*, October 1981(c).

[14] ———, AND DAVID B. WORTMAN. "SDL: The Simulation Data Language A Database Management for Modelers." *Simulation Today* in *Simulation*, August 1981.

[15] WIEDERHOLD, GIO. *Database Design*. New York: McGraw-Hill, 1977.

QUESTIONS

1. What is the difference between an MIS and a DSS? Discuss.

2. Explain how this application substitutes empirical distributions for theoretical distributions, and why.

3. In what ways do the deterministic and probabilistic models of operating this model assist the production manager? Discuss.

chapter eleven

THE MANAGEMENT OF MODELING

The management of modeling does not differ greatly from the management of information systems or, for that matter, the management of any organizational subsystem. While this book has focused on the technical aspects of modeling, the success of a modeling venture is probably even more dependent on the human factors.

THE TEAM APPROACH

Although members of the operations research staff of a firm are inclined by disposition and commonality of background to form an organizational unit just as do data processing personnel, it is essential that a team approach to project management be used to tackle each modeling venture. The project team should be composed of at least one individual with expertise in the details of modeling, at least one person thoroughly familiar with the computer system to be used for model development and execution (these may be the same person), at least one member of the department or departments whose problems are the subject of the modeling effort, and one representative of top management with comprehensive knowledge of the organization and the authority to carry through or discontinue the modeling at any point. Without the constant monitoring and support of top management, the tedious minutiae of data collection and ultimate implementation of solutions suggested by the modeling will often meet with fatal resistance. If the computer expert on the team is conversant with standard systems analysis methods, friction between the modeling team and users, or between the modeling team and management, is likely to be minimized. The user department representative on the team is charged with ensuring that the modeling effort is directed toward a real, important problem; that the model accurately defines the major aspects of the system; and that proposed alternative solutions are credible. The management representative must verify that the cost of the modeling venture does not exceed the expected benefits from improved system performance and must make sure that the team receives all needed cooperation within the organization.

PROJECT MANAGEMENT

There are numerous ways of making sure that a project such as the design, development, and implementation of a simulation model will be completed within reasonable bounds of time and cost. Project management tools such as GANTT charts and PERT and CPM software packages force the project team initially to divide the modeling process into small increments, assign time and cost estimates to each, and define the order in which the steps must be completed. Actual time and cost expenditures are supplied regularly, and comparisons are made with initial projections of time and cost at that point in the project. If the project is behind schedule or above cost estimates, the problem can be identified and resolved before it creates a disaster.

One common pitfall which may be minimized or avoided entirely by the team approach and careful project management is the underestimate of time and cost or overestimate of benefits by eager user personnel and operations research specialists. Another similar hazard is the pursuit of the elegant and optimal solution rather than the simple and satisfactory one that management would regard as preferable. A poor impression of the operations research department may thwart the approval of future project requests.

THE FUTURE OF MODELING

Since simulation and its statistical support activities depend heavily on computer technology for successful implementation, certain trends in hardware and software carry clear implications for simulation modeling. The wide availability and acceptance of decision support systems, of which simulation is a part, have increased management access to simulation at larger organizations. Even among small-computer users such popular packages as the VISICALC financial modeling program for microcomputers have enabled many to understand and appreciate simulation who might never have been exposed to it previously.

Enhancements are constantly being made to the simulation programming languages, and new languages are regularly appearing with features which consolidate desirable properties of various older languages with features which were previously unavailable. Packages specific to certain areas of simulation which require inputs more akin to data entry than to programming also tend to place simulation within the grasp of the user rather than within the esoteric realm of the operations researcher or computer scientist.

Data-base management systems offer the potential user of simulation access to a pool of information about target systems which helps to obviate the need for tedious, time-consuming, and error-prone special-purpose data collection.

In hardware, CPU, main memory, and disk speeds are constantly improving. These improvements increase the likelihood that large-scale simulations will be feasible for smaller and smaller computers. Many owners of microcomputers have adapted them to serve as intelligent terminals which can communicate with large time-sharing networks offering simulation and statistical packages.

Recent graduates of colleges with majors in business are all acquainted with simulation through management gaming and through introduction to operations research methods.

PROBLEMS YET TO BE RESOLVED

The American National Standards Institute (ANSI) has developed versions of several languages such as COBOL and FORTRAN which have common subsets that can be run almost without modification on a wide variety of computer systems. Such standardization is highly desirable for the major simulation programming languages but has not been forthcoming, probably because the relative number of users is not large enough to offer incentives to the providers of software. As simulation grows more and more popular and accessible, there will be even more pressure to standardize languages.

Perhaps a more troublesome lack of standardization is in the credentials of professional simulation modelers. Some come from backgrounds in engineering, others from computer science, others from mathematics and the physical sciences, others from humanities and social sciences. All have been attracted by problems in their subject matter areas which are amenable to analysis by simulation and perhaps in no other feasible way. While these individuals approach the modeling task with good intentions and an intensive knowledge of the problem area, many lack the methodological tools to perform efficient, effective, credible simulations. The tools which have been discussed in this book are probably the least that a simulation professional should know about the subject. Matters such as validation and reduction of simulation run time are state-of-the-art issues which have barely been mentioned. In the future, it may be possible to become certified as a simulation professional in much the same fashion as an accountant or a data processing professional becomes certified, by passing a comprehensive examination over a recognized body of material and by taking steps to keep up with the rapid progress of the field.

CONCLUSION

While in the past simulation was described as a method of last resort for solving problems, the future will most likely see simulation as the method of preference. The trend is toward more cost-effective, more reliable, easier-to-use and understand, more comprehensive modeling methods. From astronomy to zoology, from business to sociology, the problems await the methodology and ultimately the solutions.

QUESTIONS

1. What types of people should serve on a modeling project team? What is the function of each?
2. What project management tools may be used to assist in the planning and control of a modeling venture? What purposes do such tools serve?
3. Discuss the probable future of modeling in terms of
   (a) computer hardware advances
   (b) computer software advances
   (c) improvements in personnel skills
4. What difficulties remain to be resolved in simulation methodology? Discuss.

SELECTED NUMERICAL ANSWERS

2-2. Mean interarrival time $= 421.75$, variance $= 120945.5$, sample size $= 24$. Mean service time $= 316.769$, variance $= 67605.69$, sample size $= 26$.

2-4. **(a)** Computer $\chi^2 = 11.00$, critical $\chi^2 = 7.815$.
    (b) Computed $\chi^2 = 1.02$, critical $\chi^2 = 3.841$.
    (c) Computed $\chi^2 = 1.33$, critical $\chi^2 = 5.991$.

2-6. Computed $\chi^2 = 2.40$, critical $\chi^2 = 3.84$.

3-2. **(a)** Expected daily wage cost $= \$68.00$.
    (b) Expected daily opportunity cost $= \$362.10$.
    (c) Expected daily wage cost of idle time $= \$16.93$.

3-4. Mean time in system $= 507.2792$, mean number in queue $= 0.22584$.

3-6. Mean time in system $= 856.7331$, mean number in queue $= 1.21411$.

3-8. Mean time in system $= 316.7763$, mean number in system $= 0.75108$.

3-10. Mean time in system $= 1116.656$, mean number in queue $= 1.89651$.

4-2. $R = 0.48$, z yields an approximate result which is not significant.

4-4. Maximum difference $= 0.39$, critical $D = 0.565$.

4-6. Answers will vary with number of classes defined.

8-2. **(b)** Computed $\chi^2 = 7.10$, critical $\chi^2 = 15.507$.

8-6. **(a)** Mean $= 953.25$, variance $= 998435.20$, sample size $= 40$.
    (b) Observed number of runs $= 24$, expected number of runs $= 26.33$, $z = -0.7036$
    (d) Conway's rule $=$ number 1; with subgroup size $= 10$ E & S above-and-below rule $=$ subgroup 4, E & S moving-average rule $=$ subgroup 2.

9-2. $z = -1.294$.

9-4. Absolute differences 8.787, 44.875, and 53.762; criterion at the 0.05 level of significance $= 18.621$.

APPENDIX TABLES

APPENDIX TABLE I AREAS UNDER THE STANDARD NORMAL PROBABILITY
DISTRIBUTION BETWEEN THE MEAN AND SUCCESSIVE VALUE OF *z*.

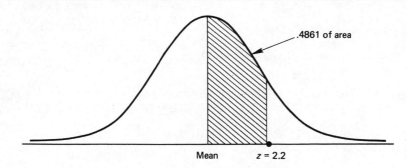

.4861 of area

Mean *z* = 2.2

EXAMPLE: To find the area under the curve between the mean and a point 2.2 standard deviations to
the right of the mean, look up the value opposite 2.2 in the table; .4861 of the area under the curve lies
between the mean and a *z* value of 2.2.

z	.00	.01	.02	.03	.04	.05	.06	.07	.08	.09
0.0	.0000	.0040	.0080	.0120	.0160	.0199	.0239	.0279	.0319	.0359
0.1	.0398	.0438	.0478	.0517	.0557	.0596	.0636	.0675	.0714	.0753
0.2	.0793	.0832	.0871	.0910	.0948	.0987	.1026	.1064	.1103	.1141
0.3	.1179	.1217	.1255	.1293	.1331	.1368	.1406	.1443	.1480	.1517
0.4	.1554	.1591	.1628	.1664	.1700	.1736	.1772	.1808	.1844	.1879
0.5	.1915	.1950	.1985	.2019	.2054	.2088	.2123	.2157	.2190	.2224
0.6	.2257	.2291	.2324	.2357	.2389	.2422	.2454	.2486	.2517	.2549
0.7	.2580	.2611	.2642	.2673	.2704	.2734	.2764	.2794	.2823	.2852
0.8	.2881	.2910	.2939	.2967	.2995	.3023	.3051	.3078	.3106	.3133
0.9	.3159	.3186	.3212	.3238	.3264	.3289	.3315	.3340	.3365	.3389

APPENDIX TABLE I (cont.)

z	.00	.01	.02	.03	.04	.05	.06	.07	.08	.09
1.0	.3413	.3438	.3461	.3485	.3508	.3531	.3554	.3577	.3599	.3621
1.1	.3643	.3665	.3686	.3708	.3729	.3749	.3770	.3790	.3810	.3830
1.2	.3849	.3869	.3888	.3907	.3925	.3944	.3962	.3980	.3997	.4015
1.3	.4032	.4049	.4066	.4082	.4099	.4115	.4131	.4147	.4162	.4177
1.4	.4192	.4207	.4222	.4236	.4251	.4265	.4279	.4292	.4306	.4319
1.5	.4332	.4345	.4357	.4370	.4382	.4394	.4406	.4418	.4429	.4441
1.6	.4452	.4463	.4474	.4484	.4495	.4505	.4515	.4525	.4535	.4545
1.7	.4554	.4564	.4573	.4582	.4591	.4599	.4608	.4616	.4625	.4633
1.8	.4641	.4649	.4656	.4664	.4671	.4678	.4686	.4693	.4699	.4706
1.9	.4713	.4719	.4726	.4732	.4738	.4744	.4750	.4756	.4761	.4767
2.0	.4772	.4778	.4783	.4788	.4793	.4798	.4803	.4808	.4812	.4817
2.1	.4821	.4826	.4830	.4834	.4838	.4842	.4846	.4850	.4854	.4857
2.2	.4861	.4864	.4868	.4871	.4875	.4878	.4881	.4884	.4887	.4890
2.3	.4893	.4896	.4898	.4901	.4904	.4906	.4909	.4911	.4913	.4916
2.4	.4918	.4920	.4922	.4925	.4927	.4929	.4931	.4932	.4934	.4936
2.5	.4938	.4940	.4941	.4943	.4945	.4946	.4948	.4949	.4951	.4952
2.6	.4953	.4955	.4956	.4957	.4959	.4960	.4961	.4962	.4963	.4946
2.7	.4965	.4966	.4967	.4968	.4969	.4970	.4971	.4972	.4973	.4974
2.8	.4974	.4975	.4976	.4977	.4977	.4978	.4979	.4979	.4980	.4981
2.9	.4981	.4982	.4982	.4983	.4984	.4984	.4985	.4985	.4986	.4986
3.0	.4987	.4987	.4987	.4988	.4988	.4989	.4989	.4989	.4990	.4990

From Robert D. Mason, *Essentials of Statistics*, © 1976, p. 307. Reprinted by permission of Prentice-Hall, Inc., Englewood Cliffs, N.J.

APPENDIX TABLE II PERCENTILE VALUES (χ_p^2) FOR THE CHI-SQUARE DISTRIBUTION, WITH ν DEGREES OF FREEDOM (SHADED AREA = p).

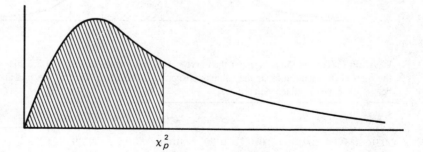

χ_p^2

ν	$\chi_{0.995}^2$	$\chi_{0.99}^2$	$\chi_{0.975}^2$	$\chi_{0.95}^2$	$\chi_{0.90}^2$
1	7.88	6.63	5.02	3.84	2.71
2	10.60	9.21	7.38	5.99	4.61
3	12.84	11.34	9.35	7.81	6.25
4	14.96	13.28	11.14	9.49	7.78
5	16.7	15.1	12.8	11.1	9.2

APPENDIX TABLE II (cont.)

ν	$\chi^2_{0.995}$	$\chi^2_{0.99}$	$\chi^2_{0.975}$	$\chi^2_{0.95}$	$\chi^2_{0.90}$
6	18.5	16.8	14.4	12.6	10.6
7	20.3	18.5	16.0	14.1	12.0
8	22.0	20.1	17.5	15.5	13.4
9	23.6	21.7	19.0	16.9	14.7
10	25.2	23.2	20.5	18.3	16.0
11	26.8	24.7	21.9	19.7	17.3
12	28.3	26.2	23.3	21.0	18.5
13	29.8	27.7	24.7	22.4	19.8
14	31.3	29.1	26.1	23.7	21.1
15	32.8	30.6	27.5	25.0	22.3
16	34.3	32.0	28.8	26.3	23.5
17	35.7	33.4	30.2	27.6	24.8
18	37.2	34.8	31.5	28.9	26.0
19	38.6	36.2	32.9	30.1	27.2
20	40.0	37.6	34.2	31.4	28.4
21	41.4	38.9	35.5	32.7	29.6
22	42.8	40.3	36.8	33.9	30.8
23	44.2	41.6	38.1	35.2	32.0
24	45.6	43.0	39.4	36.4	33.2
25	49.6	44.3	40.6	37.7	34.4
26	48.3	45.6	41.9	38.9	35.6
27	49.6	47.0	43.2	40.1	36.7
28	51.0	48.3	44.5	41.3	37.9
29	52.3	49.6	45.7	42.6	39.1
30	53.7	50.9	47.0	43.8	40.3
40	66.8	63.7	59.3	55.8	51.8
50	79.5	76.2	71.4	67.5	63.2
60	92.0	88.4	83.3	79.1	74.4
70	104.2	100.4	95.0	90.5	85.5
80	116.3	112.3	106.6	101.9	96.6
90	128.3	124.1	118.1	113.1	107.6
100	140.2	135.8	129.6	124.3	118.5

From Robert E. Shannon, *Systems Simulation: The Art and Science*, ©1975, p. 372.
Reprinted by permission of Prentice-Hall, Inc., Englewood Cliffs, N.J.

APPENDIX TABLE III   KOLMOGOROV-SMIRNOV
CRITICAL VALUES

Degrees of Freedom (N)	One-Sample Test*		
	$D_{0.10}$	$D_{0.05}$	$D_{0.01}$
1	0.950	0.975	0.995
2	0.776	0.842	0.929
3	0.642	0.708	0.828

APPENDIX TABLE III (cont.)

Degrees of Freedom (N)	*One-Sample Test**		
	$D_{0.10}$	$D_{0.05}$	$D_{0.01}$
4	0.564	0.624	0.733
5	0.510	0.565	0.669
6	0.470	0.521	0.618
7	0.438	0.486	0.577
8	0.411	0.457	0.543
9	0.388	0.432	0.514
10	0.368	0.410	0.490
11	0.352	0.391	0.468
12	0.338	0.375	0.450
13	0.325	0.361	0.433
14	0.314	0.349	0.418
15	0.304	0.338	0.404
16	0.295	0.328	0.392
17	0.286	0.318	0.381
18	0.278	0.309	0.371
19	0.272	0.301	0.363
20	0.264	0.294	0.356
25	0.24	0.27	0.32
30	0.22	0.24	0.29
35	0.21	0.23	0.27
Over 35	$\frac{1.22}{\sqrt{N}}$	$\frac{1.36}{\sqrt{N}}$	$\frac{1.63}{\sqrt{N}}$

*Used for testing goodness of fit of a sample to a theoretical distribution where N = sample size.

Adapted from Robert E. Shannon, *Systems Simulation: The Art and Science*, © 1975, p. 380, Reprinted by permission of Prentice-Hall, Inc., Englewood Cliffs, N.J.

APPENDIX TABLE IV POISSON PROBABILITIES

	ρ									
x	*0.1*	*0.2*	*0.3*	*0.4*	*0.5*	*0.6*	*0.7*	*0.8*	*0.9*	*1.0*
0	.9048	.8187	.7408	.6703	.6065	.5488	.4966	.4493	.4066	.3679
1	.0905	.1637	.2222	.2681	.3033	.3293	.3476	.3595	.3659	.3679
2	.0045	.0164	.0333	.0536	.0758	.0988	.1217	.1438	.1647	.1839
3	.0002	.0011	.0033	.0072	.0126	.0198	.0284	.0383	.0494	.0613
4	.0000	.0001	.0002	.0007	.0016	.0030	.0050	.0077	.0111	.0153
5	.0000	.0000	.0000	.0001	.0002	.0004	.0007	.0012	.0020	.0031
6	.0000	.0000	.0000	.0000	.0000	.0000	.0001	.0002	.0003	.0005
7	.0000	.0000	.0000	.0000	.0000	.0000	.0000	.0000	.0000	.0001

APPENDIX TABLE IV (cont.)

					ρ					
x	1.1	1.2	1.3	1.4	1.5	1.6	1.7	1.8	1.9	2.0
0	.3329	.3012	.2725	.2466	.2231	.2019	.1827	.1653	.1496	.1353
1	.3662	.3614	.3543	.3452	.3347	.3230	.3106	.2975	.2842	.2707
2	.2014	.2169	.2303	.2417	.2510	.2584	.2640	.2678	.2700	.2707
3	.0738	.0867	.0998	.1128	.1255	.1378	.1496	.1607	.1710	.1804
4	.0203	.0260	.0324	.0395	.0471	.0551	.0636	.0723	.0812	.0902
5	.0045	.0062	.0084	.0111	.0141	.0176	.0216	.0260	.0309	.0361
6	.0008	.0012	.0018	.0026	.0035	.0047	.0061	.0078	.0098	.0120
7	.0001	.0002	.0003	.0005	.0008	.0011	.0015	.0020	.0027	.0034
8	.0000	.0000	.0001	.0001	.0001	.0002	.0003	.0005	.0006	.0009
9	.0000	.0000	.0000	.0000	.0000	.0000	.0001	.0001	.0001	.0002

					ρ					
x	2.1	2.2	2.3	2.4	2.5	2.6	2.7	2.8	2.9	3.0
0	.1225	.1108	.1003	.0907	.0821	.0743	.0672	.0608	.0550	.0498
1	.2572	.2438	.2306	.2177	.2052	.1931	.1815	.1703	.1596	.1494
2	.2700	.2681	.2652	.2613	.2565	.2510	.2450	.2384	.2314	.2240
3	.1890	.1966	.2033	.2090	.2138	.2176	.2205	.2225	.2237	.2240
4	.0992	.1082	.1169	.1254	.1336	.1414	.1488	.1557	.1622	.1680
5	.0417	.0476	.0538	.0602	.0668	.0735	.0804	.0872	.0940	.1008
6	.0146	.0174	.0206	.0241	.0278	.0319	.0362	.0407	.0455	.0504
7	.0044	.0055	.0068	.0083	.0099	.0118	.0139	.0163	.0188	.0216
8	.0011	.0015	.0019	.0025	.0031	.0038	.0047	.0057	.0068	.0081
9	.0003	.0004	.0005	.0007	.0009	.0011	.0014	.0018	.0022	.0027
10	.0001	.0001	.0001	.0002	.0002	.0003	.0004	.0005	.0006	.0008
11	.0000	.0000	.0000	.0000	.0000	.0001	.0001	.0001	.0002	.0002
12	.0000	.0000	.0000	.0000	.0000	.0000	.0000	.0000	.0000	.0001

					ρ					
x	3.1	3.2	3.3	3.4	3.5	3.6	3.7	3.8	3.9	4.0
0	.0450	.0408	.0369	.0334	.0302	.0273	.0247	.0224	.0202	.0183
1	.1397	.1304	.1217	.1135	.1057	.0984	.0915	.0850	.0789	.0733
2	.2165	.2087	.2008	.1929	.1850	.1771	.1692	.1615	.1539	.1465
3	.2237	.2226	.2209	.2186	.2158	.2125	.2087	.2046	.2001	.1954
4	.1734	.1781	.1823	.1858	.1888	.1912	.1931	.1944	.1951	.1954
5	.1075	.1140	.1203	.1264	.1322	.1377	.1429	.1477	.1522	.1563
6	.0555	.0608	.0662	.0716	.0771	.0826	.0881	.0936	.0989	.1042
7	.0246	.0278	.0312	.0348	.0385	.0425	.0466	.0508	.0551	.0595
8	.0095	.0111	.0129	.0148	.0169	.0191	.0215	.0241	.0269	.0298
9	.0033	.0040	.0047	.0056	.0066	.0076	.0089	.0102	.0116	.0132
10	.0010	.0013	.0016	.0019	.0023	.0028	.0033	.0039	.0045	.0053
11	.0003	.0004	.0005	.0006	.0007	.0009	.0011	.0013	.0016	.0019
12	.0001	.0001	.0001	.0002	.0002	.0003	.0003	.0004	.0005	.0006
13	.0000	.0000	.0000	.0000	.0001	.0001	.0001	.0001	.0002	.0002
14	.0000	.0000	.0000	.0000	.0000	.0000	.0000	.0000	.0000	.0001

APPENDIX TABLE IV (cont.)

					ρ					
x	4.1	4.2	4.3	4.4	4.5	4.6	4.7	4.8	4.9	5.0
0	.0166	.0150	.0136	.0123	.0111	.0101	.0091	.0082	.0074	.0067
1	.0679	.0630	.0583	.0540	.0500	.0462	.0427	.0395	.0365	.0337
2	.1393	.1323	.1254	.1188	.1125	.1063	.1005	.0948	.0894	.0842
3	.1904	.1852	.1798	.1743	.1687	.1631	.1574	.1517	.1460	.1404
4	.1951	.1944	.1933	.1917	.1898	.1875	.1849	.1820	.1789	.1755
5	.1600	.1633	.1662	.1687	.1708	.1725	.1738	.1747	.1753	.1755
6	.1093	.1143	.1191	.1237	.1281	.1323	.1362	.1398	.1432	.1462
7	.0640	.0686	.0732	.0778	.0824	.0869	.0914	.0959	.1002	.1044
8	.0328	.0360	.0393	.0428	.0463	.0500	.0537	.0575	.0614	.0653
9	.0150	.0168	.0188	.0209	.0232	.0255	.0280	.0307	.0334	.0363
10	.0061	.0071	.0081	.0092	.0104	.0118	.0132	.0147	.0164	.0181
11	.0023	.0027	.0032	.0037	.0043	.0049	.0056	.0064	.0073	.0082
12	.0008	.0009	.0011	.0014	.0016	.0019	.0022	.0026	.0030	.0034
13	.0002	.0003	.0004	.0005	.0006	.0007	.0008	.0009	.0011	.0013
14	.0001	.0001	.0001	.0001	.0002	.0002	.0003	.0003	.0004	.0005
15	.0000	.0000	.0000	.0000	.0001	.0001	.0001	.0001	.0001	.0002

					ρ					
x	5.1	5.2	5.3	5.4	5.5	5.6	5.7	5.8	5.9	6.0
0	.0061	.0055	.0050	.0045	.0041	.0037	.0033	.0030	.0027	.0025
1	.0311	.0287	.0265	.0244	.0225	.0207	.0191	.0176	.0162	.0149
2	.0793	.0746	.0701	.0659	.0618	.0580	.0544	.0509	.0477	.0446
3	.1348	.1293	.1239	.1185	.1133	.1082	.1033	.0985	.0938	.0892
4	.1719	.1681	.1641	.1600	.1558	.1515	.1472	.1428	.1383	.1339
5	.153	.1748	.1740	.1728	.1714	.1697	.1678	.1656	.1632	.1603
6	.1490	.1515	.1537	.1555	.1571	.1584	.1594	.1601	.1605	.1605
7	.1086	.1125	.1163	.1200	.1234	.1267	.1298	.1326	.1353	.1377
8	.0692	.0731	.0771	.0810	.0849	.0887	.0925	.0962	.0998	.1033
9	.0392	.0423	.0454	.0486	.0519	.0552	.0586	.0620	.0654	.0688
10	.0200	.0220	.0241	.0262	.0285	.0309	.0334	.0359	.0386	.0413
11	.0093	.0104	.0116	.0129	.0143	.0157	.0173	.0190	.0207	.0225
12	.0039	.0045	.0051	.0058	.0065	.0073	.0082	.0092	.0102	.0113
13	.0015	.0018	.0021	.0024	.0028	.0032	.0036	.0041	.0046	.0052
14	.0006	.0007	.0008	.0009	.0011	.0013	.0015	.0017	.0019	.0022
15	.0002	.0002	.0003	.0003	.0004	.0005	.0006	.0007	.0008	.0009
16	.0001	.0001	.0001	.0001	.0001	.0002	.0002	.0002	.0003	.0003
17	.0000	.0000	.0000	.0000	.0000	.0001	.0001	.0001	.0001	.0001

					ρ					
x	6.1	6.2	6.3	6.4	6.5	6.6	6.7	6.8	6.9	7.0
0	.0022	.0020	.0018	.0017	.0015	.0014	.0012	.0011	.0010	.0009
1	.0137	.0126	.0116	.0106	.0098	.0090	.0082	.0076	.0070	.0064
2	.0417	.0390	.0364	.0340	.0318	.0296	.0276	.0258	.0240	.0223
3	.0848	.0806	.0765	.0726	.0688	.0652	.0617	.0584	.0552	.0521
4	.1294	.1249	.1205	.1162	.1118	.1076	.1034	.0992	.0952	.0912
5	.1579	.1549	.1519	.1487	.1454	.1420	.1385	.1349	.1314	.1277

APPENDIX TABLE IV (cont.)

x	6.1	6.2	6.3	6.4	6.5	6.6	6.7	6.8	6.9	7.0
6	.1605	.1601	.1595	.1586	.1575	.1562	.1546	.1529	.1511	.1490
7	.1399	.1418	.1435	.1450	.1462	.1472	.1480	.1486	.1489	.1490
8	.1066	.1099	.1130	.1160	.1188	.1215	.1240	.1263	.1284	.1304
9	.0723	.0757	.0791	.0825	.0858	.0891	.0923	.0954	.0985	.1014
10	.0441	.0469	.0498	.0528	.0558	.0588	.0618	.0649	.0679	.0710
11	.0245	.0265	.0285	.0307	.0330	.0353	.0377	.0401	.0426	.0452
12	.0124	.0137	.0150	.0164	.0179	.0194	.0210	.0227	.0245	.0264
13	.0058	.0065	.0073	.0081	.0089	.0098	.0108	.0119	.0130	.0142
14	.0025	.0029	.0033	.0037	.0041	.0046	.0052	.0058	.0064	.0071
15	.0010	.0012	.0014	.0016	.0018	.0020	.0023	.0026	.0029	.0033
16	.0004	.0005	.0005	.0006	.0007	.0008	.0010	.0011	.0013	.0014
17	.0001	.0002	.0002	.0002	.0003	.0003	.0004	.0004	.0005	.0006
18	.0000	.0001	.0001	.0001	.0001	.0001	.0001	.0002	.0002	.0002
19	.0000	.0000	.0000	.0000	.0000	.0000	.0000	.0001	.0001	.0001

x	7.1	7.2	7.3	7.4	7.5	7.6	7.7	7.8	7.9	8.0
0	.0008	.0007	.0007	.0006	.0006	.0005	.0005	.0004	.0004	.0003
1	.0059	.0054	.0049	.0045	.0041	.0038	.0035	.0032	.0029	.0027
2	.0208	.0194	.0180	.0167	.0156	.0145	.0134	.0125	.0116	.0107
3	.0492	.0464	.0438	.0413	.0389	.0366	.0345	.0324	.0305	.0286
4	.0874	.0836	.0799	.0764	.0729	.0696	.0663	.0632	.0602	.0573
5	.1241	.1204	.1167	.1130	.1094	.1057	.1021	.0986	.0951	.0916
6	.1468	.1445	.1420	.1394	.1367	.1339	.1311	.1282	.1252	.1221
7	.1489	.1486	.1481	.1474	.1465	.1454	.1442	.1428	.1413	.1396
8	.1321	.1337	.1351	.1363	.1373	.1382	.1388	.1392	.1395	.1396
9	.1042	.1070	.1096	.1121	.1144	.1167	.1187	.1207	.1224	.1241
10	.0740	.0770	.0800	.0829	.0858	.0887	.0914	.0941	.0967	.0993
11	.0478	.0504	.0531	.0558	.0585	.0613	.0640	.0667	.0695	.0722
12	.0283	.0303	.0323	.0344	.0366	.0388	.0411	.0434	.0457	.0481
13	.0154	.0168	.0181	.0196	.0211	.0227	.0243	.0260	.0278	.0296
14	.0078	.0086	.0095	.0104	.0113	.0123	.0134	.0145	.0157	.0169
15	.0037	.0041	.0046	.0051	.0057	.0062	.0069	.0075	.0083	.0090
16	.0016	.0019	.0021	.0024	.0026	.0030	.0033	.0037	.0041	.0045
17	.0007	.0008	.0009	.0010	.0012	.0013	.0015	.0017	.0019	.0021
18	.0003	.0003	.0004	.0004	.0005	.0006	.0006	.0007	.0008	.0009
19	.0001	.0001	.0001	.0002	.0002	.0002	.0003	.0003	.0003	.0004
20	.0000	.0000	.0001	.0001	.0001	.0001	.0001	.0001	.0001	.0002
21	.0000	.0000	.0000	.0000	.0000	.0000	.0000	.0000	.0001	.0001

x	8.1	8.2	8.3	8.4	8.5	8.6	8.7	8.8	8.9	9.0
0	.0003	.0003	.0002	.0002	.0002	.0002	.0002	.0002	.0001	.0001
1	.0025	.0023	.0021	.0019	.0017	.0016	.0014	.0013	.0012	.0011
2	.0100	.0092	.0086	.0079	.0074	.0068	.0063	.0058	.0054	.0050
3	.0269	.0252	.0237	.0222	.0208	.0195	.0183	.0171	.0160	.0150
4	.0544	.0517	.0491	.0466	.0443	.0420	.0398	.0377	.0357	.0337

APPENDIX TABLE IV (cont.)

					ρ					
x	8.1	8.2	8.3	8.4	8.5	8.6	8.7	8.8	8.9	9.0
5	.0882	.0849	.0816	.0784	.0752	.0722	.0692	.0663	.0635	.0607
6	.1191	.1160	.1128	.1097	.1066	.1034	.1003	.0972	.0941	.0911
7	.1378	.1358	.1338	.1317	.1294	.1271	.1247	.1222	.1197	.1171
8	.1395	.1392	.1388	.1382	.1375	.1366	.1356	.1344	.1332	.1318
9	.1256	.1269	.1280	.1290	.1299	.1306	.1311	.1315	.1317	.1318
10	.1017	.1040	.1063	.1084	.1104	.1123	.1140	.1157	.1172	.1186
11	.0749	.0776	.0802	.0828	.0853	.0878	.0902	.0925	.0948	.0970
12	.0505	.0530	.0555	.0579	.0604	.0629	.0654	.0679	.0703	.0728
13	.0315	.0334	.0354	.0374	.0395	.0416	.0438	.0459	.0481	.0504
14	.0182	.0196	.0210	.0225	.0240	.0256	.0272	.0289	.0306	.0324
15	.0098	.0107	.0116	.0126	.0136	.0147	.0158	.0169	.0182	.0194
16	.0050	.0055	.0060	.0066	.0072	.0079	.0086	.0093	.0101	.0109
17	.0024	.0026	.0029	.0033	.0036	.0040	.0044	.0048	.0053	.0058
18	.0011	.0012	.0014	.0015	.0017	.0019	.0021	.0024	.0026	.0029
19	.0005	.0005	.0006	.0007	.0008	.0009	.0010	.0011	.0012	.0014
20	.0002	.0002	.0002	.0003	.0003	.0004	.0004	.0005	.0005	.0006
21	.0001	.0001	.0001	.0001	.0001	.0002	.0002	.0002	.0002	.0003
22	.0000	.0000	.0000	.0000	.0001	.0001	.0001	.0001	.0001	.0001

					ρ					
x	9.1	9.2	9.3	9.4	9.5	9.6	9.7	9.8	9.9	10
0	.0001	.0001	.0001	.0001	.0001	.0001	.0001	.0001	.0001	.0000
1	.0010	.0009	.0009	.0008	.0007	.0007	.0006	.0005	.0005	.0005
2	.0046	.0043	.0040	.0037	.0034	.0031	.0029	.0027	.0025	.0023
3	.0140	.0131	.0123	.0115	.0107	.0100	.0093	.0087	.0081	.0076
4	.0319	.0302	.0285	.0269	.0254	.0240	.0226	.0213	.0201	.0189
5	.0581	.0555	.0530	.0506	.0483	.0460	.0439	.0418	.0398	.0378
6	.0881	.0851	.0822	.0793	.0764	.0736	.0709	.0682	.0656	.0631
7	.1145	.1118	.1091	.1064	.1037	.1010	.0982	.0955	.0928	.0901
8	.1302	.1286	.1269	.1251	.1232	.1212	.1191	.1170	.1148	.1126
9	.1317	.1315	.1311	.1306	.1300	.1293	.1284	.1274	.1263	.1251
10	.1198	.1210	.1219	.1228	.1235	.1241	.1245	.1249	.1250	.1251
11	.0991	.1012	.1031	.1049	.1067	.1083	.1098	.1112	.1125	.1137
12	.0752	.0776	.0799	.0822	.0844	.0866	.0888	.0908	.0928	.0948
13	.0526	.0549	.0572	.0594	.0617	.0640	.0662	.0685	.0707	.0729
14	.0342	.0361	.0380	.0399	.0419	.0439	.0459	.0479	.0500	.0521
15	.0208	.0221	.0235	.0250	.0265	.0281	.0297	.0313	.0330	.0347
16	.0118	.0127	.0137	.0147	.0157	.0168	.0180	.0192	.0204	.0217
17	.0063	.0069	.0075	.0081	.0088	.0095	.0103	.0111	.0119	.0128
18	.0032	.0035	.0039	.0042	.0046	.0051	.0055	.0060	.0065	.0071
19	.0015	.0017	.0019	.0021	.0023	.0026	.0028	.0031	.0034	.0037
20	.0007	.0008	.0009	.0010	.0011	.0012	.0014	.0015	.0017	.0019
21	.0003	.0003	.0004	.0004	.0005	.0006	.0006	.0007	.0008	.0009
22	.0001	.0001	.0002	.0002	.0002	.0002	.0003	.0003	.0004	.0004
23	.0000	.0001	.0001	.0001	.0001	.0001	.0001	.0001	.0002	.0002
24	.0000	.0000	.0000	.0000	.0000	.0000	.0000	.0001	.0001	.0001

APPENDIX TABLE IV (cont.)

					ρ					
x	11	12	13	14	15	16	17	18	19	20
0	.0000	.0000	.0000	.0000	.0000	.0000	.0000	.0000	.0000	.0000
1	.0002	.0001	.0000	.0000	.0000	.0000	.0000	.0000	.0000	.0000
2	.0010	.0004	.0002	.0001	.0000	.0000	.0000	.0000	.0000	.0000
3	.0037	.0018	.0008	.0004	.0002	.0001	.0000	.0000	.0000	.0000
4	.0102	.0053	.0027	.0013	.0006	.0003	.0001	.0001	.0000	.0000
5	.0224	.0127	.0070	.0037	.0019	.0010	.0005	.0002	.0001	.0001
6	.0411	.0255	.0152	.0087	.0048	.0026	.0014	.0007	.0004	.0002
7	.0646	.0437	.0281	.0174	.0104	.0060	.0034	.0018	.0010	.0005
8	.0888	.0655	.0457	.0304	.0194	.0120	.0072	.0042	.0024	.0013
9	.1085	.0874	.0661	.0473	.0324	.0213	.0135	.0083	.0050	.0029
10	.1194	.1048	.0859	.0663	.0486	.0341	.0230	.0150	.0095	.0058
11	.1194	.1144	.1015	.0844	.0663	.0496	.0355	.0245	.0164	.0106
12	.1094	.1144	.1099	.0984	.0829	.0661	.0504	.0368	.0259	.0176
13	.0926	.1056	.1099	.1060	.0956	.0814	.0658	.0509	.0378	.0271
14	.0728	.0905	.1021	.1060	.1024	.0930	.0800	.0655	.0514	.0387
15	.0534	.0724	.0885	.0989	.1024	.0992	.0906	.0786	.0650	.0516
16	.0367	.0543	.0719	.0866	.0960	.0992	.0963	.0884	.0772	.0646
17	.0237	.0383	.0550	.0713	.0847	.0934	.0963	.0936	.0863	.0760
18	.0145	.0256	.0397	.0554	.0706	.0830	.0909	.0936	.0911	.0844
19	.0084	.0161	.0272	.0409	.0557	.0699	.0814	.0887	.0911	.0888
20	.0046	.0097	.0177	.0286	.0418	.0559	.0692	.0798	.0866	.0888
21	.0024	.0055	.0109	.0191	.0299	.0426	.0560	.0684	.0783	.0846
22	.0012	.0030	.0065	.0121	.0204	.0310	.0433	.0560	.0676	.0769
23	.0006	.0016	.0037	.0074	.0133	.0216	.0320	.0438	.0559	.0669
24	.0003	.0008	.0020	.0043	.0083	.0144	.0226	.0328	.0442	.0557
25	.0001	.0004	.0010	.0024	.0050	.0092	.0154	.0237	.0336	.0446
26	.0000	.0002	.0005	.0013	.0029	.0057	.0101	.0164	.0246	.0343
27	.0000	.0001	.0002	.0007	.0016	.0034	.0063	.0109	.0173	.0254
28	.0000	.0000	.0001	.0003	.0009	.0019	.0038	.0070	.0117	.0181
29	.0000	.0000	.0001	.0002	.0004	.0011	.0023	.0044	.0077	.0125
30	.0000	.0000	.0000	.0001	.0002	.0006	.0013	.0026	.0049	.0083
31	.0000	.0000	.0000	.0000	.0001	.0003	.0007	.0015	.0030	.0054
32	.0000	.0000	.0000	.0000	.0001	.0001	.0004	.0009	.0018	.0034
33	.0000	.0000	.0000	.0000	.0000	.0001	.0002	.0005	.0010	.0020
34	.0000	.0000	.0000	.0000	.0000	.0000	.0001	.0002	.0006	.0012
35	.0000	.0000	.0000	.0000	.0000	.0000	.0000	.0001	.0003	.0007
36	.0000	.0000	.0000	.0000	.0000	.0000	.0000	.0001	.0002	.0004
37	.0000	.0000	.0000	.0000	.0000	.0000	.0000	.0000	.0001	.0002
38	.0000	.0000	.0000	.0000	.0000	.0000	.0000	.0000	.0000	.0001
39	.0000	.0000	.0000	.0000	.0000	.0000	.0000	.0000	.0000	.0001

From Herbert Moskowitz and Gordon P. Wright, *Operations Research Techniques for Management*, © 1979, pp. 761–766. Reprinted by permission of Prentice-Hall, Inc., Englewood Cliffs, N.J.

APPENDIX TABLE V   VALUES OF $e^{-\lambda}$

λ	$e^{-\lambda}$	λ	$e^{-\lambda}$	λ	$e^{-\lambda}$	λ	$e^{-\lambda}$
0.1	0.90484	2.6	0.07427	5.1	0.00610	7.6	0.00050
0.2	0.81873	2.7	0.06721	5.2	0.00552	7.7	0.00045
0.3	0.74082	2.8	0.06081	5.3	0.00499	7.8	0.00041
0.4	0.67032	2.9	0.05502	5.4	0.00452	7.9	0.00037
0.5	0.60653	3.0	0.04979	5.5	0.00409	8.0	0.00034
0.6	0.54881	3.1	0.04505	5.6	0.00370	8.1	0.00030
0.7	0.49659	3.2	0.04076	5.7	0.00335	8.2	0.00027
0.8	0.44933	3.3	0.03688	5.8	0.00303	8.3	0.00025
9.0	0.40657	3.4	0.03337	5.9	0.00274	8.4	0.00022
1.0	0.36788	3.5	0.03020	6.0	0.00248	8.5	0.00020
1.1	0.33287	3.6	0.02732	6.1	0.00224	8.6	0.00018
1.2	0.30119	3.7	0.02472	6.2	0.00203	8.7	0.00017
1.3	0.27253	3.8	0.02237	6.3	0.00184	8.8	0.00015
1.4	0.24660	3.9	0.02024	6.4	0.00166	8.9	0.00014
1.5	0.22313	4.0	0.01832	6.5	0.00150	9.0	0.00012
1.6	0.20190	4.1	0.01657	6.6	0.00136	9.1	0.00011
1.7	0.18268	4.2	0.01500	6.7	0.00123	9.2	0.00010
1.8	0.16530	4.3	0.01357	6.8	0.00111	9.3	0.00009
1.9	0.14957	4.4	0.01228	6.9	0.00101	9.4	0.00008
2.0	0.13534	4.5	0.01111	7.0	0.00091	9.5	0.00007
2.1	0.12246	4.6	0.01005	7.1	0.00083	9.6	0.00007
2.2	0.11080	4.7	0.00910	7.2	0.00075	9.7	0.00006
2.3	0.10026	4.8	0.00823	7.3	0.00068	9.8	0.00006
2.4	0.09072	4.9	0.00745	7.4	0.00061	9.9	0.00005
2.5	0.08208	5.0	0.00674	7.5	0.00055	10.0	0.00005

From Richard I. Levin, *Statistics for Management*, 2nd ed., © 1981, p. 757. Reprinted by permission of Prentice-Hall, Inc., Englewood Cliffs, N.J.

APPENDIX TABLE VI   FORMULAS FOR SELECTED *M/M/*1 AND *M/M/c* MODELS

Definition of Symbols

n	=	Number of units in system.
λ	=	Mean arrival rate (units per time period).
μ	=	Mean service rate (units per time period).
ρ	=	Traffic intensity (λ/μ).
N	=	Maximum number allowed in system.
m	=	Number of units in finite population.
c	=	Number of servers.
P_n	=	Probability of n units in system.
L_s	=	Steady-state mean number of units in system.
L_q	=	Steady-state mean number of units in queue.
L_b	=	Steady-state mean number of units in queue for busy system.
W_s	=	Steady-state mean time in system.
W_q	=	Steady-state mean time in queue.
W_b	=	Steady-state mean time in queue for busy system.

APPENDIX TABLE VI (cont.)

Finite Queue M/M/1 Model[1]

$$P_o = \begin{cases} \dfrac{1 - \rho}{1 - \rho^{N+1}} & \text{for } \lambda \neq \mu \\[2mm] \dfrac{1}{N + 1} & \text{for } \lambda \neq \mu \end{cases}$$

$$P(n > 0) = 1 - P_o$$

$$P_n = P_o \rho^n \quad \text{for } n \leq N$$

$$L_s = \begin{cases} \dfrac{\rho}{1 - \rho} - \dfrac{(N + 1)\rho^{N+1}}{1 - \rho^{N+1}} & \text{for } \lambda \neq \mu \\[3mm] \dfrac{N}{2} & \text{for } \lambda \neq \mu \end{cases}$$

$$L_q = L_s - (1 - P_o)$$

$$L_b = \dfrac{L_q}{1 - P_o}$$

$$W_s = \dfrac{L_q}{\lambda(1 - P_N)} + \dfrac{1}{\mu}$$

$$W_q = W_s - \dfrac{1}{\mu}$$

$$W_b = \dfrac{W_q}{1 - P_o}.$$

Finite Population M/M/1 Model[2]

$$P_o = \dfrac{1}{\displaystyle\sum_{i=0}^{m} \left[\dfrac{m!}{(m - i)!} \cdot \rho^i \right]}$$

$$P(n > 0) = 1 - P_o$$

$$P_n = \dfrac{m!}{(m - n)!} \rho^n P_o, \quad n \leq m$$

APPENDIX TABLE VI (cont.)

$$\textit{Finite Population M/M/1 Model}^2$$

$$L_s = m - \frac{1}{\rho}(1 - P_o)$$

$$L_q = m - \frac{(\lambda + \mu)(1 - P_o)}{\lambda}$$

$$L_b = \frac{L_q}{1 - P_o}$$

$$W_s = \frac{m}{\mu(1 - P_o)} - \frac{1}{\lambda}$$

$$W_q = \frac{1}{\mu}\left[\frac{m}{1 - P_o} - \frac{\lambda + \mu}{\lambda}\right]$$

$$W_b = \frac{W_q}{1 - P_o}.$$

$$\textit{Standard M/M/c Model}^3$$

$$P_o = \frac{1}{\left[\displaystyle\sum_{i=0}^{c-1}\frac{\rho^i}{i!}\right] + \dfrac{\rho^c}{c!\left[1 - \dfrac{\rho}{c}\right]}}$$

$$P_n = \begin{cases} \dfrac{\rho^n}{n!} \cdot P_o & \text{for } 0 \le n \le c \\[2ex] \left[\dfrac{\rho^n}{c!\,c^{n-c}}\right] \cdot P_o & \text{for } n \ge c \end{cases}$$

$$P(n \ge c) = \frac{\rho^c \mu c}{c!(\mu c - \lambda)} \cdot P_o$$

$$L_s = \frac{\rho^{c+1}}{(c-1)!(c-\rho)^2} \cdot P_o + \rho$$

$$L_q = L_s - \rho$$

$$L_b = \frac{L_q}{P(n \ge c)}$$

APPENDIX TABLE VI (cont.)

Standard M/M/c Model[3]

$$W_s = \frac{L_q}{\lambda} + \frac{1}{\mu}$$

$$W_q = \frac{L_q}{\lambda}$$

$$W_b = \frac{W_q}{P(n \geq c)}$$

Finite Queue M/M/c Model[4]

$$P_o = \cfrac{1}{\left[\displaystyle\sum_{i=0}^{c} \frac{\rho^i}{i!}\right] + \left[\frac{1}{c!}\right] \cdot \left[\displaystyle\sum_{i=c+1}^{N} \frac{\rho^i}{c^{i-c}}\right]}$$

$$P_n = \begin{cases} \dfrac{\rho^n}{n!} \cdot P_o & \text{for } 0 \leq n \leq c \\[3mm] \dfrac{\rho^n}{c!c^{n-c}} \cdot P_o & \text{for } c \leq n \leq N \end{cases}$$

$$P(n \geq c) = 1 - P_o \sum_{i=0}^{c-1} \frac{\rho^i}{i!}$$

$$L_s = \frac{P_o \rho^{c+1}}{(c-1)!(c-\rho)^2} \left[1 - \left[\frac{\rho}{c}\right]^{N-c} - (N-c)\left[\frac{\rho}{c}\right]^{N-c} \cdot \left[1 - \frac{\rho}{c}\right]\right]$$

$$+ \rho(1 - P_N)$$

$$L_q = L_s - \rho(1 - P_N)$$

$$L_b = \frac{L_q}{P(n \geq c)}$$

$$W_s = \frac{L_q}{\lambda(1 - P_N)} + \frac{1}{\mu}$$

$$W_q = W_s - \frac{1}{\mu}$$

$$W_b = \frac{W_q}{P(n \geq c)}$$

APPENDIX TABLE VI (cont.)

Finite Population M/M/c Model[5]

$$P_o = \frac{1}{\left[\displaystyle\sum_{i=0}^{c} \frac{m!}{(m-i)!i!} \cdot \rho^i\right] + \left[\displaystyle\sum_{i=c+1}^{m} \frac{m!}{(m-i)!c!c^{i-c}} \cdot \rho^i\right]}$$

$$P_n = \begin{cases} \dfrac{m!P_o\rho^n}{(m-n)!n!} & \text{for } 0 \leq n \leq c \\[2ex] \dfrac{m!P_o\rho^n}{(m-n)!c!c^{n-c}} & \text{for } c \leq n \leq m \end{cases}$$

$$P(n \geq c) = 1 - P_o \sum_{i=0}^{c-1} \frac{m!}{(m-i)!i!} \cdot \rho^i$$

$$L_s = \frac{Lq + m\rho}{1 + \rho}$$

$$L_q = \sum_{n=c+1}^{m} (n-c)P_n$$

$$L_b = \frac{L_q}{P(n \geq c)}$$

$$W_s = \frac{L_s}{\lambda(m - L_s)}$$

$$W_q = \frac{L_q}{\lambda(m - L_s)}$$

$$W_b = \frac{W_q}{P(n \geq c)}$$

Self-service Model[6]

$$P_n = \frac{e^{-\rho}}{n!} \cdot \rho^n \qquad \text{for } n \geq 0$$

$$L_s = \rho$$

$$W_s = \frac{1}{\mu}$$

From F. S. Budnick, R. Mojena, and T. E. Vollmann, *Principles of Operations Research for Management*, © 1977, pp. 472–474. Reprinted by permission of Richard D. Irwin, Inc., Homewood, Ill.
[1]Note: $0 < \rho < \infty$. [2]Note: $0 < \rho < \infty$. [3]Note: $0 < \rho < c$. [4]Note: $0 < \rho < \infty$.
[5]Note: $0 < \rho < \infty$. [6]Note: $0 < \rho < \infty$.

APPENDIX TABLE VII STANDARD NUMERICAL ATTRIBUTES (SNAs)*

Entity	SNA	Definition
Blocks	N_j	The count of the total number of transactions to enter block j.
	W_j	The count of the number of transactions currently waiting at block j.
Facilities	F_j	The in-use status of facility j (0 if not in use, 1 if in use).
	FR_j	Utilization in parts per thousand of facility j (range is 0 to 999).
	FC_j	Total number of transactions to enter facility j.
	FT_j	Average transaction utilization time for facility j.
Functions	FN_j	The computed value of function j.
Groups	G_j	The current number of members of group j.
Savevalues	X_j or XF_j	The current contents of full-word savevalue j.
	XH_j	The current contents of half-word savevalue j.
	XB_j	The current contents of byte savevalue j.
	XL_j	The current contents of floating-point savevalue j.
Matrix savevalues	$MX_j(a, b)$	The current contents of full-word matrix savevalue j, row a, column b.
	$MH_j(a, b)$	The current contents of half-word matrix savevalue j, row a, column b.
	$MB_j(a, b)$	The current contents of byte matrix savevalue j, row a, column b.
	$ML_j(a, b)$	The current contents of floating-point matrix savevalue j, row a, column b.
Storages	S_j	The current contents of storage j.
	R_j	Number of available units or capacity remaining of storage j.
	SR_j	Utilization in parts per thousand of storage j (range is 0 to 999).
	SA_j	Average contents of storage j.
	SM_j	Maximum contents of storage j.
	SC_j	Total number of units to enter storage j.
	ST_j	Average utilization per unit of storage j.
Queues	Q_j	The current length or number of units in queue j.
	QA_j	Average length or number of units in queue j.
	QM_j	Maximum length or contents of queue j.
	QC_j	Total number of units to enter queue j.
	QZ_j	Number of units spending zero time in queue j.

APPENDIX TABLE VII (cont.)

	QT_j	Average time each unit (including zero time units) spent in queue j.
	QX_j	Average time each unit (excluding zero time units) spent in queue j.
System attributes	$C\,1$	The current value of the relative simulator clock. Clock time relative to last RESET CLEAR operation.
	$AC\,1$	The current value of the absolute simulator clock. Clock time since start of run or last CLEAR operation.
	RN_j	A computed random number with range 0 to 999, unless used as the argument of a function (in that case, a fractional value between 0 and 0.999999 inclusive).
	$TG\,1$	The number of terminations remaining in the model to satisfy the START count.
Tables	TB_j	The mean value of table j.
	TC_j	Total number of entries in table j.
	TD_j	Standard deviation of table j.
Transactions	PF_j	The current contents of full-word parameter j of the transaction currently being processed.
	PH_j	The current contents of half-word parameter j of the transaction currently being processed.
	PB_j	The current contents of byte parameter j of the transaction currently being processed.
	PL_j	The current contents of floating-point parameter j of the transaction currently being processed.
	$M\,1$	The transit time of the transaction currently being processed.
	MP_j	The intermediate transit time of the transaction currently being processed.
	PR	The priority of the transaction currently being processed.
Variables and Boolean variables	V_j	The computed value of variable j.
	BV_j	The computed value of Boolean variable j.
User chains	CH_j	The current count of the number of transactions on user chain j.
	CA_j	The average number of transactions on user chain j.
	CM_j	The maximum number of transactions on user chain j.
	CC_j	The total number of transactions on user chain j.
	CT_j	The average time per transaction on user chain j.

Adapted from *General Purpose Simulation System V User's Manual*, SH20-0866-1, pp. 90–110, including updates by TNLs SN20-3582, SN20-2473, and SN20-3045, by permission of IBM Corp., White Plains, N.Y.

*Not all versions of GPSS contain every SNA shown. In some versions the SNA may have a different grammatical structure. Check your system's user's manual for further information.

APPENDIX TABLE VII BLOCK STATEMENT FORMATS

Operation	A	B	C	D	E	F	G	H	I
ADVANCE	Mean time	Spread							
ALTER G GE L LE E NE MIN MAX	Group no.	Count	Member attribute to be altered	Value to replace attribute	Matching member attribute	Matching SNA	Alternate exit		
ASSEMBLE	No. of transactions to assemble								
ASSIGN	Parameter no. or range	SNA value to be assigned	No. of function modifier	Parameter type	Note: The parameter type operand may be coded as the C operand if a function modifier is not specified.				
BUFFER									
CHANGE	"From" block no.	"To" block no.							
COUNT G,GE L,LE E,NE U,NU I,NI SNE,SE SNF,SF LR,LS	Parameter in which to place count	Lower limit	Upper limit	Comparison value if conditional operator is specified	Mnemonic of SNA to be counted				
DEPART	Queue no.	No. of units							
ENTER	Storage no.	No. of units							

[ ] indicates optional operand. { } indicates that one of the items within the braces must be chosen.

APPENDIX TABLE VIII (cont.)

Operation	A	B	C	D	E	F	G	H	I
EXAMINE	Group no.	Numeric value— numeric mode	Alternate exit						
EXECUTE	Block no.								
FAVAIL	Facility no. or range								
FUNAVAIL	Facility no. or range	Remove or continue option	Alternate block no.	Parameter no.	Remove or continue option	Alternate block no.	Remove or continue option	Alternate block no.	
			Options for controlling transactions		Options for preempted transactions		Options for delayed transactions		
GATE LS LR	Logic switch no.	Next block if condition is false							
GATE NI I NU U FV FNV	Facility no.	Next block if condition is false							
GATE SE SF SNE SNF SV SNV	Storage no.	Next block if condition is false							

APPENDIX TABLE VIII (cont.)

Operation	A	B	C	D	E	F	G	H	I
GATE M NM	Match block no.	Next block if condition is false							
GATHER	No. of transactions to be gathered								
GENERATE	Mean time	Spread	Initialization interval	Creation limit	Priority level				
HELP HELPA HELPB HELPC HELPAPL1 HELPBPL1 HELPCPL1	Help routine name	B – G operands SNA values to be passed to help routine							Full-word, half-word, byte, and floating-point parameters in any sequence
INDEX	Parameter no.	Increment							
JOIN	Group no.	Numeric value – numeric mode							
LEAVE	Storage no.	No. of units							
LINK	User chain no.	Ordering of chain, LIFO, FIFO, or parameter no.	Alternate block exit						
LOGIC S R I	Logic switch no.								

391

APPENDIX TABLE VIII (cont.)

Operation	A	B	C	D	E	F	G	H	I
LOOP	Parameter no.	Next block if $P_x \neq 0$							
MARK	Parameter no.								
MATCH	Conjugate MATCH block no.								
MSAVEVALUE	Matrix no. or range	Row no. or range	Column no. or range	SNA value to be saved	MSAVEVALUE type				
PREEMPT	Facility no.	Priority option	Block no. for preempted transaction	Parameter no. of preempted transaction	Remove option				
PRINT	Lower limit	Upper limit	Entity mnemonic	Paging indicator					
PRIORITY	Priority no.	Buffer option							
QUEUE	Queue no.	No. of units							
RELEASE	Facility no.								
REMOVE G,GE L,LE E,NE MIN MAX	Group no.	Count no. of members to be removed—transaction mode	Numeric value to be removed—numeric mode	Member attribute for comparison—transaction mode	Comparison SNA	Alternate exit for entering transaction			
RETURN	Facility no.								
SAVAIL	Storage no. or range								
SAVEVALUE	Savevalue no. or range	SNA value to be saved	Savevalue type						

APPENDIX TABLE VIII (cont.)

Operation	A	B	C	D	E	F	G	H	I
SCAN G GE L LE E NE MIN MAX	Group no.	Member attribute for comparison	Comparison value for B operand		Member attribute to be obtained if match is made	Entering transaction parameter no. in which to place D operand value		Alternate exit	
SEIZE	Facility no.								
SELECT G,GE L,LE E,NE U,NU I,NI SE,SNE SF,SNF LR,LS MIN,MAX	Parameter in which to place entity no.	Lower limit	Upper limit		Comparison value if conditional operator is specified	SN mnemonic to be examined if conditional		Alternate exit	
SPLIT	No. of copies	Next block for copies	Parameter for serial numbering	No. of full-word, half-word, byte, and floating-point parameters in any order					
SUNAVAIL	Storage no. or range								
TABULATE	Table no.	Weighting factor							
TERMINATE	Termination count								
TEST E,NE G,GE L,LE	First SNA	Second SNA	Next block if relation is false						
TRACE									

APPENDIX TABLE VIII (cont.)

Operation	A	B	C	D	E	F	G	H	I
TRANSFER	Selection mode	Next block A	Next block B	Indexing factor					
UNLINK G,GE L,LE E,NE	User chain no.	Next block for the unlinked transaction	Transaction unlink count	Parameter no.	Match argument	Next block for unlinking transaction if no transactions unlinked			
UNTRACE									
WRITE	Jobtape no.								
AUXILIARY	Entity mnemonic	Total entity allocation	No. of entity type to reside in core	No. of entities constituting each direct access record	Bytes in excess of basic bytes				
BVARIABLE	Combinations of elements, attributes, and operators								
CLEAR	Savevalues or ranges not to be cleared	Delimiter if multiple entries [,]							
END									
FUNCTION	Function argument	Function type and no. of points							
$x_1,y_1/x_2,y_2/$etc., for function follower points									
INITIAL	Entity or range	Value	Delimiter if multiple entries [/]						
JOB									
JOBTAPE	Jobtape no.	Next block for jobtape transactions	Transaction offset time	Scaling factor					

APPENDIX TABLE VIII (cont.)

Operation	A	B	C	D	E	F	G	H	I
LOAD	GPSS module or user-written HELP routine to be loaded	Delimiter if multiple entries [,]							
MATRIX	Matrix type	No. of matrix rows	No. of matrix columns						
NOXREF									
QTABLE	Queue no.	Upper limit of lowest frequency class	Frequency class size	No. of frequency classes					
READ	No. of files to be skipped								
REALLOCATE	Entity mnemonic to be reallocated	Total no. of that entity	Delimiter if multiple entries [,]						
RESET	Entity or range not to be reset	Delimiter if multiple entries [,]							
REWIND	Jobtape no.								
RMULT	Initial multiplier for RN1	Initial multiplier for RN2	Initial multiplier for RN3	Initial multiplier for RN4	Initial multiplier for RN5	Initial multiplier for RN6	Initial multiplier for RN7	Initial multiplier for RN8	
SAVE	Reposition option								
SIMULATE	Max. run length in minutes	Time expiration option							

APPENDIX TABLE VIII (cont.)

Operation	A	B	C	D	E	F	G	H	I
START	Run termination count	Printout suppression	Snap interval	Standard transaction printout					
STORAGE	Storage no. or range	Capacity	Delimiter if multiple entries [/]						
TABLE	Table argument	Upper limit of lowest-frequency class	Frequency class size	No. of frequency classes	Arrival rate time interval for RT mode table				
VARIABLE	Combinations of elements and arithmetic operators								
FVARIABLE	Combinations of elements and arithmetic operators								

[ ] Indicates optional operand.
{ } Indicates that one of the items within the braces must be chosen.

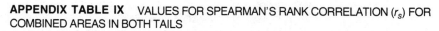

APPENDIX TABLE IX   VALUES FOR SPEARMAN'S RANK CORRELATION (r_s) FOR COMBINED AREAS IN BOTH TAILS

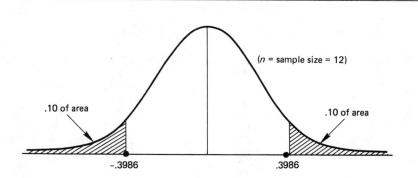

EXAMPLE: For a two-tailed test of significance at the .20 level, with $n = 12$, the appropriate value for r_s can be found by looking under the .20 column and proceeding down to the 12 row; there we find the appropriate r_s value to be .3986.

N	.20	.10	.05	.02	.01	.002
4	.8000	.8000				
5	.7000	.8000	.9000	.9000		
6	.6000	.7714	.8286	.8857	.9429	
7	.5357	.6786	.7450	.8571	.8929	.9643
8	.5000	.6190	.7143	.8095	.8571	.9286
9	.4667	.5833	.6833	.7667	.8167	.9000
10	.4424	.5515	.6364	.7333	.7818	.8667
11	.4182	.5273	.6091	.7000	.7455	.8364
12	.3986	.4965	.5804	.6713	.7273	.8182
13	.3791	.4780	.5549	.6429	.6978	.7912
14	.3626	.4593	.5341	.6220	.6747	.7670
15	.3500	.4429	.5179	.6000	.6536	.7464
16	.3382	.4265	.5000	.5824	.6324	.7265
17	.3260	.4118	.4853	.5637	.6152	.7083
18	.3148	.3994	.4716	.5480	.5975	.6904
19	.3070	.3895	.4579	.5333	.5825	.6737
20	.2977	.3789	.4451	.5203	.5684	.6586
21	.2909	.3688	.4351	.5078	.5545	.6455
22	.2829	.3597	.4241	.4963	.5426	.6318
23	.2767	.3518	.4150	.4852	.5306	.6186
24	.2704	.3435	.4061	.4748	.5200	.6070
25	.2646	.3362	.3977	.4654	.5100	.5962
26	.2588	.3299	.3894	.4564	.5002	.5856
27	.2540	.3236	.3822	.4481	.4915	.5757
28	.2490	.3175	.3749	.4401	.4828	.5660
29	.2443	.3113	.3685	.4320	.4744	.5567
30	.2400	.3059	.3620	.4251	.4665	.5479

From W. J. Conover, *Practical Nonparametric Statistics*, John Wiley & Sons, Inc., New York, 1971. Reprinted by permission.

APPENDIX TABLE X NATURAL LOGARITHMS
FOR 0 TO 1 RANDOM NUMBERS

RN	ln(RN)	RN	ln(RN)
0.010	−4.605	0.510	−0.673
0.020	−3.912	0.520	−0.654
0.030	−3.507	0.530	−0.635
0.040	−3.219	0.540	−0.616
0.050	−2.996	0.550	−0.598
0.060	−2.813	0.560	−0.580
0.070	−2.659	0.570	−0.562
0.080	−2.526	0.580	−0.545
0.090	−2.408	0.590	−0.528
0.100	−2.303	0.600	−0.511
0.110	−2.207	0.610	−0.494
0.120	−2.120	0.620	−0.478
0.130	−2.040	0.630	−0.462
0.140	−1.966	0.640	−0.446
0.150	−1.897	0.650	−0.431
0.160	−1.833	0.660	−0.416
0.170	−1.772	0.670	−0.400
0.180	−1.715	0.680	−0.386
0.190	−1.661	0.690	−0.371
0.200	−1.609	0.700	−0.357
0.210	−1.561	0.710	−0.342
0.220	−1.514	0.720	−0.329
0.230	−1.470	0.730	−0.315
0.240	−1.427	0.740	−0.301
0.250	−1.386	0.750	−0.288
0.260	−1.347	0.760	−0.274
0.270	−1.309	0.770	−0.261
0.280	−1.273	0.780	−0.248
0.290	−1.238	0.790	−0.236
0.300	−1.204	0.800	−0.223
0.310	−1.171	0.810	−0.211
0.320	−1.139	0.820	−0.198
0.330	−1.109	0.830	−0.186
0.340	−1.079	0.840	−0.174
0.350	−1.050	0.850	−0.163
0.360	−1.022	0.860	−0.151
0.370	−0.994	0.870	−0.139
0.380	−0.968	0.880	−0.128
0.390	−0.942	0.890	−0.117
0.400	−0.916	0.900	−0.105
0.410	−0.892	0.910	−0.094
0.420	−0.868	0.920	−0.083
0.430	−0.844	0.930	−0.073
0.440	−0.821	0.940	−0.062
0.450	−0.799	0.950	−0.051
0.460	−0.777	0.960	−0.041
0.470	−0.755	0.970	−0.030
0.480	−0.734	0.980	−0.020
0.490	−0.713	0.990	−0.010
0.500	−0.693	1.000	0.000

BIBLIOGRAPHY

TEXTBOOKS COVERING SIMULATION AND RELATED DISCIPLINES

SIMULATION APPLICATIONS

FORRESTER, JAY W. *Industrial Dynamics*. Cambridge, Mass.: M.I.T. Press, 1961.

HOLST, PER. *Computer Simulation 1951–1976, An Index to the Literature*. London: Mansell, 1979.

NAYLOR, T. H. *Computer Simulation Experiments with Models of Economic Systems*. New York: John Wiley, 1971.

SIMULATION GAMES

BOOCOCK, SARANE S., AND E. O. SCHILD, eds. *Simulation Games in Learning*. Beverly Hills, Calif.: Sage Publications, 1968.

DUKE, RICHARD D., AND CATHY S. GREENBLAT. *Game-Generating Games*. Beverly Hills, Calif.: Sage Publications, 1979.

GREENBLAT, CATHY STEIN, AND RICHARD D. DUKE. *Principles and Practices of Gaming-Simulation*. Beverly Hills, Calif.: Sage Publications, 1981.

HORN, ROBERT E., AND ANNE CLEAVES, eds. *The Guide to Simulations/Games for Education and Training*. Beverly Hills, Calif.: Sage Publications, 1980.

GENERAL SIMULATION

BULGREN, WILLIAM G. *Discrete System Simulation*. Englewood Cliffs, N.J.: Prentice-Hall, 1982.

EMSHOFF, JAMES R., AND ROGER L. SISSON. *Design and Use of Computer Simulation Models*. New York: Macmillan, 1970.

GORDON, GEOFFREY. *System Simulation*, 2nd ed. Englewood Cliffs, N.J.: Prentice-Hall, 1978.

GRAYBEAL, WAYNE J., AND UDO W. POOCH. *Simulation: Principles and Methods*. Cambridge, Mass.: Winthrop, 1980.

LAW, AVERILL J., AND W. DAVID KELTON. *Simulation Modeling and Analysis*. New York: McGraw-Hill, 1982.

LEWIS, T. G., AND B. J. SMITH. *Computer Principles of Modeling and Simulation*. Boston: Houghton-Mifflin, 1979.

MAISEL, HERBERT, AND GIULIANO GNUGNOLI. *Simulation of Discrete Stochastic Systems*. Chicago: Science Research Associates, 1972.

MARYANSKI, FRED J. *Digital Computer Simulation*. Rochelle Park, N.J.: Hayden, 1980.

MCMILLAN, CLAUDE, AND RICHARD F. GONZALEZ. *Systems Analysis*, 3rd ed. Homewood, Ill.: Richard D. Irwin, 1973.

MEIER, ROBERT C., WILLIAM T. NEWELL, AND HAROLD L. PAZER. *Simulation in Business and Economics*. Englewood Cliffs, N.J.: Prentice-Hall, 1969.

MIZE, JOE H., AND J. GRADY COX. *Essentials of Simulation*. Englewood Cliffs, N.J.: Prentice-Hall, 1968.

NAYLOR, THOMAS, ET AL. *Computer Simulation Techniques*. New York: John Wiley, 1966.

ORD-SMITH, R. J., AND J. STEPHENSON. *Computer Simulation of Continuous Systems*. Cambridge: Cambridge University Press, 1975.

PAYNE, JAMES A. *Introduction to Simulation*. New York: McGraw-Hill, 1982.

RUBENSTEIN, REUVEN Y. *Simulation and the Monte Carlo Method*. New York: John Wiley, 1981.

SCHMIDT, J. W., AND R. E. TAYLOR. *Simulation and Analysis of Industrial Systems*. Homewood, Ill.: Richard D. Irwin, 1970.

SHANNON, ROBERT E. *Systems Simulation: The Art and Science*. Englewood Cliffs, N.J.: Prentice-Hall, 1975.

WATSON, HUGH J. *Computer Simulation in Business*. New York: John Wiley, 1981.

SIMULATION PHILOSOPHY AND METHODOLOGY

COLELLA, A. M., M. J. O'SULLIVAN, AND D. J. CARLINO. *Systems Simulation*. Lexington, Mass.: Lexington Books, 1974.

FARRELL, W., C. H. MCCALL, AND E. C. RUSSELL. *Optimization Techniques for Computerized Simulation Models*. New York: C.A.C.I., 1975.

FISHMAN, GEORGE S. *Concepts and Methods in Discrete Event Digital Simulation*. New York: John Wiley, 1973.

———. *Principles of Discrete Event Simulation*. New York: John Wiley, 1978.

MIRHAM, G. A. *Simulation: Statistical Foundations and Methodology*. New York: Academic Press, 1972.

TOCHER, K. D. *The Art of Simulation*. New York: Van Nostrand, 1963.

ZEIGLER, BERNARD P. *Theory of Modeling and Simulation*. New York: John Wiley, 1976.

SIMULATION PROGRAMMING

BIRTWISTLE, G., O.-J. DAHL, B. MYHRHAUG, AND K. NYGAARD. *Simula Begin*. New York: Auerbach, 1973.

BOBILLIER, P. A., B. C. KAHAN AND A. R. PROBST. *Simulation with GPSS and GPSS V*. Englewood Cliffs, N.J.: Prentice-Hall, 1976.

DUNNING, KENNETH A. *Getting Started in GPSS*. San Jose, Calif.: Engineering Press, 1981.

GORDON, GEOFFREY. *The Application of GPSS V to Discrete Systems Simulation*. Englewood Cliffs, N.J.: Prentice-Hall, 1976.

GREENBERG, STANLEY. *GPSS Primer*. New York: John Wiley, 1972.

KIVIAT, P. J., R. VILLANUEVA, AND H. M. MARKOWITZ, *The SIMSCRIPT II.5 Programming Language*. New York: C.A.C.I., 1975.

KOSY, DONALD W. *The ECSS II Language for Simulating Computer Systems,* Report R-1895-GSA. Santa Monica, Calif.: Rand, December 1976.

PRITSKER, A. A. B. *The GASP IV Simulation Language*. New York: John Wiley, 1974.

———. *Modeling and Analysis Using Q-GERT Networks*. New York: Halsted, 1977.

———, AND CLAUDE DENNIS PEGDEN. *Introduction to Simulation and SLAM*. New York: Halsted, 1979.

PUGH, A., III. *DYNAMO II User's Manual*. Cambridge, Mass.: M.I.T. Press, 1973.

SCHRIBER, THOMAS J. *Simulation Using GPSS*. New York: John Wiley, 1974.

WYMAN, FORREST PAUL. *A Guide to Using SIMSCRIPT*. New York: John Wiley, 1970.

QUEUING MODELS

BUDNICK, FRANK S., RICHARD MOJENA, AND THOMAS E. VOLLMANN. *Principles of Operations Research for Management*. Homewood, Ill.: Richard D. Irwin, 1977.

GROSS, D., AND C. N. HARRIS. *Fundamentals of Queuing Theory*. New York: John Wiley, 1974.

HILLIER, FREDERICK S., AND GERALD J. LIEBERMAN. *Operations Research*, 2nd ed. San Francisco: Holden-Day, 1974.

KLEINROCK, LEONARD. *Queuing Systems*. New York: John Wiley, 1975.

PANICO, J. A. *Queuing Theory: A Study of Waiting Lines for Business, Economics, and Science*. Englewood Cliffs, N.J.: Prentice-Hall, 1968.

SAATY, THOMAS L. *Elements of Queueing Theory*. New York: McGraw-Hill, 1961.

WHITE, J. A., J. W. SCHMIDT, AND G. K. BENNETT. *Analysis of Queuing Systems*. New York: Academic Press, 1975.

STATISTICS APPLICABLE TO SIMULATION

DUDEWICZ, EDWARD J., AND THOMAS G. RALLEY. *The Handbook of Random Number Generation and Testing with TESTRAND Computer Code*. Columbus, Ohio: American Sciences Press, 1981.

GIBBONS, JEAN DICKINSON. *Nonparametric Methods for Quantitative Analysis*. New York: Holt, Rinehart and Winston, 1976.

KLEIJNEN, JACK P. C. *Statistical Techniques in Simulation*. New York, Marcel Dekker, 1974.

LEWIS, T. G. *Distribution Sampling for Computer Simulation*. Lexington, Mass.: Lexington Books, 1975.

ROSCOE, JOHN T. *Fundamental Research Statistics for the Behavioral Sciences*, 2nd ed. New York: Holt, Rinehart and Winston, 1975.

THESEN, ARNE. *Computer Methods in Operations Research*. New York: Academic Press, 1978.

PERIODICALS AND CONFERENCE PROCEEDINGS COVERING SIMULATION

There are many sources of current research and application reports about simulation. Some of which are especially appropriate for people with an interest in business are

Decision Sciences. Journal of the American Institute for Decision Sciences, University Plaza, Atlanta, Georgia 30303.

Interfaces. Journal of the Practice of Management Science, co-sponsored by the Operations Research Society of America and The Institute of Management Sciences; available from the Institute of Management Sciences, 146 Westminster Street, Providence, Rhode Island 02903.

Management Science. Journal of The Institute of Management Sciences, 146 Westminster Street, Providence, Rhode Island 02903.

Proceedings of the Summer Computer Simulation Conferences (annually since about 1975). Available from the Society for Computer Simulation, P.O. Box 2228, La Jolla, California 92038.

Proceedings of the Winter Simulation Conferences (annually since about 1970). Available from the Association for Computing Machinery, P.O. Box 12115, Church Street Station, New York, New York 10249.

Simulation. Journal of the Society for Computer Simulation, P.O. Box 2228, La Jolla, California 92038

Simuletter. Journal of the Special Interest Group on Simulation, a subdivision of the Association for Computing Machinery. P.O. Box 12115, Church Street Station, New York, New York 10249.

The Society for Computer Simulation also publishes a group of nonserial volumes on various aspects of simulation. The Institute of Management Sciences has a subgroup called the College on Simulation and Gaming, which publishes a periodical newsletter containing items of interest to simulation professionals. A private consulting firm, C.A.C.I., Inc., 12011 San Vicente Boulevard, Los Angeles, California 90049, publishes an excellent newsletter called *SIMSNIPS*, containing article reviews and announcements of simulation conferences and courses.

Interfaces is perhaps the most practical and the most readable of the group, followed by *Decision Sciences*, *Simuletter*, the various conference proceedings, *Simulation*, and Management Science.

A list of fairly recent articles about simulation in *Decision Sciences*, *Interfaces*, and *Management Science* categorized by subtopic within simulation follows:

1. Corporate and financial modeling
2. Data processing and computer system modeling
3. Simulation of flow processes (e.g., oil)
4. Gaming and educational modeling
5. Governmental, urban planning, and societal modeling
6. Health care systems modeling
7. Simulation methodology
8. Production and operations modeling
9. Philosophy of modeling
10. Queuing formula applications
11. Marketing and sales modeling
12. Other simulation topics

CORPORATE AND FINANCIAL MODELING

BACON, PETER W., AND ROBERT W. HAESSLAER. "Simulation and the Capital Asset Pricing Model." *Decision Sciences*, Vol. 6, no. 1, January 1975, pp. 202−204.

MOORE, LAURENCE J., AND BERNARD W. TAYLOR III. "Experimental Investigation of Priority Scheduling in a Bank Check Processing Operation." *Decision Sciences*, Vol. 8, no. 4, October 1977, pp. 692−710.

————, "Multiteam, Multiproject Research and Development Planning with GERT." *Management Science,* Vol. 24, no. 4, December 1977, pp. 401–410.

NAYLOR, THOMAS H., AND HORST SCHAULAND. "A Survey of Users of Corporate Planning Models." *Management Science*, Vol. 22, no. 9, May 1976, pp. 927−937.

ONSI, MOHAMED. "Simulation of the Economic Factors Affecting Organizational Slack: A Factorial Design." *Decision Sciences*, Vol. 6, no. 1, January 1975, pp. 78−91.

STRATTON, WILLIAM O. "Accounting Systems: The Reliability Approach to Internal Control Evaluation." *Decision Sciences*, Vol. 12, no. 1, January 1981, pp. 51−67.

TAYLOR, BERNARD W., III, AND LAURENCE J. MOORE. "R & D Project Planning with Q-GERT Network Modeling and Simulation." *Management Science*, Vol. 26, no. 1, January 1980, pp. 44−59.

WINKOFSKY, E. P., N. R. BAKER, AND D. J. SWEENEY. "A Decision Process Model of R & D Resource Allocation in Hierarchical Organizations." *Management Science*, Vol. 27, no. 5, March 1981, pp. 268−283.

DATA PROCESSING AND COMPUTER SYSTEM MODELING

FELLINGHAM, JOHN C., THEODORE J. MOCK, AND MIKLOS A. VASARHELY. "Simulation of Information Choice." *Decision Sciences*, Vol. 7, no. 2, April 1976, pp. 219–234.

MELLICHAMP, JOSEPH M., AND NEAL M. BENGSTON. "Designing Spacelab's Data Management System with Simulation." *Interfaces*, Vol. 9, no. 3, May 1979, pp. 87–93.

There are many articles on the simulation of computer systems in the *Proceedings of the Symposia on the Simulation of Computer Systems*, held annually from 1973 to 1976, and in the *Performance Evaluation Review*, journal of the Special Interest Group on Computer Performance Measurement and Evaluation (SIGMETRICS), a subdivision of the Association for Computer Machinery. These are available from the Association for Computing Machinery, P.O. Box 12115, Church Street Station, New York, New York 10249.

SIMULATION OF FLOW PROCESSES

BAR-LEV, D., AND M. A. POLLATSCHEK. "Simulation as an Aid in Decision Making at Israel Fertilizers and Chemicals." *Interfaces*, Vol. 11, no. 2, April 1981, pp. 17–21.

CUNNINGHAM, A. A., AND J. SWIRLES. "Modeling the Great Canadian Oil Sands Operation: The Politics of Implementation." *Interfaces*, Vol. 10, no. 5, October 1980, pp. 55–62.

MELLICHAMP, JOSEPH M., AND CHARLES P. WEAVER. "Simulation and Sewage." *Decision Sciences*, Vol. 8, no. 3, July 1977, pp. 584–597.

ROTHKOPF, M. H., J. K. MCCARRON, AND S. FROMOVITZ. "A Weather Model for Simulating Offshore Construction Alternatives." *Management Science*, Vol. 20, no. 10, June 1974, pp. 1345–1349.

GAMING AND TRAINING

BURNES, ALVIN C. "A Computer Simulation Approach to the Teaching of Mail Survey Strategy Alternatives and Choice." *Decision Sciences*, Vol. 9, no. 1, January 1978, pp. 156–168.

HAND, HERBERT H., AND HENRY P. SIMS, JR. "Statistical Evaluation of Computer Gaming Performance." *Management Science*, Vol. 21, no. 6, February 1975, pp. 708–717.

TAYLOR, BERNARD W., III, AND LAURENCE J. MOORE. "Analysis of a Ph.D. Program via GERT Modeling and Simulation." *Decision Sciences*, Vol. 9, no. 4, October 1978, pp. 725–737.

WOLFE, JOSEPH. "Effective Performance Behaviors in a Simulated Policy and Decision Making Environment." *Management Science*, Vol. 21, no. 8, April 1975, pp. 872–882.

GOVERNMENTAL, URBAN PLANNING, AND SOCIETAL MODELING

AYAL, IGAL, AND DONALD J. HEMPEL. "Simulating Housing Market Processes at the Micro Level." *Management Science*, Vol. 25, no. 6, June 1979, pp. 565–576.

DOWNING, PAUL B., AND WILLIAM D. WATSON, JR. "A Simulation Study of Alternative Pollution Control Enforcement Systems." *Management Science*, Vol. 22, no. 5, January 1976, pp. 558–569.

FERGUSON, CARL E., JR., J. BARRY MASON, AND J. B. WILKINS. "Simulating Food Shoppers' Economic Losses as a Result of Supermarket Unavailability." *Decision Sciences*, Vol. 11, no. 3, July 1980, pp. 535–556.

GRANT, FLOYD H., III. "Reducing Voter Waiting Time." *Interfaces*, Vol. 10, no. 5, October 1980, pp. 19–25.

MACON, MAX RODERICK, AND EFRAIM TURBAN. "Energy Audit Program Simulation" *Interfaces*, Vol. 11, no. 1, February 1981, pp. 13–19.

MONARCHI, DAVID E., THOMAS E. HENDRICK, AND DONALD R. PLANE. "Simulation for Fire Department Deployment Policy Analysis." *Decision Sciences*, Vol. 8, no. 1, January 1977, pp. 211–227.

SEXTON, DONALD E. "Evaluating Urban Growth Policies with a Systems Simulation." *Management Science*, Vol. 25, no. 1, January 1979, pp. 43–53.

SHEPHERD, K. W. "Applying Simulation Techniques to Legislative Analysis: The Veterans and Survivors Pension Reform Act." *Interfaces*, Vol. 7, no. 1, November 1976, pp. 31–40.

SMITH, V. KERRY, DAVID B. WEBSTER, AND NORMAN A. HECK. "The Management of Wilderness Areas: A Simulation Model." *Decision Sciences*, Vol. 7, no. 3, July 1976, pp. 524–537.

HEALTH CARE SYSTEMS MODELING

FRERICHS, RALPH R., AND JUAN PRAWDA. "A Computer Simulation Model for the Control of Rabies in an Urban Area of Colombia." *Management Science*, Vol. 22, no. 4, December 1975, pp. 411–421.

HERSHEY, JOHN C., WILLIAM J. ABERNATHY, AND NICHOLAS BALOFF. "Comparison of Nurse Allocation Policies—A Monte Carlo Mode." *Decision Sciences*, Vol. 5, no. 1, January 1974, pp. 58–72.

KWAK, N. K., P. J. KUZDRALL, AND HOMER H. SCHMITZ. "The GPSS Simulation of Scheduling Policies for Surgical Patients." *Management Science*, Vol. 22, no. 9, May 1976, pp. 982–989.

REISMAN, A., W. CULL, H. EMMONS, B. DEAN, C. LIN, J. RASMUSSEN, P. DARUKHANA-VALA, AND T. GEORGE. "On the Design of Alternative Obstetric Anesthesia Team Configurations." *Management Science*, Vol. 23, no. 6, February 1977, pp. 545–556.

UYENO, DEAN H. "Health Manpower Systems: An Application of Simulation to the Design of Primary Health Care Teams." *Management Science*, Vol. 20, no. 6, February 1974, pp. 981–989.

SIMULATION METHODOLOGY

AHLUND, MIKAEL C., HIRAM C. BARKSDALE, AND JIMMY E. HILLIARD. "Multivariate Spectral Analysis—An Illustration." *Decision Sciences*, Vol. 8, no. 4, October 1977, pp. 734−752.

CARTER, GRACE, AND EDWARD J. IGNALL. "Virtual Measures: A Variance Reduction Technique for Simulation." *Management Science*, Vol. 21, no. 6, February 1975, pp. 607−616.

COOLEY, BELVA J., AND JOHN W. COOLEY. "Data Analysis for Simulation Experiments: Application of a Distribution-Free Multiple Comparisons Procedure." *Decision Sciences*, Vol. 11, no. 3, July 1980, pp. 482−492.

ENGLEBRECHT-WIGGANS, RICHARD, AND WILLIAM L. MAXWELL. "Analysis of the Time Indexed List Procedure for Synchronization of Discrete Event Simulations." *Management Science*, Vol. 24, no. 13, September 1978, pp. 1417−1427.

HSU, D. A., AND J. S. HUNTER. "Analysis of Simulation-Generated Responses Using Autoregressive Models." *Management Science*, Vol. 24, no. 2, October 1977, pp. 181−190.

KING, WILLIAM R. "Methodological Analysis Through Systems Simulation." *Decision Sciences*, Vol. 5, no. 1, January 1974, pp. 1−9.

KLEIJNEN, JACK P. C. "Antithetic Variates, Common Random Numbers, and Optimal Computer Time Allocation in Simulation." *Management Science*, Vol. 21, no. 10, June 1975, pp. 1176−1185.

LAVENBERG, S. S., AND P. D. WELCH. "A Perspective on the Use of Control Variables to Increase the Efficiency of Monte Carlo Simulations." *Management Science*, Vol. 27, no. 3, March 1981, pp. 322−335.

SARIN, RAKESH K. "Experimental Results of an Approach for Evaluating Multiattribute Alternatives." *Decision Sciences*, Vol. 8, no. 4, October 1977, pp. 722−733.

SCHRUVEN, LEE W. "A Coverage Function for Interval Estimators of Simulation Response." *Management Science*, Vol. 26, no. 1, January 1980, pp. 18−27.

SIMS, HENRY P., JR., AND DAVID A. WILKERSON. "Time-Lags in Cross-Lag Correlation Studies: A Computer Simulation." *Decision Sciences*, Vol. 8, no. 3, July 1977, pp. 630−644.

WRIGHT, R. D., AND T. E. RAMSAY, JR. "On the Effectiveness of Common Random Numbers." *Management Science*. Vol. 25, no. 7, July 1979, pp. 649−656.

PHILOSOPHY OF MODELING

SOLOMON, SUSAN L. "Building Modelers: Teaching the Art of Simulation." *Interfaces*, Vol. 10, no. 2, April 1980, pp. 65−72.

WOOLSEY, GENE. "Whatever Happened to Simple Simulation? A Question and Answer." *Interfaces*, Vol. 9, no. 4, August 1979, pp. 9−11.

PRODUCTION AND OPERATIONS MODELING

ADAM, NABIL R., AND JULIUS SURKIS. "Priority Update Intervals and Anomalies in Dynamic Ratio Type Job Shop Scheduling Rules." *Management Science*, Vol. 26, no. 12, December 1980, pp. 1227−1237.

BERRY, WILLIAM L., AND VITTAL RAO. "Critical Ratio Scheduling: An Experimental Analysis." *Management Science*, Vol. 22, no. 2, October 1975, pp. 192−201.

FRYER, JOHN S. "Organizational Structure of Dual-Constraint Job Shops." *Decision Sciences*, Vol. 5, no. 1, January 1974, pp. 45−57.

―――. "Labor Flexibility in Multiechelon Dual-Constraint Job Shops." *Management Science*, Vol. 20, no. 7, March 1974, pp. 1073−1080.

―――. "Organizational Segmentation and Labor Transfer Policies in Labor and Machine Limited Production Systems." *Decision Sciences*, Vol. 7, no. 4, October 1976, pp. 725−738.

GOLOVIN, LEWIS. "Product Blending: A Simulation Case Study in Double-Time." *Interfaces*, Vol. 9, no. 5, November 1979, pp. 64−76.

HARDY, S. T., AND L. J. KRAJEWSKI. "A Simulation of Interactive Maintenance Decisions." *Decision Sciences,* Vol. 6, no. 1, January 1975, pp. 92–105.

HERSHAUER, JAMES C., AND RONALD J. EBERT. "Research and Simulation Selection of a Job-Shop Sequencing Rule." *Management Science*, Vol. 21, no. 8, April 1975, pp. 872−882.

HOLLOWAY, C. A., AND R. T. NELSON. "Job Shop Scheduling with Due Dates and Variable Processing Times." *Management Science*, Vol. 20, no. 9, May 1974, pp. 1264−1272.

JAIN, SURESH K. "A Simulation-Based Scheduling and Management Information System for a Machine Shop." *Interfaces,* Vol. 6, no. 1, November 1975, pp. 81−96.

LEE, WILLIAM B., AND BASHEER M. KHUMAWALA. "Simulation Testing of Aggregate Production Models in an Implementation Methodology." *Management Science*, Vol. 20, no. 6, February 1974, pp. 903−911.

MOORE, JACK. "A Computer Modeling Approach to Manufacturing and Distribution Strategy." *Interfaces*, Vol. 10, no. 4, August 1980, pp. 16−21.

WEEKS, JAMES K. "A Simulation Study of Predictable Due-Dates." *Management Science*, Vol. 25, no. 4, April 1979, pp. 363−373.

―――, AND TONY R. WINGLER. "A Stochastic Dominance Ordering of Scheduling Rules." *Decision Sciences*, Vol. 10, no. 2, April 1979, pp. 245−257.

QUEUING FORMULA APPLICATIONS

BYRD, JACK, JR. "The Value of Queueing Theory." *Interfaces*, Vol. 8, no. 3, March 1978, pp. 22−26.

COSMETATOS, GEORGE P. "The Value of Queueing Theory—A Case Study." *Interfaces*, Vol. 9, no. 3, May 1979, pp. 20−25.

DEUTSCH, HOWARD, AND VINCENT A. MABERT. "Queuing Theory and Teller Staffing: A Successful Implementation." *Interfaces*, Vol. 10, no. 5, October 1980, pp. 63−67.

FOOTE, B. L. "A Queuing Case Study of Drive-in Banking." *Interfaces*, Vol. 6, no. 4, August 1976, pp. 31–37.

KOLESAR, PETER. "A Quick and Dirty Response to the Quick and Dirty Crowd: Particularly to Jack Byrd's 'The Value of Queueing Theory.' " *Interfaces*, Vol. 9, no. 2, February 1979, pp. 77–82.

MCKEOWN, PATRICK G. "An Application of Queueing Analysis to the New York State Child Abuse and Maltreatment Register Telephone Reporting System." *Interfaces*, Vol. 9, no. 3, May 1979, pp. 20–25.

VAZSONYI, ANDREW. "To Queue or Not to Queue: A Rejoinder." *Interfaces*, Vol. 9, no. 2, February 1979, pp. 83–86.

MARKETING AND SALES MODELING

CRASK, MELVIN R. "A Simulation Model of Patronage Behavior Within Shopping Centers." *Decision Sciences*, Vol. 10, no. 1, January 1979, pp. 1–15.

CURRY, DAVID J., JORDAN J. LOUVIERE, AND MICHAEL J. AUGUSTINE. "On the Sensitivity of Brand-Choice Simulations to Attribute Importance Weights." *Decision Sciences*, Vol. 12, no. 3, July 1981, pp. 502–516.

GREEN, PAUL E., WAYNE S. DESARBO, AND PRADEGS K. KEDIA. "On the Insensitivity of Brand-Choice Simulations to Attribute Importance Weights." *Decision Sciences*, Vol. 11, no. 3, July 1980, pp. 439–450.

TRANSPORTATION MODELING

COOK, THOMAS M., AND ROBERT A. RUSSELL. "A Simulation and Statistical Analysis of Stochastic Vehicle Routing with Timing Constraints." *Decision Sciences*, Vol. 9, no. 4, October 1978, pp. 673–687.

GAITHER, NORMAN. "A Stochastic Constrained Optimization Model for Determining Commercial Fishing Seasons." *Management Science*, Vol. 26, no. 2, February 1980, pp. 143–154.

SCHECHTER, MORDECHAI, AND ROBERT L. LUCAS. "Validating a Large-Scale Simulation Model of Wilderness Recreational Travel." *Interfaces*, Vol. 10, no. 5, October 1980, pp. 11–18.

WYMAN, F. PAUL. "Simulation of Tar Sands Mining Operations." *Interfaces*, Vol. 8, no. 1, November 1977, pp. 6–20.

OTHER SIMULATION TOPICS

GUNTHER, RICHARD E. "Dual-Resource Parallel Queues with Server Transfer and Information Access Delays." *Decision Sciences*, Vol. 12, no. 1, January 1981, pp. 97-111.

HANNAN, EDWARD L., AND LARRY A. SMITH. "A Simulation of the Effects of Alternative Rule Systems for Jai Alai." *Decision Sciences*, Vol. 12, no. 1, January 1981, pp. 75–84.

HOLZMAN, ALBERT G., AND DONALD B. JOHNSON. "A Simulation Model of the College Admission Process." *Interfaces*, Vol. 5, no. 3, May 1975, pp. 55–64.

SIMSTAT PACKAGE PROGRAM LISTINGS

```
        DATCOL PROGRAM
        PROGRAM DATCOL
C THIS PROGRAM PROVIDES ANALYSIS OF OBSERVED DATA FOR SIMULATION
C INPUT.  SOURCE DATA FOR EACH OBSERVATION (CUSTOMER) ARE TIME
C ARRIVED, TIME SERVICE BEGAN AND TIME SERVICE ENDED.  SIMULATION
C INPUTS COMPUTED ARE INTERARRIVAL TIME AND SERVICE TIME.
C PERFORMANCE MEASURES COMPUTED ARE NUMBER AND TIME IN QUEUE,
C TRANSIT TIME, AND SERVER IDLE TIME.  AVERAGES OF THESE PERFORMANCE
C MEASURES ARE CALCULATED OVER ALL OBSERVATIONS.  TIMES
C OVERLAPPING MIDNIGHT(ZERO HOURS) MAY NOT BE USED.  TIMES
C OVERLAPPING NOON SHOULD USE 24-HOUR MILITARY CLOCK TIMES.
C DATA ITEMS ARE TO BE ENTERED USING THE FIRST TWO DIGITS AS HOURS,
C THE SECOND TWO DIGITS AS MINUTES, AND THE THIRD TWO DIGITS AS
C SECONDS.  FRACTIONS OF A SECOND ARE PERMISSIBLE.  THE THREE FIELDS
C FOR EACH CUSTOMER--ARRIVAL TIME, SERVICE START TIME AND SERVICE
C COMPLETION TIME--SHOULD BE SEPARATED BY COMMAS ON A SINGLE LINE.
        DIMENSION ARSTD(201),SSTSTD(201),SNDSTD(201),ARTIM(201)
        DIMENSION SVSTTM(201),SVNDTM(201)
        DIMENSION LQ(201),STIME(201),TRTIME(201),WTIME(201),SIT(201)
        DIMENSION ORIGSS(15),ORIGSE(15),ORST(15),ORISST(15),ORIEST(15)
        REAL INTARV(201)
        TLTRT=0
        TLIAT=0
        TLSIT=0
        TLWT=0
        TLLQ=0
        TLST=0
        IOUT=2
        DO 3 I=1,15
        ORIGSS(I)=0
        ORIGSE(I)=0
        ORISST(I)=0
        ORIEST(I)=0
    3   ORST(15)=0
        DO 5 I=1,201
        LQ(I)=0
```

```
        STIME(I)=0
        TRTIME(I)=0
        WTIME(I)=0
        SIT(I)=0
        ARSTD(I)=0
        SSTSTD(I)=0
        INTARV(I)=0
5       SNDSTD(I)=0
        WRITE(IOUT,10)
10      FORMAT('0THIS PROGRAM COMPUTES PERFORMANCE MEASURES FROM ',
       1'SIMULATION INPUT DATA'/' AND WRITES THE INFORMATION ON DISK.',
       2' TABLE OUTPUT IS GIVEN IN SECONDS'/' OR IN STANDARD',
       3' UNITS. IF THE DATA ARE ON DISK, TYPE A 1.'/' OTHERWISE',
       4', TYPE A 0.')
        READ, IDISK
        WRITE(IOUT,25)
25      FORMAT('0IF DATA ARE IN HRS, MIN AND SEC, TYPE A 1.'/
       1' IF DATA ARE IN STANDARD UNITS STARTING WITH 0, TYPE A 0.')
        READ, ISTD
        IF(ISTD .EQ. 1)WRITE(IOUT,30)
30      FORMAT('0AT WHAT TIME DID OBSERVATION BEGIN?')
        IF(ISTD .EQ. 1)READ, BEGIN
        IF(ISTD .EQ. 0)BEGST=0
        IF(ISTD .EQ. 0)GO TO 35
        IHR=BEGIN/10000.
        IMIN=BEGIN/100.-IHR*100
        SEC=BEGIN-(IHR*10000+IMIN*100)
        BEGST1=IHR*3600+IMIN*60+SEC
        BEGST=0
35      WRITE(IOUT,40)
40      FORMAT('0HOW MANY CUSTOMERS WERE IN THE QUEUE INITIALLY? DO',
       1' NOT COUNT'/' THE CUSTOMER WHO MAY HAVE BEEN BEING SERVED.')
        READ, INITQ
        IF(INITQ .EQ. 0)GO TO 95
        IF(IDISK .EQ. 0)WRITE(IOUT,85)
85      FORMAT('0ENTER THE SERVICE START AND END TIMES FOR THE '/
       1' CUSTOMERS IN THE INITIAL QUEUE, SEPARATED BY COMMAS,'/
       2' ONE CUSTOMER PER LINE.')
        DO 90 I=1,INITQ
        IF(IDISK .EQ.0)READ, ORIGSS(I),ORIGSE(I)
        IF(IDISK .EQ. 1)READ(20,*)ORIGSS(I),ORIGSE(I)
        IF(ISTD .EQ. 0)GO TO 88
        IHR=ORIGSS(I)/10000.
        IMIN=ORIGSS(I)/100.-IHR*100.
        SEC=ORIGSS(I)-(IHR*10000+IMIN*100)
        ORISST(I)=IHR*3600+IMIN*60+SEC-BEGST1
        IHR=ORIGSE(I)/10000.
        IMIN=ORIGSE(I)/100.-IHR*100.
        SEC=ORIGSE(I)-(IHR*10000.+IMIN*100)
        ORIEST(I)=IHR*3600+IMIN*60+SEC-BEGST1
88      IF(ISTD .EQ. 0)ORIEST(I)=ORIGSE(I)
        IF(ISTD .EQ. 0)ORISST(I)=ORIGSS(I)
        ORST(I)=ORIEST(I)-ORISST(I)
90      TLST=TLST+ORST(I)
95      I=0
        IF(IDISK .EQ. 1)GO TO 70
        WRITE(IOUT,45)
45      FORMAT('0ENTER ARRIVAL TIME, SERVICE START TIME, AND SERVICE',
       1' END TIME, SEPARATED BY'/' COMMAS, FOR ALL CUSTOMERS ',
       2'WHO ARRIVED AFTER THE OBSERVATION '/' PERIOD BEGAN.',
       3' THERE SHOULD BE ONE LINE PER CUSTOMER.  TO INDICATE '/
       4' THE END OF DATA TYPE A LINE CONTAINING ONLY 999999,0,0  .')
50      I=I+1
        READ,ARTIM(I),SVSTTM(I),SVNDTM(I)
        IF(ARTIM(I) .NE. 999999) GO TO 50
        GO TO 100
```

```
70      READ(20,*,END=100)(ARTIM(I),SVSTTM(I),SVNDTM(I),I=1,201)
100     IMAX=I-1
        IF(ISTD .EQ. 0)GO TO 140
        DO 120 J=1,IMAX
        IHR=ARTIM(J)/10000.
        IMIN=ARTIM(J)/100.-IHR*100
        SEC=ARTIM(J)-(IHR*10000+IMIN*100)
        ARSTD(J)=IHR*3600+IMIN*60+SEC-BEGST1
        IHR=SVSTTM(J)/10000.
        IMIN=SVSTTM(J)/100.-IHR*100
        SEC=SVSTTM(J)-(IHR*10000+IMIN*100)
        SSTSTD(J)=IHR*3600+IMIN*60+SEC-BEGST1
        IHR=SVNDTM(J)/10000.
        IMIN=SVNDTM(J)/100.-IHR*100
        SEC=SVNDTM(J)-(IHR*10000+IMIN*100)
120     SNDSTD(J)=IHR*3600+IMIN*60+SEC-BEGST1
        GO TO 150
140     DO 145 I=1,IMAX
        ARSTD(I)=ARTIM(I)
        SSTSTD(I)=SVSTTM(I)
145     SNDSTD(I)=SVNDTM(I)
150     DO 170 I=1,IMAX
        IF(I .EQ. 1)GO TO 160
        L=I-1
        DO 155 M=1,L
        N=L-M+1
        IF(SNDSTD(N) .LE. ARSTD(I))GO TO 170
155     LQ(I)=LQ(I)+1
160     IF(INITQ .EQ. 0)GO TO 170
        DO 165 J=1,INITQ
        K=INITQ-J+1
        IF(ORIEST(K) .LE. ARSTD(I))GO TO 170
165     LQ(I)=LQ(I)+1
170     CONTINUE
        INTARV(1)=-99999.9
        STIME(1)=SNDSTD(1)-SSTSTD(1)
        TRTIME(1)=SNDSTD(1)-ARSTD(1)
        WTIME(1)=SSTSTD(1)-ARSTD(1)
        IF(INITQ .EQ. 0)SIT(1)=SSTSTD(1)-BEGST
        IF(INITQ .NE. 0)SIT(1)=SSTSTD(1)-ORIEST(INITQ)
        TLLQ=TLLQ+LQ(1)
        TLST=TLST+STIME(1)
        TLSIT=TLSIT+SIT(1)
        TLTRT=TLTRT+TRTIME(1)
        TLWT=TLWT+WTIME(1)
        DO 180 I=2,IMAX
        TRTIME(I)=SNDSTD(I)-ARSTD(I)
        WTIME(I)=SSTSTD(I)-ARSTD(I)
        SIT(I)=SSTSTD(I)-SNDSTD(I-1)
        STIME(I)=SNDSTD(I)-SSTSTD(I)
        INTARV(I)=ARSTD(I)-ARSTD(I-1)
        TLST=TLST+STIME(I)
        TLSIT=TLSIT+SIT(I)
        TLIAT=TLIAT+INTARV(I)
        TLTRT=TLTRT+TRTIME(I)
        TLLQ=TLLQ+LQ(I)
180     TLWT=TLWT+WTIME(I)
245     WRITE(IOUT,250)
250     FORMAT('0OBS.   TIME     I/A  QUEUE SERVICE  SERVICE  SRVICE',
       1' TRANSIT WAITING SERVER')
        WRITE(IOUT,255)
255     FORMAT(' NO.  ARRIVED   TIME  LNGTH START T. END TIME  TIME ',
       1'  TIME    TIME IDLE T.'/)
        IF(INITQ .EQ. 0)GO TO 264
        DO 260 I=1,INITQ
        WRITE(30,263)I,ORISST(I),ORIEST(I),ORST(I)
```

```
260      WRITE(IOUT,263)I,ORISST(I),ORIEST(I),ORST(I)
263      FORMAT(I4,24X,2F9.1,F7.1)
         WRITE(IOUT,263)
264      DO 265 I=1,IMAX
         WRITE(30,270)I,ARSTD(I),INTARV(I),LQ(I),SSTSTD(I),SNDSTD(I),
        1STIME(I),TRTIME(I),WTIME(I),SIT(I)
265      WRITE(IOUT,270)I,ARSTD(I),INTARV(I),LQ(I),SSTSTD(I),SNDSTD(I),
        1STIME(I),TRTIME(I),WTIME(I),SIT(I)
270      FORMAT(I4,F9.1,F8.1,2X,I5,F9.1,F9.1,F7.1,F8.1,F8.1,F7.1)
         AVSIT=TLSIT/SNDSTD(IMAX)*100.
         AVIAT=TLIAT/(IMAX-1)
         AVTRT=TLTRT/IMAX
         AVWT=TLWT/IMAX
         AVST=TLST/(IMAX+INITQ)
         AVLQ=TLLQ/IMAX
         WRITE(IOUT,300)AVIAT,AVLQ,AVST,AVTRT,AVWT,AVSIT
300      FORMAT('0AVERAGE INTERARRIVAL TIME=',F15.3/
        1'0AVERAGE QUEUE LENGTH=',F20.3/
        2'0AVERAGE SERVICE TIME=',F20.3/'0AVERAGE TRANSIT TIME=',
        3F20.3/'0AVERAGE WAITING TIME=',F20.3/'0AVERAGE SERVER ',
        4'IDLE TIME PERCENT=',F8.3)
         STOP
         END

         PROGRAM DESCST
C        THIS PROGRAM COMPUTES THE MEAN, VARIANCE AND STANDARD
C        DEVIATION OF A DATA SET, PRINTS THE DATA, SORTS AND REPRINTS
C        IT IF DESIRED TO FACILITATE CATEGORIZATION.
C        IT WILL ALSO COMPUTE THE MEANS OF M ADJACENT OBSERVATIONS OR
C        SELECT EVERY N-TH OBSERVATION.
C        IF INPUT FROM A DISK FILE IS DESIRED, THE FILE SHOULD BE
C        DESIGNATED AS DEVICE NUMBER 20.
C        IF OUTPUT TO A DISK FILE IS DESIRED, THE FILE SHOULD BE
C        DESIGNATED AS DEVICE NUMBER 30.
         DIMENSION Y(5000),M(5000)
         REAL MEAN,M
         DO 2 I=1,5000
2        M(I)=0.
         N=0
         IOUT=2
         SUMX=0
         INDX=0
         SUMSQ=0
         WRITE(IOUT,3)
3        FORMAT('0DESCRIPTIVE STATISTICS')
         WRITE(IOUT,5)
5        FORMAT('0DO YOU WANT INSTRUCTIONS? TYPE 1 FOR YES, 0 FOR NO.')
         READ, INSTR
         IF(INSTR .EQ. 0) GO TO 14
         WRITE(IOUT,10)
10       FORMAT('0UNDER OPTION 1 THIS PROGRAM COMPUTES DESCRIPTIVE '/
        1' STATISTICS, INCLUDING THE MEAN, VARIANCE AND STANDARD '/
        2' DEVIATION OF A DATA SET, AND CREATES A SORTED LIST OF '/
        3' INPUTS.  UNDER OPTION 2 THE PROGRAM COMPUTES THE '/
        4'/ MEANS OF EACH M ADJACENT OBSERVATIONS OR SELECTS EVERY'/
        5' N-TH OBSERVATION.')
14       WRITE(IOUT,15)
15       FORMAT('0WHICH OPTION? TYPE 1 OR 2?')
         READ, NOP
         WRITE(IOUT,22)
22       FORMAT('0IF INPUT IS TO BE PROVIDED FROM A DISK '/
        1' FILE,TYPE A 1.  OTHERWISE TYPE A 0.')
         READ, IINFL
         IF(IINFL .EQ. 0) GO TO 19
21       READ(20,*,END=50)X
         N=N+1
         SUMX=SUMX+X
         Y(N)=X
```

```
          GO TO 21
19        WRITE(IOUT,20)
20        FORMAT('0ENTER THE DATA ONE OBSERVATION PER LINE.')
          WRITE(IOUT,30)
30        FORMAT('0THE LAST OBSERVATION SHOULD BE GIVEN AS -99999.')
40        READ, X
          IF(X .EQ. -99999)GO TO 50
          N=N+1
          SUMX=SUMX+X
          Y(N)=X
          GO TO 40
50        IF(NOP .EQ. 2) GO TO 110
          WRITE(IOUT,89)
89        FORMAT('0DO YOU WANT TO SEE THE RAW DATA?'/
         1' IF YES, TYPE A 1.  IF NO, TYPE A 0.')
          READ, IDATA
          IF(IDATA .EQ. 1)WRITE(IOUT,90)(Y(I),I=1,N)
          NM1=N-1
          MEAN=SUMX/N
          DO 55 I=1,N
55        SUMSQ=SUMSQ+(Y(I)-MEAN)**2
          VAR=SUMSQ/NM1
          STDEV=SQRT(VAR)
          WRITE(IOUT,60)MEAN,VAR,STDEV,N
60        FORMAT('0THE MEAN IS ',F8.4/' THE VARIANCE IS ',F10.2/
         1' THE STANDARD DEVIATION IS ',F8.4/' THE SAMPLE SIZE IS ',I5)
          WRITE(IOUT,70)
70        FORMAT('0WOULD YOU LIKE TO SEE A SORTED DATA LIST?'/
         1' IF YES, TYPE 1.  IF NO, TYPE 0.')
          READ, ISORT
          IF(ISORT .EQ. 0)GO TO 100
          DO 80 I=1,NM1
          K=I+1
          DO 80 J=K,N
          IF(Y(I) .LE. Y(J))GO TO 80
          W=Y(I)
          Y(I)=Y(J)
          Y(J)=W
80        CONTINUE
          WRITE(IOUT,90)(Y(I),I=1,N)
90        FORMAT(1X,10F7.1)
100       STOP
110       WRITE(IOUT,115)
115       FORMAT('0IF YOU WANT TO AVERAGE ADJACENT OBSERVATIONS,',
         1' TYPE A 1.'/' IF YOU WANT TO SELECT EVERY N-TH ',
         2'OBSERVATION, TYPE A 0.')
          READ, ISEL
          IF(ISEL .EQ. 0)GO TO 200
112       WRITE(IOUT,120)
120       FORMAT('0HOW MANY ADJACENT OBSERVATIONS DO',
         1' YOU WANT TO AVERAGE?')
          READ, NADJ
          NTIMES=N/NADJ
          DO 140 I1=1,NTIMES
          I=1+NADJ*(I1-1)
          M(I1)=0
          DO 130 J=1,NADJ
          K=I+J-1
130       M(I1)=M(I1)+Y(K)
140       M(I1)=M(I1)/NADJ
160       WRITE(IOUT,170)
170       FORMAT('0IF YOU WANT YOUR OUTPUT WRITTEN ON PAPER ONLY,',
         1' TYPE A 1.'/' IF YOU WANT YOUR OUTPUT WRITTEN ON DISK ',
         2'ONLY, TYPE A 2.'/' IF YOU WANT YOUR OUTPUT WRITTEN BOTH',
         3' ON PAPER AND ON DISK, TYPE A 3.')
          READ, ICOPY
          IF(ICOPY .EQ. 1 .OR. ICOPY .EQ. 3)WRITE(IOUT,180)(M(I),I=1,NTIMES)
```

```
180     FORMAT(1X,10F7.1)
        IF(ICOPY .EQ. 2 .OR. ICOPY .EQ. 3)WRITE(30,190)(M(I),I=1,NTIMES)
190     FORMAT(F7.2)
        STOP
200     WRITE(IOUT,210)
210     FORMAT('0YOU CHOSE SELECTION OF EVERY N-TH OBSERVATION.'/
       1' TYPE THE VALUE OF N.')
        READ, NSEL
        NTIMES=0
        DO 220 K=1,N,NSEL
        NTIMES=NTIMES+1
220     M(NTIMES)=Y(K)
        GO TO 160
        END
        GO TO 610
620     WRITE(IOUT,521)
  610   CONTINUE
 1000   FORMAT(1X,25HIN SYSTEM=                            ,F15.5)
 1001   FORMAT(1X,25HIN QUEUE=                             ,F15.5)
 1002   FORMAT(1X,25HIN QUEUE FOR BUSY SYSTEM=             ,F15.5)
 1003   FORMAT(1X,25HIN SYSTEM=                            ,F15.5)
 1004   FORMAT(1X,25HIN QUEUE=                             ,F15.5)
 1005   FORMAT(1X,25HIN QUEUE FOR BUSY SYSTEM=             ,F15.5)
 1006   FORMAT(1X,40HSTEADY-STATE MEAN NUMBER OF UNITS:         )
 1007   FORMAT(1X,40HSTEADY-STATE MEAN TIME:                    )
 1008   FORMAT(1X,40HIN QUEUE FOR BUSY SYSTEMS IS               )
 1009   FORMAT(1X,45HUNDEFINED FOR P(N> OR =C) APPROACHING ZERO.          )
        RETURN
        END
        FUNCTION SUMM (R,NUT)
        IF(NUT.GT.0) GO TO 500
        SUMM=1
        GO TO 1100
  500   CONTINUE
        IF(NUT.GT.1) GO TO 800
        SUMM=1+R
        GO TO 1100
  800   CONTINUE
        SUMM=1
        CONST=1
        DO 1000 I=1,NUT
        CONST=CONST*R/I
        SUMM=SUMM+CONST
        IF(CONST.GT. 0.00000001) GO TO 1000
        CONST=0
 1000   CONTINUE
 1100   CONTINUE
        RETURN
        END


        PROGRAM GOFFIT
C       TEST THE GOODNESS OF FIT OF EITHER THE NEGATIVE EXPONENTIAL
C       DISTRIBUTION, THE POISSON DISTRIBUTION, THE NORMAL DISTRIBUTION,
C       THE UNIFORM DISTRIBUTION, OR AN EMPIRICALLY DERIVED
C       DISTRIBUTION.  THE USER MUST PROVIDE THE MEAN AND/OR STANDARD
C       DEVIATION IF APPROPRIATE AS WELL AS CLASSES AND
C       CLASS FREQUENCIES.  THE USER MAY SELECT EITHER A CHI SQUARE
C       TEST OR A KOLMOGOROV-SMIRNOV TEST.
        IOUT=2
        REAL MEAN,II1,II2
        DIMENSION WTH(100),OBSFRQ(100),ULIM(100),CCHISQ(100),CHI(25)
        DIMENSION CRLFRQ(100),CTHFRQ(100),EXPFRQ(100),RELFRQ(100)
        DIMENSION ADIFF(100),ORLFRQ(100),XKS(35),PR(100),BOUND(100)
        DATA CHI(1),CHI(2),CHI(3),CHI(4)/3.841,5.991,7.815,9.488/
        DATA CHI(5),CHI(6),CHI(7),CHI(8)/11.070,12.592,14.067,15.507/
        DATA CHI(9),CHI(10),CHI(11),CHI(12)/16.919,18.307,19.675,21.026/
        DATA CHI(13),CHI(14),CHI(15),CHI(16)/22.362,23.685,24.996,26.296/
```

```
        DATA CHI(17),CHI(18),CHI(19)/27.587,28.869,30.144/
        DATA CHI(21),CHI(22),CHI(23)/32.671,33.924,35.172/
        DATA CHI(24),CHI(25),CHI(20)/36.415,37.652,31.410/
        DATA XKS(1),XKS(2),XKS(3),XKS(4)/.975,.842,.708,.624/
        DATA XKS(6),XKS(7),XKS(8),XKS(9)/.521,.486,.457,.432/
        DATA XKS(11),XKS(12),XKS(13),XKS(14)/.391,.375,.361,.349/
        DATA XKS(16),XKS(17),XKS(18),XKS(19)/.328,.318,.309,.301/
        DATA XKS(21),XKS(22),XKS(23),XKS(24)/.287,.281,.275,.269/
        DATA XKS(26),XKS(27),XKS(28),XKS(29)/.259,.254,.250,.246/
        DATA XKS(31),XKS(32),XKS(33),XKS(34)/.238,.234,.231,.227/
        DATA XKS(5),XKS(10),XKS(15),XKS(20)/.565,.410,.338,.294/
        DATA XKS(25),XKS(30),XKS(35)/.264,.242,.224/
1       TCHISQ=0
        TOBSFR=0
        TORLFR=0.
        TTFRQ=0.
        DIFF=0.
        TEXPFR=0
        TRELFR=0
        DO 3 I=1,100
        WTH(I)=0.
        OBSFRQ(I)=0.
        ORLFRQ(I)=0.
        ADIFF(I)=0.
        ULIM(I)=0.
        CCHISQ(I)=0.
        EXPFRQ(I)=0.
        CRLFRQ(I)=0.
        CTHFRQ(I)=0.
        BOUND(I)=0.
        PR(I)=0.
3       RELFRQ(I)=0.
        WRITE(IOUT,2)
2       FORMAT('0GOODNESS OF FIT TESTS')
        WRITE(IOUT,4)
4       FORMAT('0DO YOU WANT INSTRUCTIONS?',
       1' TYPE 1 FOR YES, 0 FOR NO.')
        READ, INST
        IF(INST .EQ. 0)GO TO 57
        WRITE(IOUT,36)
36      FORMAT('0YOU MAY CHOOSE EITHER A CHI SQUARE OR A '/
       1' KOLMOGOROV-SMIRNOV GOODNESS OF FIT TEST.')
        WRITE(IOUT,5)
5       FORMAT('0WHICH DISTRIBUTION DO YOU WANT TO TEST '
       1/' FOR GOODNESS OF FIT?'/' YOUR ALTERNATIVES ARE:'
       2/' 1. NEGATIVE EXPONENTIAL DISTRIBUTION'
       3/' 2. POISSON DISTRIBUTION'/' 3. NORMAL DISTRIBUTION'
       4/' 4. UNIFORM DISTRIBUTION'/' 5. EMPIRICALLY DEFINED '
       5'DISTRIBUTION')
        WRITE(IOUT,40)
40      FORMAT('0***WARNING***BE SURE THAT THE CLASSES YOU '/
       1' DEFINED INCLUDE A MUTUALLY EXCLUSIVE,COLLECTIVELY '/
       2' EXHAUSTIVE SET OF POSSIBLE OUTCOMES, SUCH AS '/
       3' "PLUS OR MINUS INFINITY", IF APPROPRIATE ')
57      WRITE(IOUT,6)
6       FORMAT('0TYPE THE NUMBER OF THE DISTRIBUTION YOU WANT TO TEST')
61      READ, IDIST
        WRITE(IOUT,7)
7       FORMAT('0YOU HAVE CHOSEN THE GOODNESS OF FIT TEST FOR:')
        IF(IDIST .EQ. 1) WRITE(IOUT,8)
8       FORMAT(' 1. NEGATIVE EXPONENTIAL DISTRIBUTION.')
        IF(IDIST .EQ. 2) WRITE(IOUT,9)
9       FORMAT(' 2. POISSON DISTRIBUTION.')
        IF(IDIST .EQ. 3)WRITE(IOUT,10)
10      FORMAT(' 3. NORMAL DISTRIBUTION.')
        IF(IDIST .EQ. 4)WRITE(IOUT,11)
11      FORMAT(' 4. UNIFORM DISTRIBUTION.')
```

```
      IF(IDIST .EQ. 5)WRITE(IOUT,12)
12    FORMAT(' 5. EMPIRICALLY DEFINED DISTRIBUTION.')
      IF(IDIST .GT. 5)WRITE(IOUT,13)
13    FORMAT('0YOU MUST CHOOSE A NUMBER FROM 1 THROUGH 5.')
      IF(IDIST .GT. 5) GO TO 61
      WRITE(IOUT,38)
38    FORMAT('0IF YOU WANT TO DO A CHI SQUARE TEST, TYPE A 1.'/
     1' IF YOU WANT TO DO A KOLMOGOROV-SMIRNOV TEST, TYPE A 0.')
      READ, ITEST
      IF(IDIST .EQ. 1)WRITE(IOUT,16)
16    FORMAT('0ENTER THE MEAN TIME')
      IF(IDIST .EQ. 2)WRITE(IOUT,18)
18    FORMAT('0ENTER THE MEAN RATE')
      IF(IDIST .EQ. 3)WRITE(IOUT,20)
20    FORMAT('0ENTER THE MEAN')
      IF(IDIST .LE. 3)READ, MEAN
      IF(IDIST .EQ. 3)WRITE(IOUT,25)
25    FORMAT('0ENTER THE STANDARD DEVIATION')
      IF(IDIST .EQ. 3)READ, STDEV
      IF(ITEST .EQ. 1)WRITE(IOUT,29)
29    FORMAT('0 YOU HAVE CHOSEN A CHI SQUARE TEST.')
      IF(ITEST .EQ. 0)WRITE(IOUT,31)
31    FORMAT('0YOU HAVE CHOSEN A KOLMOGOROV-SMIRNOV TEST.')
      WRITE(IOUT,30)
30    FORMAT('0ENTER THE NUMBER OF CLASSES')
      READ, NCLASS
      IF(IDIST .EQ. 2) GO TO 300
      IF(IDIST .EQ. 3) GO TO 400
      IF(IDIST .EQ. 4) GO TO 500
      IF(IDIST .EQ. 5) GO TO 600
      IF(IDIST .NE. 1) GO TO 1
      WRITE(IOUT,705)
705   FORMAT('0IF YOU WANT TO DO A GOODNESS OF FIT TEST,',
     1' TYPE A 1.'/' IF YOU WANT TO COMPUTE BOUNDARIES,',
     2' TYPE A 0.')
      READ, IPRO
      IF(IPRO .EQ. 1) GO TO 790
      XINC=1./NCLASS
      MEAN=1./MEAN
      PR(1)=XINC
      WRITE(IOUT,710)
710   FORMAT('0CLASS',3X,'CUMULATIVE GT',5X,'CLASS',5X,'CLASS'/
     1' IDENT',3X,'PROBABILITY',5X,'BOUNDARY',4X,'WIDTH'/)
      DO 750 IJ=1,NCLASS
      BOUND(IJ)=ALOG(PR(IJ))/(-MEAN)
750   PR(IJ+1)=PR(IJ)+XINC
      DO 760 IJ=1,NCLASS
      IF(IJ .NE. NCLASS)WTH(IJ)=BOUND(NCLASS-IJ)-BOUND(NCLASS-IJ+1)
      IF(IJ .EQ. NCLASS)WTH(IJ)=9999.99
760   WRITE(IOUT,720)IJ,PR(NCLASS-IJ+1),BOUND(NCLASS-IJ+1),WTH(IJ)
720   FORMAT(1X,I5,6X,F6.4,6X,F8.2,2X,F8.2)
      WRITE(IOUT,770)
770   FORMAT('0IF YOU WANT TO PROCEED TO A CHI SQUARE GOODNESS '/
     1' OF FIT TEST OF THIS NEGATIVE EXPONENTIAL DISTRIBUTION, '/
     2' TYPE A 1.  OTHERWISE, TYPE A 0.')
      READ, NEGF
      IF(NEGF .EQ. 0) GO TO 178
      IF(NEGF .EQ. 1)GO TO 55
790   WTH(NCLASS)=9999.
      NXTLST=NCLASS-1
      WRITE(IOUT,54)
54    FORMAT('0','IF ALL CLASSES EXCEPT THE LAST ARE OF EQUAL WIDTH,'/
     1' ENTER THAT WIDTH.',' OTHERWISE,ENTER 999.99')
      READ, X
      IF(X .EQ. 999.99) GO TO 59
      DO 53 I=1,NXTLST
```

```
53       WTH(I)=X
55       DO 58 I=1,NCLASS
         NXTLST=NCLASS-1
         WRITE(IOUT,56)I
56       FORMAT('0ENTER THE OBSERVED FREQUENCY FOR CLASS ',I3)
         READ,N
         OBSFRQ(I)=N
58       TOBSFR=TOBSFR+OBSFRQ(I)
         GO TO 98
59       DO 95 I=1,NXTLST
         WRITE(IOUT,60)I
60       FORMAT('0ENTER THE CLASS WIDTH FOR CLASS ',I3)
         READ, WTH(I)
         WRITE(IOUT,80)I
80       FORMAT('0ENTER THE OBSERVED FREQUENCY FOR CLASS ',I3)
         READ, N
         OBSFRQ(I)=N
95       TOBSFR=TOBSFR+OBSFRQ(I)
         WRITE(IOUT,96)
96       FORMAT('0','ENTER **ONLY** THE OBSERVED FREQUENCY,'
        1'NOT THE CLASS WIDTH, FOR THE LAST CLASS')
         READ, N
         OBSFRQ(NCLASS)=N
         TOBSFR=TOBSFR+OBSFRQ(NCLASS)
98       ULIM(1)=WTH(1)
         IF(IPRO .EQ. 1)MEAN=1./MEAN
         EXPON1=(2.71828**(-1.*MEAN*ULIM(1)))
         RELFRQ(1)=1.-EXPON1
         CRLFRQ(1)=RELFRQ(1)
         DO 100 I=2,NXTLST
         ULIM(I)=ULIM(I-1)+WTH(I)
         CRLFRQ(I)=1.-2.71828**(-1.*MEAN*ULIM(I))
100      RELFRQ(I)=CRLFRQ(I)-CRLFRQ(I-1)
         RELFRQ(NCLASS)=2.71828**(-1.*MEAN*ULIM(NXTLST))
         ULIM(NCLASS)=9999
         GO TO 700
C  BEGIN POISSON CALCULATIONS
300      NXTLST=NCLASS-1
         WRITE(IOUT,303)
303      FORMAT('0**NOTE**ONLY SEQUENTIAL INTEGER VALUES BEGINNING '/
        1' WITH ZERO MAY BE USED.')
         DO 320 I=1,NXTLST
         WRITE(IOUT,305)I
305      FORMAT('0ENTER THE CLASS MIDPOINT FOR CLASS ',I3)
         READ, MPT
         ULIM(I)=MPT
         WRITE(IOUT,310)I
310      FORMAT('0ENTER THE OBSERVED FREQUENCY FOR CLASS ',I3)
         READ, OBSFRQ(I)
         TOBSFR=TOBSFR+OBSFRQ(I)
         RELFRQ(I)=MEAN**MPT*2.71828**(-MEAN)/FCTRL(MPT)
         IF(I .EQ. 1)CRLFRQ(I)=RELFRQ(I)
320      IF(I .NE. 1)CRLFRQ(I)=CRLFRQ(I-1)+RELFRQ(I)
         WRITE(IOUT,330)NCLASS
330      FORMAT('0ENTER**ONLY**THE OBSERVED FREQUENCY FOR CLASS ',I3)
         READ, OBSFRQ(NCLASS)
         RELFRQ(NCLASS)=1.-CRLFRQ(NXTLST)
         TOBSFR=TOBSFR+OBSFRQ(NCLASS)
         ULIM(NCLASS)=9999
         GO TO 700
C  BEGIN NORMAL CALCULATIONS
400      NXTLST=NCLASS-1
         WRITE(IOUT,405)
405      FORMAT('0ENTER THE UPPER CLASS BOUNDARY FOR CLASS   1')
         READ, ULIM(1)
         WRITE(IOUT,408)
```

```
408    FORMAT('0ENTER THE OBSERVED FREQUENCY FOR CLASS    1')
       READ, OBSFRQ(1)
       TOBSFR=TOBSFR+OBSFRQ(1)
       XLOW=MEAN-4.*STDEV
       XHI=ULIM(1)
       NN=10
       P1=1./(2.*3.1415926)**(0.5)
       A=(XLOW-MEAN)/STDEV
       B=(XHI-MEAN)/STDEV
       NN=2*NN
       D=(B-A)/NN
       II1=0
       II2=0
       II3=NN-1
       DO 425 II=1,II3,2
       Z=A+II*D
       P=P1*2.71828**(-Z**2/2.)
       II1=II1+P
       Z=A+(II+1)*D
       P=P1*2.71828**(-Z**2/2.)
425    II2=II2+P
       ZA=P1*2.71828**(-A**2/2.)
       ZB=P1*2.71828**(-B**2/2.)
       S=(4./3.)*D*II1+(2./3.)*D*II2+(1./3.)*D*(ZA-ZB)
       RELFRQ(1)=S
       CRLFRQ(1)=RELFRQ(1)
       IF(NCLASS .EQ. 2)GO TO 460
       DO 450 I=2,NXTLST
       WRITE(IOUT,410)I
410    FORMAT('0ENTER THE UPPER CLASS BOUNDARY FOR CLASS ',I3)
       READ, ULIM(I)
       WRITE(IOUT,420)I
420    FORMAT('0ENTER THE OBSERVED FREQUENCY FOR CLASS ',I3)
       READ, OBSFRQ(I)
       TOBSFR=TOBSFR+OBSFRQ(I)
       XLOW=ULIM(I-1)
       XHI=ULIM(I)
       A=(XLOW-MEAN)/STDEV
       B=(XHI-MEAN)/STDEV
       D=(B-A)/NN
       II1=0
       II2=0
       DO 435 II=1,II3,2
       Z=A+II*D
       P=P1*2.71828**(-Z**2/2.)
       II1=II1+P
       Z=A+(II+1)*D
       P=P1*2.71828**(-Z**2/2.)
435    II2=II2+P
       ZA=P1*2.71828**(-A**2/2.)
       ZB=P1*2.71828**(-B**2/2.)
       S=(4./3.)*D*II1+(2./3.)*D*II2+(1./3.)*D*(ZA-ZB)
       RELFRQ(I)=S
450    CRLFRQ(I)=CRLFRQ(I-1)+RELFRQ(I)
460    WRITE(IOUT,470)NCLASS
470    FORMAT('0ENTER***ONLY***THE OBSERVED FREQUENCY FOR CLASS ',I3)
       READ, OBSFRQ(NCLASS)
       ULIM(NCLASS)=9999.
       TOBSFR=TOBSFR+OBSFRQ(NCLASS)
       RELFRQ(NCLASS)=1.-CRLFRQ(NXTLST)
       GO TO 700
C  BEGIN UNIFORM CALCULATIONS
500    DO 520 I=1,NCLASS
       WRITE(IOUT,505)I
505    FORMAT('0ENTER THE CLASS MIDPOINT FOR CLASS ',I3)
       READ, ULIM(I)
```

```
         WRITE(IOUT,510)I
510      FORMAT('0ENTER THE OBSERVED FREQUENCY FOR CLASS NUMBER ',I3)
         READ, OBSFRQ(I)
520      TOBSFR=TOBSFR+OBSFRQ(I)
         DO 530 I=1,NCLASS
530      RELFRQ(I)=1./NCLASS
         GO TO 700
C BEGIN EMPIRICAL CALCULATIONS
600      DO 630 I=1,NCLASS
         WRITE(IOUT,610)I
610      FORMAT('0ENTER THE OBSERVED FREQUENCY FOR CLASS ',I3)
         READ, OBSFRQ(I)
         TOBSFR=TOBSFR+OBSFRQ(I)
         WRITE(IOUT,615)I
615      FORMAT('0ENTER THE EMPIRICAL RELATIVE FREQUENCY FOR CLASS ',I3/
        1' AS A DECIMAL FRACTION')
         JLIM(I)=I
630      READ, RELFRQ(I)
         GO TO 700
700      DO 110 I=1,NCLASS
         IF(ITEST .EQ. 0)TORLFR=TORLFR+OBSFRQ(I)/TOBSFR
         IF(ITEST .EQ. 0)ORLFRQ(I)=TORLFR
         IF(ITEST .EQ. 0)TTFRQ=TTFRQ+RELFRQ(I)
         IF(ITEST .EQ. 0)CTHFRQ(I)=TTFRQ
         IF(ITEST .EQ. 0)ADIFF(I)=ABS(ORLFRQ(I)-CTHFRQ(I))
         EXPFRQ(I)=TOBSFR*RELFRQ(I)
         IF(EXPFRQ(I) .LT. 0.0001) CCHISQ(I)=999.99
         IF(EXPFRQ(I) .LT. 0.0001) GO TO 110
         CCHISQ(I)=((OBSFRQ(I)-EXPFRQ(I))**2)/EXPFRQ(I)
110      CONTINUE
         IF(ITEST .EQ. 0)GO TO 900
         WRITE(IOUT,120)
120      FORMAT('0',8X,'CLASS',3X,'OBSERVED',3X,'RELATIVE',
        13X,'EXPECTED',3X,'  CHI')
         WRITE(IOUT,130)
130      FORMAT(1X,'CLASS',3X,'IDENT',3X,'  FREQ. ',3X,'  FREQ. ',
        13X,'  FREQ. ',3X,'SQUARE'/)
         DO 140 I=1,NCLASS
         TCHISQ=TCHISQ+CCHISQ(I)
         TEXPFR=TEXPFR+EXPFRQ(I)
140      TRELFR=TRELFR+RELFRQ(I)
         DO 150 I=1,NCLASS
150      WRITE(IOUT,160)I,ULIM(I),OBSFRQ(I),RELFRQ(I),EXPFRQ(I),CCHISQ(I)
160      FORMAT(1X,I5,2X,F6.1,2X,F8.2,3X,F8.3,3X,F8.2,3X,F6.2)
         WRITE(IOUT,170)TOBSFR,TRELFR,TEXPFR,TCHISQ
170      FORMAT('0','**TOTALS***',3X,F9.2,2X,F9.3,2X,F9.2,
        13X,F6.2/)
         IF(IDIST .EQ. 1 .OR. IDIST .EQ. 2)GO TO 210
         IF(IDIST .EQ. 3)GO TO 220
         NDF=NCLASS-1
         WRITE(IOUT,171)NDF
171      FORMAT('0DEGREES OF FREEDOM EQUALS NUMBER OF CLASSES '/
        1' MINUS ONE, OR ',I3,'.')
         MIN1=NDF
         MIN2=NDF
         GO TO 195
190      MIN1=NCLASS-NDF1
         MIN2=NCLASS-NDF2
         WRITE(IOUT,172)NDF1,MIN1
172      FORMAT('0','IF YOU USED THESE OBSERVED DATA TO ESTIMATE '
        1'HYPOTHESIS PARAMETERS,'/' DEGREES OF FREEDOM EQUALS '
        2'NUMBER OF CLASSES MINUS ',I3,' OR ',I3,' .')
         WRITE(IOUT,173)NDF2,MIN2
173      FORMAT('0','IF YOU USED AN INDEPENDENT ESTIMATE OF THE '
        1'HYPOTHESIS PARAMETERS, '/' DEGREES OF FREEDOM EQUALS ',
        2'NUMBER OF CLASSES MINUS ',I3,' OR ',I3,'.')
```

```
195     IF(MIN1 .LE. 0 .OR. MIN2 .LE. 0)WRITE(IOUT,200)
200     FORMAT('0DEGREES OF FREEDOM ARE DUBIOUS.')
        IF(MIN1 .GT. 0 .AND. MIN2 .GT. 0 .AND. IDIST .LE. 3
       1.AND. MIN2 .LE. 20)WRITE(IOUT,201)CHI(MIN1),MIN1,CHI(MIN2),MIN2
        IF(MIN1 .GT. 0 .AND. MIN2 .GT. 0 .AND. IDIST .GT. 3
       1.AND. MIN1 .LE. 20)WRITE(IOUT,202)CHI(MIN1),MIN1
201     FORMAT('0CRITICAL CHI-SQUARE VALUES AT THE .05 LEVEL OF',
       1' SIGNIFICANCE ARE '/F9.3,' FOR ',I3,' DEGREES OF FREEDOM ',
       2'AND ',F8.3,' FOR ',I3,' DEGREES OF FREEDOM.')
202     FORMAT('0CRITICAL CHI-SQUARE VALUE AT THE .05 LEVEL OF',
       1' SIGNIFICANCE IS '/F9.3,' FOR ',I3,' DEGREES OF FREEDOM.')
203     DO 174 I=1,NCLASS
        EXPFRQ(I)=EXPFRQ(I)+.001
        IF(EXPFRQ(I) .LT. 5)GO TO 175
174     CONTINUE
        GO TO 178
175     WRITE(IOUT,176)
176     FORMAT('0AT LEAST ONE EXPECTED FREQUENCY IS LESS THAN 5.',
       1' YOU MIGHT COMBINE CLASSES.')
178     WRITE(IOUT,180)
180     FORMAT('0IF YOU WANT TO SOLVE ANOTHER PROBLEM, TYPE A 1.',
       1' OTHERWISE, TYPE A 0.')
        READ, J
        IF(J .EQ. 1) GO TO 1
        STOP
210     NDF1=2
        NDF2=1
        GO TO 190
220     NDF1=3
        NDF2=1
        GO TO 190
900     WRITE(IOUT,910)
910     FORMAT('0',26X,'OBSERVED',3X,'THEORETICAL')
        WRITE(IOUT,915)
915     FORMAT(8X,'CLASS',2X,'OBSERVED',3X,'CUMULATIVE'2X,
       1'CUMULATIVE',4X,'ABSOLUTE')
        WRITE(IOUT,920)
920     FORMAT(1X,'CLASS',2X,'IDENT',2X,'FREQUENCY',2X,
       1'PROPORTION',2X,'PROPORTION',3X,'DIFFERENCE'/)
        DO 940 I=1,NCLASS
        WRITE(IOUT,930)I,ULIM(I),OBSFRQ(I),ORLFRQ(I),CTHFRQ(I),ADIFF(I)
930     FORMAT(1X,I5,1X,F6.1,4X,F5.0,5X,F6.4,6X,F6.4,8X,F6.4)
        IF(ADIFF(I) .GT. DIFF) DIFF=ADIFF(I)
940     CONTINUE
        WRITE(IOUT,945)TOBSFR
945     FORMAT('0',' ***TOTAL***',3X,F6.0)
        WRITE(IOUT,950)DIFF
950     FORMAT('0THE MAXIMUM SAMPLE DIFFERENCE IS ',F8.4)
        IF(TOBSFR .LE. 35)CXKS=XKS(TOBSFR)
        IF(TOBSFR .GT. 35)CXKS=1.36/SQRT(TOBSFR)
        WRITE(IOUT,960)CXKS
960     FORMAT('0THE CRITICAL VALUE OF D IS ',F8.4,
       1' AT THE .05 LEVEL OF SIGNIFICANCE.')
        GO TO 178
        END
        FUNCTION FCTRL(NUM)
        FCTRL = 1
        IF(NUM .EQ. 0 .OR. NUM .EQ. 1)GO TO 9001
        DO 9000 L=1,NUM
9000    FCTRL=FCTRL*L
9001    RETURN
        END

        PROGRAM QMODEL
C       THIS PROGRAM COMPUTES PERFORMANCE CHARACTERISTICS SUCH
C       AS LENGTH OF QUEUE, NUMBER IN QUEUE, TIME IN SYSTEM, AND
C       NUMBER IN SYSTEM FOR SOME COMMON WAITING LINE SITUATIONS
```

```
C      SUCH AS M/M/C AND M/G/1.
       DIMENSION P(1000)
       IOUT=2
       WRITE(IOUT,47)
47     FORMAT('0QUEUING MODELS')
49     WRITE(IOUT,51)
51     FORMAT('0',1X,'DO YOU WANT INSTRUCTIONS?',
      1/' ','TYPE 1 FOR YES, 0 FOR NO')
       READ, INS
       IF(INS .EQ. 0) GO TO 416
  100  WRITE(IOUT,200)
200    FORMAT(1X,'THIS PROGRAM WORKS INTERACTIVELY WITH THE ')
       WRITE(IOUT,300)
  300  FORMAT(1X,37HUSER TO PERFORM QUEUING CALCULATIONS.      )
   50  CONTINUE
       WRITE(IOUT,400)
  400  FORMAT(/1X,36HTHE FOLLOWING MODELS ARE AVAILABLE:       )
       WRITE(IOUT,401)
  401  FORMAT(/1X,30H1. FINITE QUEUE M/M/1 MODEL.       )
       WRITE(IOUT,402)
  402  FORMAT(1X,40H2. FINITE POPULATION M/M/1 MODEL.        )
       WRITE(IOUT,403)
  403  FORMAT(1X,40H3. STANDARD M/M/C MODEL.                 )
       WRITE(IOUT,404)
  404  FORMAT(1X,40H4. FINITE QUEUE M/M/C MODEL.             )
       WRITE(IOUT,405)
  405  FORMAT(1X,40H5. FINITE POPULATION M/M/C MODEL.        )
       WRITE(IOUT,406)
  406  FORMAT(1X,40H6. SELF-SERVICE MODEL.                   )
       WRITE(IOUT,411)
411    FORMAT(1X,38H7. POLLACZEK-KHINTCHINE (P-K) FORMULA.)
       WRITE(IOUT,412)
412    FORMAT(1X,16H   M/G/1 SYSTEM.)
       WRITE(IOUT,408)
  408  FORMAT(1X,40HTYPE THE APPROPRIATE NUMBER.             )
416    WRITE(IOUT,407)
  407  FORMAT('0',1X,40HWHICH MODEL DO YOU WISH TO USE?          )
   40  CONTINUE
       READ, I
  409  FORMAT('0',1X,30HYOU HAVE CHOSEN                          )
       IF(I.LE.7) GO TO 500
       WRITE(IOUT,450)
  450  FORMAT(1X,40HYOU BLEW IT  TRY AGAIN.                  )
       WRITE(IOUT,451)
  451  FORMAT(1X,40HBE SURE TO PICK BETWEEN 1 AND 7          )
       GO TO 40
  500  CONTINUE
       WRITE(IOUT,409)
       IF(I.EQ.1) WRITE(IOUT,401)
       IF(I.EQ.2) WRITE(IOUT,402)
       IF(I.EQ.3) WRITE(IOUT,403)
       IF(I.EQ.4) WRITE(IOUT,404)
       IF(I.EQ.5) WRITE(IOUT,405)
       IF(I.EQ.6) WRITE(IOUT,406)
       IF(I .EQ. 7)WRITE(IOUT,411)
  550  CONTINUE
       WRITE(IOUT,800)
       READ, ALAMDA
  800  FORMAT(/1X,40HWHAT IS THE MEAN ARRIVAL RATE?            )
       WRITE(IOUT,801)
       READ, AMU
  801  FORMAT(/1X,40HWHAT IS THE MEAN SERVICE RATE?            )
  802  FORMAT(/1X,40HWHAT IS THE NUMBER OF SERVERS?           )
  803  FORMAT(/1X,46HWHAT IS THE MAXIMUM NUMBER ALLOWED IN SYSTEM?   )
  804  FORMAT(/1X,46HHOW MANY UNITS IN THE FINITE POPULATION?        )
       RHO=ALAMDA/AMU
```

```
          IF(I.EQ.1) GO TO 1000
          IF(I.EQ.2) GO TO 2000
          IF(I.EQ.3) GO TO 3000
          IF(I.EQ.4) GO TO 4000
          IF(I.EQ.5) GO TO 5000
          IF(I.EQ.6) GO TO 6000
          IF(I.EQ.7)GO TO 7000
     1000 CONTINUE
          IF(ALAMDA.LT.AMU) GO TO 1010
          GO TO 1100
     1014 WRITE(IOUT,1011)
     1011 FORMAT(1X,45HTHIS METHOD WILL NOT WORK UNLESS THE           )
     1012 FORMAT(1X,45HMEAN SERVICE RATE EXCEEDS THE MEAN             )
     1013 FORMAT(1X,45HARRIVAL RATE. TRY AGAIN                        )
          WRITE(IOUT,1012)
          WRITE(IOUT,1013)
          GO TO 550
     1010 CONTINUE
     1100 IF(I.EQ.1) WRITE(IOUT,803)
          READ, N
          IF(RHO.EQ.1) GO TO 1500
          PNOT=(1-RHO)/(1-RHO**(N+1))
          ALS=(RHO/(1-RHO))-((N+1)*(RHO**(N+1)))/((1-RHO)/PNOT)
          GO TO 1600
     1500 CONTINUE
          PNOT=1./(N+1)
          ALS=N/2.
     1600 CONTINUE
     1601 FORMAT(1X,7HP(  0)=,3X,F15.5)
          PNGTO=1-PNOT
          WRITE(IOUT,1602) PNGTO
          WRITE(IOUT,1601) PNOT
     1602 FORMAT(1X,12HP(N> OR =0)=,3X,F15.5)
          DO 1650 J=1,N
          P(J)=PNOT*(RHO**J)
          XX=RHO**J
          IF(XX.GT. 0.00000001) GO TO 1640
          P(J)=0
     1640 CONTINUE
          WRITE(IOUT,1400) J,P(J)
     1400 FORMAT(1X,2HP(,I3,2H)=,3X,F15.5)
     1650 CONTINUE
          ALQ=ALS-1+PNOT
          ALB=ALQ/PNGTO
          WS=ALQ/(ALAMDA*(1-P(N))) +1/AMU
          WQ=WS-(1/AMU)
          WB=WQ/(1-PNOT)
          CALL GA(ALS,ALQ,ALB,WS,WQ,WB)
          GO TO 8000
     2000 CONTINUE
     2010 IF(I.NE.2) GO TO 3000
          WRITE(IOUT,804)
          WRITE(IOUT,1611)
          WRITE(IOUT,1612)
     1611 FORMAT(1X,45HDUE TO THE CONSTRAINTS OF THE OPERATING        )
     1612 FORMAT(1X,45HSYSTEM, PICK A NUMBER LESS THAN 51.            )
          READ, M
          IF(M.LE.50) GO TO 2100
     1613 FORMAT(1X,45H***ERROR***MGT-50 NOTE FOLLOWS                 )
     1614 FORMAT(1X,45HIF THE NUMBER OF UNITS IN THE FINITE           )
     1615 FORMAT(1X,45HPOPULATION IS MORE THAN FIFTY, CONSIDER        )
     1616 FORMAT(1X,45HUSING THE FINITE QUEUE MODEL M/M/1             )
     1617 FORMAT(1X,45HWHICH APPROXIMATES THIS MODEL FOR FINITE       )
     1618 FORMAT(1X,45HPOPULATIONS MORE THAN THIRTY.                  )
          WRITE(IOUT,1613)
          WRITE(IOUT,1614)
```

```
            WRITE(IOUT,1615)
            WRITE(IOUT,1616)
            WRITE(IOUT,1617)
            WRITE(IOUT,1618)
            GO TO 2010
      2100  CONTINUE
C           COMPUTE PNOT
            SUM=0
            M1=M+1
            DO 2200 LIKE=1,M1
            SUM=SUM+(FCTRL(M)*RHO**(LIKE-1)/FCTRL(M-LIKE+1))
      2200  CONTINUE
            PNOT=1/SUM
       820  FORMAT(1X,7HP(0)=         ,F15.5)
            PNGTO=1-PNOT
            WRITE(IOUT,1602) PNGTO
            WRITE(IOUT,1601) PNOT
            DO 2250 KIS=1,M
            P(KIS)=(FCTRL(M)*(RHO**(KIS))*PNOT)/FCTRL(M-KIS)
            IF(P(KIS) .GE. 0.01)WRITE(IOUT,1400) KIS,P(KIS)
      2250  CONTINUE
      2260  ALS=M-(1-PNOT)/RHO
            ALQ=M-(ALAMDA+AMU)*(1-PNOT)/ALAMDA
            ALB=ALQ/(1-PNOT)
            WS=((M/AMU)/(1-PNOT))-1/ALAMDA
            WQ=(1/AMU)*(M/(1-PNOT)-(ALAMDA+AMU)/ALAMDA)
            WB=WQ/(1-PNOT)
            CALL GA (ALS,ALQ,ALB,WS,WQ,WB)
            GO TO 8000
      3000  CONTINUE
            WRITE(IOUT,802)
            READ, C
            IF(RHO.LT.C) GO TO 3010
      3001  FORMAT(1X,45HTHE ASSUMPTIONS FOR THIS MODEL ARE            )
      3002  FORMAT(1X,45HNOT VALID IF THE TRAFFIC INTENSITY EQUALS OR  )
      3003  FORMAT(1X,45HEXCEEDS THE NUMBER OF SERVERS.                )
      3004  FORMAT(1X,25HTRY AGAIN                         )
            WRITE(IOUT,3001)
            WRITE(IOUT,3002)
            WRITE(IOUT,3003)
            WRITE(IOUT,3004)
            GO TO 3000
      3010  CONTINUE
            CONST=1/(1-RHO/C)
            NIKE=C
            B=1
            IF(NIKE.GT.2) GO TO 3018
            IF(NIKE.GT.1) GO TO 3016
            K=0
            PNOT=1/(SUMM(RHO,K)+RHO*CONST)
            P(1)=RHO*PNOT
            WRITE(IOUT,1601) PNOT
            I=1
            WRITE(IOUT,1400) I,P(I)
            GO TO 3026
      3016  CONTINUE
            K=1
            PNOT=1/(SUMM(RHO,K)+(RHO**2)*CONST/2)
            IF(K .EQ. 1) WRITE(IOUT,3014)PNOT
      3014  FORMAT(1X,7HP(  0)=,3X,F15.5)
            DO 3017 I=1,2
            P(I)=PNOT*(RHO**I)/FCTRL(I)
            IF(P(I) .GE. 0.01)WRITE(IOUT,1400) I,P(I)
            IF(P(I) .LT. 0.0001) GO TO 3026
      3017  CONTINUE
            GO TO 3026
```

```
3018 CONTINUE
     DO 3019 I=1,NIKE
     CONST=CONST*(RHO/I)
     IF(CONST.GT. 0.00000001) GO TO 3019
     CONST=0
3019 CONTINUE
     IC=C-1
     PNOT=1/(SUMM(RHO,IC)+CONST)
     WRITE(IOUT,1601) PNOT
     P(1)=PNOT*RHO
     I=1
     WRITE(IOUT,1400) I,P(I)
     DO 3025 I=2,NIKE
     P(I)=P(I-1)*(RHO/I)
     IF(P(I).GT. 0.00000001) GO TO 3024
     IF(I.LT.RHO) GO TO 3024
     P(I)=0
     B=0
3024 CONTINUE
     WRITE(IOUT,1400) I,P(I)
3025 CONTINUE
3026 CONTINUE
     PC=P(NIKE)
     IF(B.EQ.0) GO TO 3030
     I=NIKE
3027 CONTINUE
     I=I+1
     P(I)=PC*((RHO/C)**(I-NIKE))
     IF(P(I).LT.  0.00000001) GO TO 3030
     IF(P(I) .GE. 0.01)WRITE(IOUT,1400) I,P(I)
     GO TO 3027
3030 CONTINUE
     PNGTC=PC*((AMU*C)/((AMU*C)-ALAMDA))
C        WRITE OUTPUT FOR P(N> OR =C)
     WRITE(IOUT,825) PNGTC
 825 FORMAT(1X,12HP(N> OR =C)=,3X,F15.5)
     ALS=PC*(RHO*C)/((C-RHO)**2) +RHO
     ALQ=ALS-RHO
     IF(PC.EQ.0) GO TO 3031
     ALB=ALQ/PNGTC
     GO TO 3032
3031 ALB=0
3032 CONTINUE
     WS=ALQ/ALAMDA +1/AMU
     WQ=ALQ/ALAMDA
     IF(PC.EQ.0) GO TO 3033
     WB=WQ/PNGTC
     GO TO 3034
3033 WB=0
3034 CONTINUE
     CALL GA(ALS,ALQ,ALB,WS,WQ,WB)
     GO TO 8000
4000 CONTINUE
     WRITE(IOUT,802)
     READ,      C
     IF(RHO.LT.C) GO TO 4005
     WRITE(IOUT,3001)
     WRITE(IOUT,3002)
     WRITE(IOUT,3003)
     WRITE(IOUT,3004)
     GO TO 550
4005 CONTINUE
     WRITE(IOUT,803)
     READ,       N
     IF(C.LT.N) GO TO 4009
     WRITE(IOUT,3001)
     WRITE(IOUT,4006)
```

```
      WRITE(IOUT,4007)
      WRITE(IOUT,3004)
      GO TO 550
 4006 FORMAT(1X,45HNOT VALID IF THE NUMBER OF SERVERS IS >OR=     )
 4007 FORMAT(1X,45HTHE MAXIMUM ALLOWED IN THE SYSTEM.             )
 4009 CONTINUE
C     TEST FOR RHO_C_N
      NIKE=C
      CC=1
      DO 4010 I=1,NIKE
      CC=CC*(RHO/I)
      IF(CC.GT. 0.00000001) GO TO 4010
      CC=0
 4010 CONTINUE
      SUM=0
      IF(CC.EQ.0) GO TO 4020
      CONST=CC
      DO 4015 I=NIKE,N
      CONST=CONST*(RHO/C)
      SUM=SUM+CONST
      IF(CONST.GT. 0.00000001) GO TO 4015
      CONST=0
 4015 CONTINUE
 4020 CONTINUE
      K=NIKE
      PNOT=1/(SUMM(RHO,K)+SUM)
      PNOT=1/(SUMM(RHO,NIKE)+SUM)
C     COMPUTE P(N) FOR N=1,N
      WRITE(IOUT,1601) PNOT
      P(1)=PNOT*RHO
      I=1
      WRITE(IOUT,1400) I,P(I)
      IF(NIKE.EQ.1) GO TO 4051
      DO 4050 I=2,NIKE
      P(I)=P(I-1)*RHO/I
      IF(P(I) .GE. 0.01)WRITE(IOUT,1400) I,P(I)
      IF(P(I).GT. 0.00000001) GO TO 4050
      IF(I.LE.RHO) GO TO 4050
      P(I)=0
 4050 CONTINUE
 4051 CONTINUE
      IF(P(NIKE).EQ.0) GO TO 4075
      J=NIKE-1
 4060 CONTINUE
      J=J+1
      I=J+1
      P(I)=P(J)*(RHO/C)
      WRITE(IOUT,1400) I,P(I)
      IF(I.EQ.N) GO TO 4075
      IF(P(I).GT. 0.00000001) GO TO 4060
      P(N)=0
 4075 CONTINUE
      MIKE=NIKE-1
      PNGTC=1-PNOT*SUMM(RHO,MIKE)
      WRITE(IOUT,825) PNGTC
      PC=P(NIKE)
      NC=N-NIKE
      A=RHO/C
      CONST=1-(A**NC)-(N-C)*(A**NC)*(1-A)
      ALS=(PC*RHO*C*CONST)/((C-RHO)**2) +RHO*(1-P(N))
      ALQ=ALS-RHO*(1-P(N))
      ALB=ALQ/PNGTC
      WS=ALQ/(ALAMDA*(1-P(N))) + 1/AMU
      WQ=WS-1/AMU
      WB=WQ/PNGTC
      CALL GA(ALS,ALQ,ALB,WS,WQ,WB)
      GO TO 8000
```

```
5000 CONTINUE
     IF(I.EQ.5) WRITE(IOUT,804)
     WRITE(IOUT,1611)
     WRITE(IOUT,1612)
     READ, M
     IF(M.GT.50) GO TO 5000
5010 CONTINUE
     WRITE(IOUT,802)
     WRITE(IOUT,860)
     WRITE(IOUT,870)
     WRITE(IOUT,880)
 860 FORMAT(1X,45HTHIS MODEL ASSUMES THE NUMBER OF SERVERS      )
 870 FORMAT(1X,45HTO BE LESS THAN THE NUMBER OF UNITS IN        )
 880 FORMAT(1X,45HTHE FINITE POPULATION.                        )
     READ, C
     IF(C.GE.M) GO TO 5010
C    COMPUTE P(0)
     NIKE=C+1
     SUM=0
     DO 5020 I=1,NIKE
     KKK=M-I+1
     SUM=SUM+FCTRL(M)*(RHO**(I-1))/(FCTRL(KKK)*FCTRL(I-1))
5020 CONTINUE
     TUM=0
     A=FCTRL(M)/FCTRL(NIKE-1)
     DO 5030 I=NIKE,M
     LLL=I-NIKE+1
     KKK=M-I
     TUM=TUM+A*(RHO**I)/(FCTRL(KKK)*(C**LLL))
5030 CONTINUE
     PNOT=1/(SUM+TUM)
     WRITE(IOUT,1601) PNOT
     LIKE=NIKE-1
     DO 5040 I=1,LIKE
     KKK=M-I
     P(I)=FCTRL(M)*PNOT*(RHO**I)/(FCTRL(KKK)*FCTRL(I))
     IF(P(I) .GE. 0.01)WRITE(IOUT,1400) I,P(I)
5040 CONTINUE
     DO 5050 I=NIKE,M
     LLL=I-LIKE
     KKK=M-I
     P(I)=A*PNOT*(RHO**I)/(FCTRL(KKK)*(C**LLL))
     IF(P(I) .GE. 0.01)WRITE(IOUT,1400) I,P(I)
5050 CONTINUE
C    COMPUTE P(N> OR =C)
     SUM=0
     MIKE=LIKE-1
     DO 5060 I=1,LIKE
     SUM=SUM+FCTRL(M)*(RHO**(I-1))/(FCTRL(M-I+1)*FCTRL(I-1))
5060 CONTINUE
     PNGTC=1-PNOT*SUM
     WRITE(IOUT,825) PNGTC
     SUM=0
     DO 5070 I=NIKE,M
     SUM=SUM+(I-LIKE)*P(I)
5070 CONTINUE
     ALQ=SUM
     ALS=(ALQ+M*RHO)/(1+RHO)
     ALB=ALQ/PNGTC
     WS=ALS/(ALAMDA*(M-ALS))
     WQ=ALQ/(ALAMDA*(M-ALS))
     WB=WQ/PNGTC
     CALL GA(ALS,ALQ,ALB,WS,WQ,WB)
     GO TO 8000
6000 CONTINUE
     X=1/(10**10)
     PNOT=1/(2.7182818284**RHO)
```

```
      WRITE(IOUT,1601) PNOT
      I=1
      P(1)=PNOT*RHO/1
      I=1
      IF(P(I) .GE. 0.01)WRITE(IOUT,1400) I,P(I)
6010 CONTINUE
      I=I+1
      P(I)=P(I-1)*RHO/I
      IF(P(I) .GE. 0.01)WRITE(IOUT,1400) I,P(I)
      IF(P(I).GT. 0.00000001) GO TO 6010
      ALS=RHO
      WS=1/AMU
      WRITE(IOUT,6100)
      WRITE(IOUT,6111) ALS
      WRITE(IOUT,6112)
      WRITE(IOUT,6113) WS
6100 FORMAT(1X,45HSTEADY STATE MEAN NUMBER OF UNITS IN          )
6111 FORMAT(1X,10HSYSTEM=          ,F15.5)
6112 FORMAT(1X,45HSTEADY STATE MEAN TIME IN SYSTEM=             )
6113 FORMAT(20X,F15.5)
      GO TO 8000
7000 IF(RHO .GE. 1.0)WRITE(IOUT,7005)
      IF(RHO .GE. 1.0)GO TO 550
7005 FORMAT(/1X,30HMEAN SERVICE RATE MUST EXCEED ,
     1/1X,18HMEAN ARRIVAL RATE )
      WRITE(IOUT,7010)
7010 FORMAT(/1X,37HWHAT IS THE VARIANCE OF SERVICE TIME?)
      READ, VT
      ALS=RHO+(RHO**2+ALAMDA**2*VT)/(2.*(1.-RHO))
      ALQ=ALS-RHO
      WQ=ALQ/ALAMDA
      WS=WQ+1./AMU
      ALB=99999.
      WB=99999.
      CALL GA(ALS,ALQ,ALB,WS,WQ,WB)
8000 CONTINUE
      WRITE(IOUT,9000)
9000 FORMAT('0', 'DO YOU WANT TO TRY ANOTHER PROBLEM?',
     1/' IF YES, TYPE 1.  IF NO, TYPE 0.')
      READ, I9010
      IF(I9010 .EQ. 1) GO TO 416
      STOP
      END
      FUNCTION FCTRL (NUM)
      IF(NUM.EQ.0) FCTRL=1
      IF(NUM.EQ.1) FCTRL=1
      IF(NUM.LE.1) GO TO 9001
      FCTRL = 1
      DO 9000 L=1,NUM
      FCTRL=FCTRL*L
9000 CONTINUE
9001 CONTINUE
      RETURN
      END
      SUBROUTINE GA(A,B,C,D,E,F)
      IOUT=2
      WRITE(IOUT,1006)
      WRITE(IOUT,1000)A
      WRITE(IOUT,1001)B
      IF(C.EQ.0) GO TO 500
      IF(C .EQ. 99999.) GO TO 520
      WRITE(IOUT,1002)C
      GO TO 510
500 CONTINUE
      WRITE(IOUT,1008)
      WRITE(IOUT,1009)
      GO TO 510
```

```
520     WRITE(IOUT,521)
521     FORMAT(1X,40HIN QUEUE FOR BUSY SYSTEM IS NOT DEFINED.)
  510 CONTINUE
        WRITE(IOUT,1007)
        WRITE(IOUT,1003)D
        WRITE(IOUT,1004)E
        IF(F.EQ.0) GO TO 600
        IF(F .EQ. 99999.) GO TO 620
        WRITE(IOUT,1005)F
        GO TO 610
  600 CONTINUE
        WRITE(IOUT,1008)
        WRITE(IOUT,1009)
        GO TO 610
620     WRITE(IOUT,521)
  610 CONTINUE
 1000 FORMAT(1X,25HIN SYSTEM=                              ,F15.5)
 1001 FORMAT(1X,25HIN QUEUE=                               ,F15.5)
 1002 FORMAT(1X,25HIN QUEUE FOR BUSY SYSTEM=               ,F15.5)
 1003 FORMAT(1X,25HIN SYSTEM=                              ,F15.5)
 1004 FORMAT(1X,25HIN QUEUE=                               ,F15.5)
 1005 FORMAT(1X,25HIN QUEUE FOR BUSY SYSTEM=               ,F15.5)
 1006 FORMAT(1X,40HSTEADY-STATE MEAN NUMBER OF UNITS:         )
 1007 FORMAT(1X,40HSTEADY-STATE MEAN TIME:                    )
 1008 FORMAT(1X,40HIN QUEUE FOR BUSY SYSTEMS IS               )
 1009 FORMAT(1X,45HUNDEFINED FOR P(N> OR =C) APPROACHING ZERO.      )
        RETURN
        END
        FUNCTION SUMM (R,NUT)
        IF(NUT.GT.0) GO TO 500
        SUMM=1
        GO TO 1100
  500 CONTINUE
        IF(NUT.GT.1) GO TO 800
        SUMM=1+R
        GO TO 1100
  800 CONTINUE
        SUMM=1
        CONST=1
        DO 1000 I=1,NUT
        CONST=CONST*R/I
        SUMM=SUMM+CONST
        IF(CONST.GT. 0.00000001) GO TO 1000
        CONST=0
 1000 CONTINUE
 1100 CONTINUE
        RETURN
        END

        PROGRAM RANGEN
        COMMON XI,XII,X,RN,A,B,ALAMBD,THETA,EX,STD,X1,X2,IROUND
C       THIS PROGRAM GENERATES A RANDOM VARIABLE WHICH IS
C       EITHER UNIFORMLY, POISSON, NEGATIVE EXPONENTIALLY,
C       OR NORMALLY DISTRIBUTED.
C       IF DESIRED, OUTPUT CAN BE WRITTEN TO DISK DEVICE 30.
        DIMENSION Y(1000)
        DO 60 I=1,1000
60      Y(I)=0.
        XI=5381
        XII=7213
        IOUT=2
        WRITE(IOUT,6)
6       FORMAT(' RANDOM NUMBER GENERATION')
        WRITE(IOUT,7)
7       FORMAT('0DO YOU WANT INSTRUCTIONS?  TYPE 1 FOR YES,',
       1'OR 0 FOR NO.')
        READ, INST
```

```
          IF(INST .EQ. 0)GO TO 12
          WRITE(IOUT,10)
10        FORMAT('1','THIS PROGRAM GENERATES RANDOM VARIATES '/
         1' FROM YOUR CHOICE OF THE FOLLOWING DISTRIBUTIONS:'/
         2' 1. UNIFORM'/' 2. POISSON'/' 3. NEGATIVE EXPO',
         3'NENTIAL'/' 4. NORMAL')
12        WRITE(IOUT,15)
15        FORMAT('0IF YOU WANT TO ENTER A RANDOM NUMBER SEED, TYPE IT AS'/
         1' A FOUR-DIGIT INTEGER. OTHERWISE, TYPE -1.')
          READ, ISEED
          IF(ISEED .NE. -1) XII=ISEED
          WRITE(IOUT,5)
5         FORMAT('0','HOW MANY RANDOM NUMBERS DO YOU WANT?')
          READ, N
          WRITE(IOUT,14)
14        FORMAT('0', 'WHICH DISTRIBUTION DO YOU WANT TO USE?')
1         READ, IDIST
          WRITE(IOUT,25)
25        FORMAT('0IF YOU WANT YOUR OUTPUT PRINTED IN LINES OF '/
         1' 10 NUMBERS EACH, TYPE A 1. IF YOU WANT YOUR OUTPUT '/
         2' PRINTED ONE NUMBER PER LINE, TYPE A 0.')
          READ, LINE
          WRITE(IOUT,40)
40        FORMAT('0IF YOU WANT YOUR RANDOM NUMBERS TO PRINT IN SORTED '/
         1' ORDER, TYPE A 1. OTHERWISE, TYPE A 0.')
          READ, ISORT
          IF(IDIST .EQ. 1) GO TO 100
          IF(IDIST .EQ. 2) GO TO 140
          IF(IDIST .EQ. 3) GO TO 300
          IF(IDIST .EQ. 4) GO TO 400
          WRITE(IOUT,30)
30        FORMAT('0','PLEASE RE-ENTER DISTRIBUTION NUMBER.')
          GO TO 1
100       WRITE(IOUT,110)
110       FORMAT('0','YOU HAVE CHOSEN THE UNIFORM DISTRIBUTION.')
          WRITE(IOUT,114)
114       FORMAT(1X,'PLEASE ENTER THE LOWER LIMIT.')
          READ, A
          WRITE(IOUT,130)
130       FORMAT(' ','PLEASE ENTER THE UPPER LIMIT.')
          READ, B
          WRITE(IOUT,150)
150       FORMAT('0IF YOU WANT YOUR UNIFORM RANDOM NUMBERS ROUNDED '/
         1' TO THE NEAREST INTEGER, TYPE A 1. OTHERWISE, TYPE A 0.')
          READ, IROUND
          DO 190 I=1,N
          CALL UNIFRM
190       Y(I)=X
          GO TO 491
140       WRITE(IOUT,210)
210       FORMAT('0YOU HAVE CHOSEN THE POISSON DISTRIBUTION.')
          WRITE(IOUT,214)
214       FORMAT(' PLEASE ENTER THE MEAN.')
          READ, ALAMBD
          DO 290 I=1,N
          CALL POISSN
290       Y(I)=X
          GO TO 491
300       WRITE(IOUT,310)
310       FORMAT('0YOU CHOSE THE NEGATIVE EXPONENTIAL DISTRIBUTION.')
          WRITE(IOUT,314)
314       FORMAT(' PLEASE ENTER THE MEAN.')
          READ, THETA
          DO 390 I=1,N
          CALL EXPONT
390       Y(I)=X
          GO TO 491
```

```
400     WRITE(IOUT,410)
410     FORMAT('OYOU HAVE CHOSEN THE NORMAL DISTRIBUTION.')
        WRITE(IOUT,414)
414     FORMAT(' PLEASE ENTER THE MEAN.')
        READ, EX
        WRITE(IOUT,430)
430     FORMAT(' PLEASE ENTER THE STANDARD DEVIATION.')
        READ, STD
        DO 490 I=1,N
        CALL NORMAL
490     Y(I)=X
491     WRITE(IOUT,550)
550     FORMAT('ODO YOU WANT OUTPUT WRITTEN ON DISK?'/
       1' IF YES, TYPE 1.  IF NO, TYPE 0.')
        READ, IDISK
        IF(IDISK .EQ. 1)WRITE(30,600)(Y(J),J=1,N)
600     FORMAT(F7.2)
        IF(ISORT .EQ. 0)GO TO 495
        NMIN1=N-1
        DO 493 I=1,NMIN1
        J=I+1
        DO 493 K=J,N
        IF(Y(I) .LE. Y(K))GO TO 493
        HOLD=Y(I)
        Y(I)=Y(K)
        Y(K)=HOLD
493     CONTINUE
495     IF(LINE .EQ. 1)WRITE(IOUT,640)(Y(J),J=1,N)
640     FORMAT('0',10F7.2)
        IF(LINE .EQ. 0)WRITE(IOUT,500)(Y(J),J=1,N)
500     FORMAT(1X,F7.2)
        STOP
        END
        SUBROUTINE RANDUM
        COMMON XI,XII,X,RN,A,B,ALAMBD,THETA,EX,STD,X1,X2,IROUND
        PROD=XI*XII
        XI=XII
        IPROD=PROD/1000000.
        DIFF=PROD-IPROD*1000000.
        IDIFF=DIFF/100.
        IF(IDIFF .EQ. 0)IDIFF=2539
        XII=IDIFF
        RN=XII/10000.
        RETURN
        END
        SUBROUTINE UNIFRM
        COMMON XI,XII,X,RN,A,B,ALAMBD,THETA,EX,STD,X1,X2,IROUND
        CALL RANDUM
        X=A+(B-A)*RN
        IF(IROUND .EQ. 0)GO TO 610
        NX=X+0.5
        X=NX
610     RETURN
        END
        SUBROUTINE POISSN
        COMMON XI,XII,X,RN,A,B,ALAMBD,THETA,EX,STD,X1,X2,IROUND
        X=0.0
        ALAMBM=(-ALAMBD)
        A=2.71828**(ALAMBM)
        S=1.0
4       CALL RANDUM
        S=S*RN
        IF(S-A)9,7,7
7       X=X+1.0
        GO TO 4
9       RETURN
        END
```

```
      SUBROUTINE EXPONT
      COMMON XI,XII,X,RN,A,B,ALAMBD,THETA,EX,STD,X1,X2,IROUND
      CALL RANDUM
      X=-THETA*ALOG(RN)
      RETURN
      END
      SUBROUTINE NORMAL
      COMMON XI,XII,X,RN,A,B,ALAMBD,THETA,EX,STD,X1,X2,IROUND
      SUM=-6.0
      DO 14 I=1,12
      CALL RANDUM
14    SUM=SUM+RN
      X=SUM*STD+EX
      RETURN
      END


      PROGRAM YULES
C     THIS PROGRAM PERFORMS YULE'S TEST OF THE DISTRIBUTION
C     OF SUMS OF 2, 3, OR 4 DIGITS.
      DIMENSION E(3,37),IR(5000),O(37),ITL(4),SUM(2500),XKS(35)
      DATA XKS(1),XKS(2),XKS(3),XKS(4)/.975,.842,.708,.624/
      DATA XKS(6),XKS(7),XKS(8),XKS(9)/.521,.486,.457,.432/.
      DATA XKS(11),XKS(12),XKS(13),XKS(14)/.391,.375,.361,.349/
      DATA XKS(16),XKS(17),XKS(18),XKS(19)/.328,.318,.309,.301/
      DATA XKS(21),XKS(22),XKS(23),XKS(24)/.287,.281,.275,.269/
      DATA XKS(26),XKS(27),XKS(28),XKS(29)/.259,.254,.250,.246/
      DATA XKS(31),XKS(32),XKS(33),XKS(34)/.238,.234,.231,.227/
      DATA XKS(5),XKS(10),XKS(15),XKS(20)/.565,.410,.338,.294/
      DATA XKS(25),XKS(30),XKS(35)/.264,.242,.224/
      DATA E(1,1),E(1,2),E(1,3),E(1,4)/.01,.02,.03,.04/
      DATA E(1,5),E(1,6),E(1,7),E(1,8)/.05,.06,.07,.08/
      DATA E(1,9),E(1,10),E(1,11),E(1,12)/.09,.1,.09,.08/
      DATA E(1,13),E(1,14),E(1,15),E(1,16)/.07,.06,.05,.04/
      DATA E(1,17),E(1,18),E(1,19)/.03,.02,.01/
      DATA E(2,1),E(2,2),E(2,3),E(2,4)/.001,.003,.006,.01/
      DATA E(2,5),E(2,6),E(2,7),E(2,8)/.015,.021,.028,.036/
      DATA E(2,9),E(2,10),E(2,11),E(2,12)/.045,.055,.063,.069/
      DATA E(2,13),E(2,14),E(2,15),E(2,16)/.073,.075,.075,.073/
      DATA E(2,17),E(2,18),E(2,19),E(2,20)/.069,.063,.055,.045/
      DATA E(2,21),E(2,22),E(2,23),E(2,24)/.036,.028,.021,.015/
      DATA E(2,25),E(2,26),E(2,27),E(2,28)/.01,.006,.003,.001/
      DATA E(3,1),E(3,2),E(3,3),E(3,4)/.0001,.0004,.0010,.0020/
      DATA E(3,5),E(3,6),E(3,7),E(3,8)/.0035,.0056,.0084,.0120/
      DATA E(3,9),E(3,10),E(3,11),E(3,12)/.0165,.022,.0282,.0348/
      DATA E(3,13),E(3,14),E(3,15),E(3,16)/.0415,.048,.054,.0592/
      DATA E(3,17),E(3,18),E(3,19),E(3,20)/.0633,.066,.067,.066/
      DATA E(3,21),E(3,22),E(3,23),E(3,24)/.0633,.0592,.054,.048/
      DATA E(3,25),E(3,26),E(3,27),E(3,28)/.0415,.0348,.0282,.022/
      DATA E(3,29),E(3,30),E(3,31),E(3,32)/.0165,.012,.0084,.0056/
      DATA E(3,33),E(3,34),E(3,35),E(3,36)/.0035,.002,.001,.0004/
      DATA E(3,37)/.0001/
      DATA ITL(1),ITL(2),ITL(3),ITL(4)/0,19,28,37/
      DATA CUME,CUMO,GDIFF/0.0,0.0,0.0/
      IIN=1
      IOUT=2
      J2=0
      DO 8 I=1,2500
8     SUM(I)=0.
      DO 5 I=1,37
5     O(I)=0.
      WRITE(IOUT,20)
20    FORMAT('OTHIS PROGRAM PERFORMS YULE''S TEST OF SUMS ',
     1'OF CONSECUTIVE DIGITS.')
      WRITE(IOUT,21)
21    FORMAT(' A KOLMOGOROV-SMIRNOV GOODNESS OF FIT TEST IS USED.')
      WRITE(IOUT,10)
10    FORMAT('0 HOW MANY RANDOM NUMBERS ARE THERE?')
```

```
        READ,N1
        WRITE(IOUT,12)
12      FORMAT('0HOW MANY DIGITS ARE IN EACH NUMBER?',
       1' YOU MAY CHOOSE 1,2,3 OR 4.')
        READ,IDIG
        N=N1*IDIG
        WRITE(IOUT,15)
15      FORMAT('0IF INPUT IS FROM DISK, TYPE A 1.',
       1' IF INPUT IS FROM A TERMINAL, TYPE A 0.')
        READ, M
        IF(M .EQ. 0) GO TO 38
        IF(IDIG .EQ. 1)READ(20,55)(IR(J),J=1,N)
        IF(IDIG .EQ. 2)READ(20,60)(IR(J),J=1,N)
        IF(IDIG .EQ. 3)READ(20,65)(IR(J),J=1,N)
        IF(IDIG .EQ. 4)READ(20,70)(IP(J),J=1,N)
        IF(IDIG .EQ. 5)READ(20,75)(IR(J),J=1,N)
55      FORMAT(I1)
60      FORMAT(2I1)
65      FORMAT(3I1)
70      FORMAT(4I1)
75      FORMAT(5I1)
        GO TO 40
38      DO 39 J=1,N,IDIG
        J1=J+IDIG-1
        IF(IDIG .EQ. 1)READ(IIN,55)(IR(I),I=J,J1)
        IF(IDIG .EQ. 2)READ(IIN,60)(IR(I),I=J,J1)
        IF(IDIG .EQ. 3)READ(IIN,65)(IR(I),I=J,J1)
        IF(IDIG .EQ. 4)READ(IIN,70)(IR(I),I=J,J1)
        IF(IDIG .EQ. 5)READ(IIN,75)(IR(I),I=J,J1)
39      CONTINUE
40      WRITE(IOUT,42)
42      FORMAT('0HOW MANY DIGITS ARE USED FOR EACH SUM?',
       1' YOU MAY CHOOSE 2,3 OR 4.')
        READ, IDSUM
        M1=ITL(IDSUM)
        DO 350 J=1,N,IDSUM
        J1=J+IDSUM-1
        IF(J1 .GT. N)GO TO 355
        J2=J2+1
        DO 350 K=J,J1
350     SUM(J2)=SUM(J2)+IR(K)
355     DO 110 J=1,J2
        DO 110 K=1,M1
        K1=K-1
        IF(SUM(J) .EQ. K1)O(K)=O(K)+1
110     CONTINUE
        DO 120 K=1,M1
120     O(K)=O(K)/J2
        IDSUM1=IDSUM-1
        WRITE(IOUT,150)
150     FORMAT('0',5X,'OBS.REL. ',2X,'EXP.REL. ',2X,'CUMULATIVE ',
       12X,'CUMULATIVE',3X,'ABSOLUTE'/1X,'SUM',2X,'FREQUENCY',
       22X,'FREQUENCY',
       22X,'OBS.REL.FR.',2X,'EXP.REL.FR.',2X,'DIFFERENCE'/)
        DO 170 I=1,M1
        CUME=CUME+E(IDSUM1,I)
        CUMO=CUMO+O(I)
        DIFF=ABS(CUME-CUMO)
        IF(DIFF .GT. GDIFF)GDIFF=DIFF
        IM1=I-1
170     WRITE(IOUT,175)IM1,O(I),E(IDSUM1,I),CUMO,CUME,DIFF
175     FORMAT(1X,I3,2X,3(F9.4,2X),2(F10.4,3X))
        WRITE(IOUT,180)GDIFF
180     FORMAT('0THE MAXIMUM DIFFERENCE IS ',F10.4)
        IF(J2 .LE. 35)CXKS=XKS(J2)
        FJ2=J2
        IF(J2 .GT. 35)CXKS=1.36/SQRT(FJ2)
```

```
          WRITE(IOUT,190)CXKS
190       FORMAT('0THE CRITICAL VALUE OF D IS ',F8.4,
        1' AT THE .05 LEVEL OF SIGNIFICANCE')
          STOP
          END

          PROGRAM GAP
C         THIS PROGRAM TESTS TO SEE WHETHER THE GAP BETWEEN
C         CONSECUTIVE OCCURRENCES OF THE SAME DIGIT IS
C         NEGATIVE BINOMIALLY DISTRIBUTED.
          DIMENSION IR(5000),IC(5000),EFR(5000),CHI(5000)
          DIMENSION TOBS(5000),TEXP(5000),XKS(35),GPR(5000)
          IIN=1
          IOUT=2
          DATA XKS(1),XKS(2),XKS(3),XKS(4)/.975,.842,.708,.624/
          DATA XKS(6),XKS(7),XKS(8),XKS(9)/.521,.486,.457,.432/
          DATA XKS(11),XKS(12),XKS(13),XKS(14)/.391,.375,.361,.349/
          DATA XKS(16),XKS(17),XKS(18),XKS(19)/.328,.318,.309,.301/
          DATA XKS(21),XKS(22),XKS(23),XKS(24)/.287,.281,.275,.269/
          DATA XKS(26),XKS(27),XKS(28),XKS(29)/.259,.254,.250,.246/
          DATA XKS(31),XKS(32),XKS(33),XKS(34)/.238,.234,.231,.227/
          DATA XKS(5),XKS(10),XKS(15),XKS(20)/.565,.410,.338,.294/
          DATA XKS(25),XKS(30),XKS(35)/.264,.242,.224/
          SW1=0
          SW2=0
          SW3=0
          ICT=0
          GDIFF=0
          TPROB=0.
          NCLAS1=0
          CCHI=0
          NGAP=0
          DO 1 I=1,5000
1         IR(I)=0
          DO 2 I=1,5000
          EFR(I)=0.
          TOBS(I)=0.
          TEXP(I)=0.
          CHI(I)=0.
2         IC(I)=0
          CGPR=0
          WRITE(IOUT,20)
20        FORMAT('0THIS PROGRAM PERFORMS THE GAP TEST',
        1' OF SPREAD BETWEEN CONSECUTIVE DIGITS.')
          WRITE(IOUT,21)
21        FORMAT(' A KOLMOGOROV-SMIRNOV GOODNESS OF FIT TEST IS USED.')
          WRITE(IOUT,10)
10        FORMAT('0','HOW MANY RANDOM NUMBERS ARE THERE?')
          READ,N1
          WRITE(IOUT, 12)
12        FORMAT('0HOW MANY DIGITS ARE IN EACH NUMBER?',
        1' YOU MAY CHOOSE 1,2,3,4 OR 5.')
          READ,IDIG
          N=N1*IDIG
14        WRITE(IOUT,15)
15        FORMAT('0','IF INPUT IS FROM DISK, TYPE A 1.',
        1' IF INPUT IS FROM A TERMINAL, TYPE A 0.')
          READ,M
          IF(M .EQ. 0)GO TO 38
          IF(IDIG .EQ. 1)READ(20,55)(IR(J),J=1,N)
          IF(IDIG .EQ. 2)READ (20,60)(IR(J),J=1,N)
          IF(IDIG .EQ. 3)READ (20,65)(IR(J),J=1,N)
          IF(IDIG .EQ. 4)READ (20,70)(IR(J),J=1,N)
          IF(IDIG .EQ. 5)READ (20,75)(IR(J),J=1,N)
55        FORMAT(I1)
60        FORMAT(2I1)
65        FORMAT(3I1)
```

```
70       FORMAT(4I1)
75       FORMAT(5I1)
         GO TO 40
38       DO 39 J=1,N,IDIG
         J1=J+IDIG-1
         IF(IDIG .EQ. 1)READ(IIN,55)(IR(I),I=J,J1)
         IF(IDIG .EQ. 2)READ(IIN,60)(IR(I),I=J,J1)
         IF(IDIG .EQ. 3)READ(IIN,65)(IR(I),I=J,J1)
         IF(IDIG .EQ. 4)READ(IIN,70)(IR(I),I=J,J1)
         IF(IDIG .EQ. 5)READ(IIN,75)(IR(I),I=J,J1)
39       CONTINUE
40       DO 53 I=1,10
         J=I-1
         ICT=0
         DO 50 K=1,N
         IF(IR(K) .EQ. J)IC(ICT+1)=IC(ICT+1)+1
         IF(IR(K) .EQ. J)ICT=0
         IF(IR(K) .NE. J)ICT=ICT+1
50       CONTINUE
53       CONTINUE
         DO 210 I=1,N
210      NGAP=NGAP+IC(I)
         NP1=NGAP+1
         WRITE(IOUT,90)
90       FORMAT('0IF YOU WANT TO SEE THE DETAILS OF THE TEST, ',
        1' TYPE A 1.  OTHERWISE, TYPE A 0.')
         READ, IDET
         IF(IDET .EQ. 1)WRITE(IOUT,490)
490      FORMAT('0',6X,'OBSERVED',2X,'CUM.OBS.',2X,'NEGATIVE',3X,
        1'EXPECTED',2X,'CUM.EXP.',5X,'ABSOLUTE')
         IF(IDET .EQ. 1)WRITE(IOUT,492)
492      FORMAT(2X,'GAP',1X,'FREQUENCY',1X,'REL.FREQ.',2X,'BIN.',
        1'PROB.',1X,'FREQUENCY',2X,'REL.FREQ.',2X,'DIFFERENCE'/)
         DO 200 L=1,NGAP
         GPR(L)=(0.9**(L-1))/10.
         IF(GPR(L) .LT. .001)NP1=L
         IF(GPR(L) .LT. .001)GO TO 202
         CGPR=CGPR+GPR(L)
200      CONTINUE
202      GPR(NP1)=1.-CGPR
         CGPR=1.0000
         DO 500 I=1,NGAP
         EFR(I)=N*GPR(I)
500      TPROB=TPROB+GPR(I)
         GPR(NP1)=1.-TPROB
         EFR(NP1)=(1.-TPROB)*N
         TEXP(1)=EFR(1)
         TOBS(1)=IC(1)
         DO 503 I=2,NP1
         TOBS(I)=TOBS(I-1)+IC(I)
503      TEXP(I)=TEXP(I-1)+EFR(I)
         DO 505 I=1,NP1
         TOBS(I)=TOBS(I)/N
         TEXP(I)=TEXP(I)/N
         DIFF=ABS(TOBS(I)-TEXP(I))
         IF(DIFF .GT. GDIFF)GDIFF=DIFF
         IF(I .EQ. 1)GO TO 504
         IF(TOBS(I) .EQ. 1.0 .AND. TOBS(I-1) .EQ. 1.0)SW1=1
         IF(TEXP(I) .EQ. 1.0 .AND. TEXP(I-1) .EQ. 1.0)SW2=1
         IF(SW1 .EQ. 1 .OR. SW2 .EQ. 1)SW3=1
504      IF(IDET .EQ. 1 .AND. SW3 .EQ. 0)WRITE(IOUT,494)I,IC(I),TOBS(I),
        1GPR(I),EFR(I),TEXP(I),DIFF
494      FORMAT(I5,I9,F10.4,3X,F8.4,2X,F8.2,3X,F8.4,3X,F9.4)
505      CONTINUE
         IF(IDET .EQ. 1)WRITE(IOUT,340)
340      FORMAT('0THE LAST GAP CATEGORY CONTAINS GREATER-THAN-',
```

```
       1'OR-EQUALS INFORMATION.'/' FREQUENCIES MAY NOT ACCUMU',
       2'LATE TO 1.0000 FOR THAT CLASS, HOWEVER.')
         WRITE(IOUT,350)GDIFF
  350    FORMAT('0THE MAXIMUM DIFFERENCE IS ',F11.4)
         IF(N .LE. 35)CXKS=XKS(N)
         FN=N
         IF(N .GT. 35)CXKS=1.36/SQRT(FN)
         WRITE(IOUT,190)CXKS
  190    FORMAT('0THE CRITICAL VALUE OF D IS ',F8.4,
       1' AT THE .05 LEVEL OF SIGNIFICANCE.')
         STOP
         END

         PROGRAM RUNUPD
C        THIS PROGRAM TESTS A STREAM OF OBSERVATIONS FOR RUNS
C        UP AND DOWN FOR CONSECUTIVE VALUES. IT WILL GIVE
C        BOTH LEFT-TAIL AND TWO-TAIL VALUES, TO TEST FOR TRANSIENT
C        STATE AS WELL AS AUTOCORRELATION. THE PROGRAM WILL
C        NOT PERFORM ACCURATELY UNLESS THE EFFECTIVE SAMPLE SIZE
C        EXCEEDS 25 AND THE NUMBER OF TIES IS SMALL.
C        IF INPUT FROM A DISK FILE IS DESIRED, THE FILE SHOULD
C        BE DESIGNATED AS DEVICE NUMBER 20.
         DIMENSION X(5000)
         IOUT=2
         DO 10 I=1,5000
  10     X(I)=0
         I=1
         N=0
         N1=0
         NRUNS=0
         UPSW=0
         WRITE(IOUT,20)
  20     FORMAT('0THIS PROGRAM PERFORMS THE RUNS-UP-AND-DOWN TEST.')
         WRITE(IOUT,25)
  25     FORMAT('0IF INPUT IS TO COME FROM A DISK FILE, TYPE A 1.'/
       1' OTHERWISE, TYPE A 0.')
         READ, IDISK
         IF(IDISK .EQ. 1)GO TO 60
         WRITE(IOUT,30)
  30     FORMAT('0ENTER AT LEAST 26 OBSERVATIONS, ONE TO A LINE.')
         WRITE(IOUT,40)
  40     FORMAT('0TERMINATE WITH A DUMMY VALUE -99999')
  50     READ, X(I)
         IF(X(I) .EQ. -99999)GO TO 100
         N=N+1
         N1=N1+1
         I=I+1
         GO TO 50
  60     READ(20,*,END=100)X(I)
         N=N+1
         N1=N1+1
         I=I+1
         GO TO 60
  100    WRITE(IOUT,105)N
  105    FORMAT('0THE ORIGINAL SAMPLE SIZE IS ',I6)
         IF(X(1) .LT. X(2))GO TO 120
         IF(X(1) .EQ. X(2))GO TO 130
         UPSW=0
         NRUNS=NRUNS+1
         GO TO 130
  120    UPSW=1
         NRUNS=NRUNS+1
  130    DO 150 I=3,N
         K=I-1
         IF(X(K) .LT. X(I) .AND. UPSW .EQ. 1)GO TO 150
         IF(X(K) .LT. X(I) .AND. UPSW .EQ. 0)GO TO 145
```

```
        IF(X(K) .GT. X(I) .AND. UPSW .EQ. 1)GO TO 140
        IF(X(K) .GT. X(I) .AND. UPSW .EQ. 0)GO TO 150
        IF(X(K) .EQ. X(I))N1=N1-1
        IF(X(K) .EQ. X(I))GO TO 150
140     UPSW=0
        NRUNS=NRUNS+1
        GO TO 150
145     UPSW=1
        NRUNS=NRUNS+1
150     CONTINUE
        N=N1
        WRITE(IOUT,110)N
110     FORMAT('0THE EFFECTIVE SAMPLE SIZE IS ',I6)
        EXMEAN=(2.*N-1.)/3.
        STDEV=SQRT((16.*N-29.)/90.)
        Z=(NRUNS-EXMEAN)/STDEV
        IF(Z .LT. 0)Z=Z+0.5/STDEV
        IF(Z .GT. 0)Z=Z-0.5/STDEV
        WRITE(IOUT,170)NRUNS,EXMEAN,STDEV,Z
170     FORMAT('0THE NUMBER OF RUNS IS ',I6/' THE EXPECTED ',
       1'NUMBER OF RUNS IS ',F10.2/' THE STANDARD DEVIATION ',
       2'IS ',F8.4/' THE VALUE OF Z IS ',F8.4)
        STOP
        END
        GO TO 190
220     NDF1=3
        NDF2=1
        GO TO 190
900     WRITE(IOUT,910)
910     FORMAT('0',26X,'OBSERVED',3X,'THEORETICAL')
        WRITE(IOUT,915)
915     FORMAT(8X,'CLASS',2X,'OBSERVED',3X,'CUMULATIVE'2X,
       1'CUMULATIVE',4X,'ABSOLUTE')
        WRITE(IOUT,920)
920     FORMAT(1X,'CLASS',2X,'IDENT',2X,'FREQUENCY',2X,
       1'PROPORTION',2X,'PROPORTION',3X,'DIFFERENCE'/)
        DO 940 I=1,NCLASS
        WRITE(IOUT,930)I,ULIM(I),OBSFRQ(I),ORLFRQ(I),CTHFRQ(I),ADIFF(I)
930     FORMAT(1X,I5,1X,F6.1,4X,F5.0,5X,F6.4,6X,F6.4,8X,F6.4)
        IF(ADIFF(I) .GT. DIFF) DIFF=ADIFF(I)
940     CONTINUE
        WRITE(IOUT,945)TOBSFR
945     FORMAT('0',' ***TOTAL***',3X,F6.0)
        WRITE(IOUT,950)DIFF
950     FORMAT('0THE MAXIMUM SAMPLE DIFFERENCE IS ',F8.4)
        IF(TOBSFR .LE. 35)CXKS=XKS(TOBSFR)
        IF(TOBSFR .GT. 35)CXKS=1.36/SQRT(TOBSFR)
        WRITE(IOUT,960)CXKS
960     FORMAT('0THE CRITICAL VALUE OF D IS ',F8.4,
       1' AT THE .05 LEVEL OF SIGNIFICANCE.')
        GO TO 178
        END
        FUNCTION FCTRL(NUM)
        FCTRL = 1
        IF(NUM .EQ. 0 .OR. NUM .EQ. 1)GO TO 9001
        DO 9000 L=1,NUM
9000    FCTRL=FCTRL*L
9001    RETURN
        END


        PROGRAM SPEAR
C       THIS PROGRAM COMPUTES THE SPEARMAN RANK CORRELATION
C       COEFFICIENT AS A METHOD OF DETECTING AUTOCORRELATION IN
C       TIME SERIES.  IF INPUT FROM A DISK FILE IS DESIRED,
C       USE DEVICE NUMBER 20.
        DIMENSION Y(500,3),X(500,3)
```

```
        IOUT=2
5       SUMD2=0
        SUMU=0
        SUMV=0
        RSUM=0
        NRSAM=0
        DIFF=0
        DO 10 I=1,500
        DO 10 J=1,3
        Y(I,J)=0
10      X(I,J)=0
        I=1
        WRITE(IOUT,2)
2       FORMAT('0SPEARMAN RANK CORRELATION')
        WRITE(IOUT,6)
6       FORMAT('0DO YOU WANT INSTRUCTIONS? TYPE 1 FOR YES, 0 FOR NO')
        READ, INDEX
        IF(INDEX .EQ. 1)WRITE(IOUT,30)
30      FORMAT('1THIS PROGRAM TESTS A SET OF DATA FOR AUTOCORRELATION '/
       1' USING THE SPEARMAN RANK CORRELATION COEFFICIENT.  IT ALSO '/
       2' GIVES A Z STATISTIC FOR THE SIGNIFICANCE OF R.'/
       3' THE SAMPLE SIZE SHOULD EXCEED 30.')
        WRITE(IOUT,20)
20      FORMAT('0IF INPUT IS FROM A DISK FILE, TYPE A 1.'/
       1' IF DATA ARE ENTERED DIRECTLY, TYPE A 0.')
        READ, IDISK
        IF(IDISK .EQ. 1)GO TO 60
        WRITE(IOUT,40)
40      FORMAT('0ENTER THE VALUES ONE TO A LINE, ENDING WITH A DUMMY ',
       1'VALUE -99999.')
45      READ, Y(1,1)
        Y(1,3)=1
50      READ, X(I,1)
        IF(X(I,1) .EQ. -99999)GO TO 100
        X(I,3)=I
        I=I+1
        Y(I,3)=I
        K=I-1
        Y(I,1)=X(K,1)
        GO TO 50
60      READ(20,*,END=100)Y(1,1)
        Y(1,3)=1
70      READ(20,*,END=100)X(I,1)
        X(I,3)=I
        I=I+1
        Y(I,3)=I
        K=I-1
        Y(I,1)=X(K,1)
        GO TO 70
100     I=I-1
150     NM1=I-1
        DO 230 K=1,NM1
        L=K+1
        DO 220 M=L,I
        IF(Y(K,1) .LE. Y(M,1))GO TO 220
        A=Y(K,1)
        Y(K,1)=Y(M,1)
        Y(M,1)=A
        B=Y(K,3)
        Y(K,3)=Y(M,3)
        Y(M,3)=B
220     CONTINUE
230     Y(K,2)=K
        Y(I,2)=I
        DO 245 J=1,NM1
        K=J+1
```

```
        IF(Y(J,1) .NE. Y(K,1))GO TO 240
        IF(NRSAM .EQ. 0) GO TO 235
        RSUM=RSUM+Y(K,2)
        NRSAM=NRSAM+1
        IF(J .NE. NM1)GO TO 245
        GO TO 242
235     RSUM=Y(J,2)+Y(K,2)
        NRSAM=2
        IF(J .NE. NM1)GO TO 245
        GO TO 242
240     IF(NRSAM .EQ. 0)GO TO 245
242     AVGRK=RSUM/NRSAM
        DO 237 L=1,NRSAM
        IF(J .NE. NM1)M=J-L+1
        IF(J .EQ. NM1 .AND. Y(J,1) .EQ. Y(K,1))M=J-L+2
        IF(J .EQ. NM1 .AND. Y(J,1) .NE. Y(K,1))M=J-L+1
237     Y(M,2)=AVGRK
        RSUM=0
        SUMU=SUMU+(NRSAM**3-NRSAM)/12.
        NRSAM=0
245     CONTINUE
        DO 260 K=1,NM1
        L=K+1
        DO 250 M=L,I
        IF(X(K,1) .LE. X(M,1))GO TO 250
        A=X(K,1)
        X(K,1)=X(M,1)
        X(M,1)=A
        B=X(K,3)
        X(K,3)=X(M,3)
        X(M,3)=B
250     CONTINUE
260     X(K,2)=K
        X(I,2)=I
        DO 300 J=1,NM1
        K=J+1
        IF(X(J,1) .NE. X(K,1))GO TO 270
        IF(NRSAM .EQ. 0)GO TO 265
        RSUM=RSUM+X(K,2)
        NRSAM=NRSAM+1
        IF(J .NE. NM1)GO TO 300
        GO TO 272
265     RSUM=X(J,2)+X(K,2)
        NRSAM=2
        IF(J .NE. NM1)GO TO 300
        GO TO 272
270     IF(NRSAM .EQ. 0)GO TO 300
272     AVGRK=RSUM/NRSAM
        DO 280 L=1,NRSAM
        IF(J .NE. NM1)M=J-L+1
        IF(J .EQ. NM1 .AND. X(J,1) .EQ. X(K,1))M=J-L+2
        IF(J .EQ. NM1 .AND. X(J,1) .NE. X(K,1))M=J-L+1
280     X(M,2)=AVGRK
        RSUM=0
        SUMV=SUMV+(NRSAM**3-NRSAM)/12.
        NRSAM=0
300     CONTINUE
        DO 310 K=1,NM1
        L=K+1
        DO 310 M=L,I
        IF(X(K,3) .LE. X(M,3))GO TO 310
        A=X(K,3)
        X(K,3)=X(M,3)
        X(M,3)=A
        B=X(K,1)
        X(K,1)=X(M,1)
        X(M,1)=B
```

```
          C=X(K,2)
          X(K,2)=X(M,2)
          X(M,2)=C
310       CONTINUE
          DO 340 K=1,NM1
          L=K+1
          DO 340 M=L,I
          IF(Y(K,3) .LE. Y(M,3)) GO TO 340
          A=Y(K,3)
          Y(K,3)=Y(M,3)
          Y(M,3)=A
          B=Y(K,1)
          Y(K,1)=Y(M,1)
          Y(M,1)=B
          C=Y(K,2)
          Y(K,2)=Y(M,2)
          Y(M,2)=C
340       CONTINUE
          WRITE(IOUT,345)
345       FORMAT('0IF YOU WANT TO SEE THE INDIVIDUAL DATA ITEMS, '/
         1' TYPE A 1.  OTHERWISE, TYPE A 0.')
          READ, IDATA
          IF(IDATA .EQ. 1)WRITE(IOUT,350)
350       FORMAT('0X VALUE  Y VALUE  X RANK  Y RANK         D        D2'/)
          DO 380 J=1,I
          D=Y(J,2)-X(J,2)
          DSQD=D**2
          SUMD2=SUMD2+DSQD
          IF(IDATA .EQ. 1)WRITE(IOUT,360)Y(J,1),X(J,1),Y(J,2),X(J,2),D,DSQD
360       FORMAT(1X,F7.2,2X,F7.2,2X,F6.2,2X,F6.2,F9.2,F13.2)
380       CONTINUE
          IORIG=I+1
          WRITE(IOUT,388)IORIG
388       FORMAT('0THE ORIGINAL SAMPLE SIZE IS ',I5)
          WRITE(IOUT,390)SUMD2
390       FORMAT('0THE SUM OF D SQUARED IS',F15.2)
          WRITE(IOUT,400)SUMU,SUMV
400       FORMAT(' SUMU=',F12.2/' SUMV=',F12.2)
          SUMX2=(I**3-I)/12.-SUMV
          SUMY2=(I**3-I)/12.-SUMU
          IF(SUMX2 .EQ. 0 .OR. SUMY2 .EQ. 0)GO TO 410
          R=(SUMX2+SUMY2-SUMD2)/(2.*SQRT(SUMX2*SUMY2))
          WRITE(IOUT,404)R
404       FORMAT('0R=',F8.2)
          Z=R*SQRT(I-1.)
          WRITE(IOUT,408)Z
408       FORMAT('0Z=',F8.2)
410       IF(SUMX2 .EQ. 0 .OR. SUMY2 .EQ. 0)WRITE(IOUT, 420)
420       FORMAT('0THE VALUE OF R IS UNDEFINED DUE TO TIES.')
          WRITE(IOUT,440)
440       FORMAT('0DO YOU WANT TO SOLVE ANOTHER PROBLEM?'/
         1' IF YES, TYPE 1. IF NO, TYPE 0.')
          READ, INDEX
          IF(INDEX .EQ. 1) GO TO 5
          STOP
          END


      PROGRAM PLOT
C     THIS PROGRAM PLOTS 70X30 GRAPHS OF DATA POINTS WHICH ARE
C     ENTERED EITHER DIRECTLY OR FROM A DISK FILE.  THE MAXIMUM
C     NUMBER OF DATA POINTS IS 1000.  A MAXIMUM OF 70 POINTS
C     WILL BE PLOTTED ON EACH GRAPH, FOR A TOTAL OF UP TO 15
C     GRAPHS.  THE PRINCIPAL PURPOSE IS TO HELP ESTABLISH
C     THE BEGINNING OF STEADY STATE.
      DIMENSION Y(30),IX(1001),IXY(1000,30),DATA(1001)
      DATA ((IXY(I,J),I=1,1000),J=1,30)/30000*'-'/
      DATA MARK/'X'/
```

```
        IIN=1
        IOUT=2
        DO 10 I=1,30
10      Y(I)=0.
        WRITE(IOUT,30)
30      FORMAT('0THIS PROGRAM PLOTS A GRAPH OF DATA POINTS.  THE',
       1' Y-AXIS CONSISTS OF THE RANGE '/' OF VALUES OF THE',
       2' SUCCESSIVE OBSERVATIONS, X, DIVIDED INTO 29 EQUAL INCRE',
       3'MENTS.'/' THE X-AXIS GIVES A SEQUENCE NUMBER FOR EACH ',
       4' OBSERVATION, UP TO THE SAMPLE'/' SIZE, N.  EACH X IS ',
       5' PLOTTED AT THE Y-AXIS POINT CLOSEST TO ITS TRUE VALUE.')
        WRITE(IOUT,35)
35      FORMAT('0IF INPUT IS FROM DISK, TYPE A 1.  OTHERWISE, TYPE A 0.')
        READ, INPUT
        IF(INPUT .EQ. 1)IIN=20
        IF(INPUT .EQ. 0)WRITE(IOUT,40)
40      FORMAT('0TYPE THE DATA POINTS ONE TO A LINE. END WITH -99999.')
        READ(IIN,*)DATA(1)
        IX(1)=0
        SMALL=DATA(1)
        BIG=DATA(1)
        DO 50 N1=2,1001
        N=N1
        IF(IIN .EQ. 20)READ(IIN,*,END=60)DATA(N)
        IF(IIN .EQ. 1)READ(IIN,*)DATA(N)
        IX(N)=N-1
        IF(DATA(N) .EQ. -99999) GO TO 60
        IF(DATA(N) .LT. SMALL)SMALL=DATA(N)
        IF(DATA(N) .GT. BIG)BIG=DATA(N)
50      CONTINUE
60      N=N-1
        IX(N+1)=N
        GRAF=FLOAT(N)/70.
        NGRAF=N/70
        IF(GRAF .GT. NGRAF)NGRAF=NGRAF+1
        DO 80 I=1,30
80      Y(I)=SMALL+(BIG-SMALL)*(I-1)/29.
        Y(30)=Y(30)+.00005
        DO 160 I=1,N
        DO 150 J=1,30
        MATCH=0
        IF(DATA(I) .EQ. Y(J))IXY(I,J)=MARK
        IF(DATA(I) .EQ. Y(J))MATCH=1
150     CONTINUE
        IF(MATCH .EQ. 1)GO TO 160
        DO 155 J=1,29
        P1=ABS(DATA(I)-Y(J))
        P2=ABS(DATA(I)-Y(J+1))
        IF(DATA(I) .GT. Y(J) .AND. DATA(I) .LT. Y(J+1)
       1 .AND. P1 .LT. P2)IXY(I,J)=MARK
        IF(DATA(I) .GT. Y(J) .AND. DATA(I) .LT. Y(J+1)
       1 .AND. P1 .GE. P2)IXY(I,J+1)=MARK
155     CONTINUE
160     CONTINUE
        DO 100 I=1,NGRAF
        DO 110 J=1,30
        K2=1+70*(I-1)
        K3=I*70+1
        K3A=I*70
        IF(K3A .GT. N)K3A=N
110     WRITE(IOUT,120)Y(30-J+1),(IXY(K,30-J+1),K=K2,K3A)
120     FORMAT(1X,F8.3,1H:,70A1)
        IF(K3 .GT. N)K3=N+1
        WRITE(IOUT,140)(IX(K1),K1=K2,K3,5)
140     FORMAT(5X,15I5/)
100     CONTINUE
        STOP
        END
```

```
          PROGRAM STEADY
C         THIS PROGRAM IMPLEMENTS THREE HEURISTICS FOR DETERMINING
C         STEADY STATE:  1) CONWAY'S TRUNCATION TECHNIQUE, 2) EMSHOFF
C         AND SISSON'S MOVING AVERAGE METHOD, AND 3) EMSHOFF AND SISSON'S
C         GREATER-LESSER METHOD.  THE MAXIMUM RUN LENGTH IS 4000
C         OBSERVATIONS.  OBSERVATIONS ARE PARTITIONED INTO SUBGROUPS OF SIZE L.
C         THE SAMPLE SIZE SHOULD EXCEED 50.
C         THE RATIONALE FOR THE ABOVE-AND-BELOW SET DECISION IS THAT THE
C         VALUE OF CRITICAL CHI-SQUARE FOR 1 D.F. AT THE .05 LEVEL IS
C         3.84.  THUS, NO MORE THAN 17 OR FEWER THAN 8 IN A SUBGROUP OF 25
C         SHOULD BE ABOVE OR BELOW THE PREVIOUS MEAN TO QUALIFY.
          DIMENSION DATA(4001),TL(161),AVG(161)
          DIMENSION HIGH(3)
          DATA HIGH(1),HIGH(2),HIGH(3)/9.,15.,21./
          IIN=1
          IOUT=2
          N=4001
          SUMX=0.
          SUMX2=0.
          PAVG=0
          CAVG=0
          NCUM=0
          IFIRST=0
          JFIRST=0
          DO 10 J=1,161
          TL(J)=0.
10        AVG(J)=0.
          WRITE(IOUT,20)
20        FORMAT('0THIS PROGRAM PERFORMS THREE HEURISTICS FOR STEADY ',
         1'STATE DETECTION.')
          WRITE(IOUT,25)
25        FORMAT('0IF YOU WANT INFORMATION, TYPE A 1.',
         1' OTHERWISE, TYPE A 0.')
          READ, INST
          IF(INST .EQ. 0)GO TO 29
          WRITE(IOUT,27)
27        FORMAT('0THIS PROGRAM COMPUTES THE POINT OF STEADY STATE',
         1' BY THESE CRITERIA:'/' 1. CONWAY''S RULE OF THE ',
         2'FIRST BEING NEITHER THE MAX NOR THE MIN OF THE REST;'/
         3' 2. EMSHOFF AND SISSON''S RULE OF THE FIRST SUBGROUP HAVING ',
         4'AS MANY ABOVE '/'    THE PREVIOUS MEAN AS BELOW IT;'/
         5' 3. EMSHOFF AND SISSON''S RULE OF THE FIRST SUBGROUP HAVING '/
         6'    A MOVING AVERAGE SIMILAR TO THAT OF THE PREVIOUS SUBGROUP.')
29        WRITE(IOUT,30)
30        FORMAT('0ACCEPTABLE SUBGROUP SIZES ARE 10, 20 OR 30.'/
         1' TYPE THE DESIRED SUBGROUP SIZE.')
          READ, L
          L1=L/10
          XHI=HIGH(L1)
          XLO=L-HIGH(L1)
          LSTSIZ=L
          WRITE(IOUT,31)
31        FORMAT('0IF DATA ARE ENTERED FROM DISK, TYPE A 1.',
         1' OTHERWISE, TYPE A 0.')
          READ, IDISK
          DO 33 I=1,N
33        IF(IDISK .EQ. 1)READ(20,*,END=32)DATA(I)
32        IF(IDISK .EQ. 1)I=I-1
          IF(IDISK .EQ. 1)GO TO 40
          WRITE(IOUT,35)
35        FORMAT('0ENTER THE DATA, ENDING WITH A DUMMY -99999.')
          DO 38 K=1,4001
          READ, VALUE
          IF(VALUE .EQ. -99999)I=K-1
          IF(VALUE .EQ. -99999)GO TO 40
38        DATA(K)=VALUE
40        GRP=I/FLOAT(L)
          IGRP=I/L
```

```
        IF(IGRP .LT. GRP)IGRP1=IGRP+1
        IF(IGRP .EQ. GRP)IGRP1=IGRP
C       THIS SECTION PERFORMS CONWAY'S TRUNCATION TECHNIQUE
100     IM1=I-1
        DO 120 K=1,IM1
        KP1=K+1
        XMIN=999999.
        XMAX=-999999.
        DO 110 J=KP1,I
        IF(DATA(J) .GT. XMAX) XMAX=DATA(J)
110     IF(DATA(J) .LT. XMIN)XMIN=DATA(J)
        IF(DATA(K) .GE. XMIN .AND. DATA(K) .LE. XMAX)GO TO 130
120     CONTINUE
        WRITE(IOUT,125)
125     FORMAT('0STEADY STATE WAS NEVER REACHED ACCORDING TO ',
       1' CONWAY''S RULE.')
        GO TO 204
130     WRITE(IOUT,150)K,DATA(K),XMIN,XMAX
150     FORMAT('0THE FIRST STEADY STATE OBSERVATION ACCORDING TO ',
       1'CONWAY''S RULE IS NUMBER ',I5,'.'/' THE VALUE OF',
       2' THAT OBSERVATION IS      ',F15.2/' THE MINIMUM VALUE IN THE ',
       3'REMAINING SET IS ',F10.2/' THE MAXIMUM VALUE IN THE ',
       4'REMAINING SET IS ',F10.2/)
C       THIS SECTION USES EMSHOFF AND SISSON'S ABOVE-AND-BELOW
C       AND MOVING-AVERAGE METHODS
204     WRITE(IOUT,205)
205     FORMAT(' IF YOU WANT TO SEE DETAILS OF THE EMSHOFF',
       1' AND SISSON TESTS,'/' TYPE A 1. OTHERWISE, TYPE A 0.')
        READ, IDET
        IF(IDET .EQ. 1)WRITE(IOUT,218)
218     FORMAT('0',7X,'*A*B*O*V*E*--*A*N*D*--*B*E*L*O*W*--*M*O*V*',
       1'I*N*G*--*A*V*E*R*A*G*E*')
        IF(IDET .EQ. 1)WRITE(IOUT,219)
219     FORMAT(9X,'PREVIOUS',23X,'AVERAGE AVERAGE ST.ERR.',
       1' ST.ERR.')
        IF(IDET .EQ. 1)WRITE(IOUT,220)
220     FORMAT(' SUBGP. CUMULATIVE NUMBER NUMBER',7X,
       1'PREVIOUS CURRENT   PREV.   CURR.')
        IF(IDET .EQ. 1)WRITE(IOUT,221)
221     FORMAT(' NUMBER    AVERAGE    ABOVE  BELOW CHISQ ',
       1'   SBGP.  SBGP.  SQRD.   SQRD.      Z')
        DO 210 J=1,L
        NCUM=L
        SUMX2=SUMX2+DATA(J)**2
210     TL(1)=TL(1)+DATA(J)
        NOP=L
        VARP=((SUMX2-TL(1)**2/NOP)/(NOP-1))/NOP
        AVG(1)=TL(1)/L
        PAVG=AVG(1)
        DO 250 J=2,IGRP1
        ILESS=0
        IMORE=0
        SUMX2=0
        X=(GRP-IGRP)*L
        IX=X
        IF(J .EQ. IGRP1 .AND. IGRP .NE. GRP .AND. X .EQ. IX)
       1LSTSIZ=IX
        IF(J .EQ. IGRP1 .AND. IGRP .NE. GRP .AND. X .NE. IX)
       1LSTSIZ=IX+1
        LL=(J-1)*L+1
        LU=(J-1)*L+LSTSIZ
        NUM=LU-LL+1
        DO 230 K=LL,LU
        IF(DATA(K) .LT. AVG(J-1))ILESS=ILESS+1
        IF(DATA(K) .GT. AVG(J-1))IMORE=IMORE+1
        TL(J)=TL(J)+DATA(K)
```

```
          SUMX2=SUMX2+DATA(K)**2
          CAVG=CAVG+DATA(K)
230       NCUM=NCUM+1
          CHISQ=2.*(IMORE-L/2.)**2/(L/2.)
          CAVG=CAVG/NUM
          VARC=((SUMX2-TL(J)**2/NUM)/(NUM-1))/NUM
          Z=(PAVG-CAVG)/SQRT(VARP+VARC)
          IF(IDET .EQ. 1)WRITE(IOUT,240)J,AVG(J-1),IMORE,ILESS,
         1CHISQ,PAVG,CAVG,VARP,VARC,Z
240       FORMAT(I6,F11.3,I7,I7,F7.2,F8.3,F9.3,F8.2,F8.2,F7.2)
          IF(IMORE.LT.XHI.AND.IMORE.GT.XLO.AND.IFIRST.EQ.0)IFIRST=J
          Z=ABS(Z)
          IF(Z .LT. 1.96 .AND. JFIRST .EQ. 0)ZF=Z
          IF(Z .LT. 1.96 .AND. JFIRST .EQ. 0)JFIRST=J
          TL(J)=TL(J-1)+TL(J)
          PAVG=CAVG
          CAVG=0.
          VARP=VARC
          NOP=NUM
250       AVG(J)=TL(J)/NCUM
300       IF(IFIRST .GT. 0)WRITE(IOUT,310)IFIRST
310       FORMAT('0THE CRITICAL VALUE OF CHI-SQUARE AT THE .05 ',
         1'LEVEL OF SIGNIFICANCE IS 3.84 .'/
         1' ACCORDING TO THE EMSHOFF AND SISSON ABOVE-AND-BELOW',
         2' CRITERION, '/' THE FIRST SUBGROUP FOR WHICH THE ',,
         3'COMPUTED CHI-SQUARE IS BELOW 3.84'/' IS SUBGROUP ',
         4'NUMBER ',I5,'.')
          IF(IFIRST .EQ. 0)WRITE(IOUT,311)
311       FORMAT('0STEADY STATE WAS NEVER REACHED ACCORDING TO ',
         1' THE ABOVE-AND-BELOW CRITERION.')
          IF(JFIRST .GT. 0)WRITE(IOUT,320)JFIRST
320       FORMAT('0THE CRITICAL VALUE OF Z AT THE .05 LEVEL OF ',
         1'SIGNIFICANCE IS 1.96 .'/' ACCORDING TO THE F&S MOVING-AVE',
         2'RAGE CRITERION, THE FIRST SUBGROUP '/' FOR WHICH THE COMPUTED',
         3' Z IS BELOW 1.96 IS SUBGROUP NUMBER ',I5,'.')
          IF(JFIRST .GT. 0 .AND. L .LT. 25)WRITE(IOUT,330)
330       FORMAT(' *WARNING*SUBGROUP SIZE TOO SMALL FOR PRECISE TEST*')
          IF(JFIRST .EQ. 0)WRITE(IOUT,321)
321       FORMAT('0STEADY STATE WAS NEVER REACHED ACCORDING TO ',
         1' THE MOVING-AVERAGE CRITERION.')
          STOP
          END

          PROGRAM SIGN
          DIMENSION X(10)
          DATA PLUS/'+'/
          IOUT=2
          WRITE(IOUT,5)
5         FORMAT('0THIS PROGRAM EXTRACTS SIGNED SAVEVALUES FROM '/
         1' THE IMAGE OF GPSS OUTPUT ON DISK, REMOVES THE '/
         2' SIGN, ADDS A DECIMAL POINT, AND WRITES THE '/
         3' RESULTING VALUES IN ANOTHER DISK FILE.')
25        READ(20,30,END=200)(X(I),I=1,10)
30        FORMAT(23X,10A1)
          DO 100 I=1,9
          J=I+1
          IF(I .EQ. 1 .AND. X(I) .EQ. PLUS)WRITE(30,40)(X(K),K=J,10)
          IF(I .EQ. 2 .AND. X(I) .EQ. PLUS)WRITE(30,45)(X(K),K=J,10)
          IF(I .EQ. 3 .AND. X(I) .EQ. PLUS)WRITE(30,50)(X(K),K=J,10)
          IF(I .EQ. 4 .AND. X(I) .EQ. PLUS)WRITE(30,55)(X(K),K=J,10)
          IF(I .EQ. 5 .AND. X(I) .EQ. PLUS)WRITE(30,60)(X(K),K=J,10)
          IF(I .EQ. 6 .AND. X(I) .EQ. PLUS)WRITE(30,65)(X(K),K=J,10)
          IF(I .EQ. 7 .AND. X(I) .EQ. PLUS)WRITE(30,70)(X(K),K=J,10)
          IF(I .EQ. 8 .AND. X(I) .EQ. PLUS)WRITE(30,75)(X(K),K=J,10)
          IF(I .EQ. 9 .AND. X(I) .EQ. PLUS)WRITE(30,80)(X(K),K=J,10)
40        FORMAT(1X,9A1,'.')
```

```
45       FORMAT(2X,8A1,'.')
50       FORMAT(3X,7A1,'.')
55       FORMAT(4X,6A1,'.')
60       FORMAT(5X,5A1,'.')
65       FORMAT(6X,4A1,'.')
70       FORMAT(7X,3A1,'.')
75       FORMAT(8X,2A1,'.')
80       FORMAT(9X,A1,'.')
100        CONTINUE
           GO TO 25
200      STOP
         END


         PROGRAM NANOVA
C        THIS PROGRAM COMPUTES THE KRUSKAL-WALLIS "H" STATISTIC
C        FOR ONE-WAY ANALYSIS OF VARIANCE IF THE NUMBER OF TREATMENT
C        GROUPS IS 3 OR MORE.  IF THE NUMBER OF TREATMENT GROUPS IS
C        2, THE PROGRAM COMPUTES THE MANN-WHITNEY-WILCOXON
C        STATISTIC.  A CORRECTION FACTOR FOR TIES IS INCLUDED.  FOR
C        SUFFICIENTLY LARGE SAMPLE SIZES THE "H" STATISTIC IS CHI
C        SQUARE DISTRIBUTED.
C        DUNN'S TEST OF MULTIPLE COMPARISONS IS ALSO PERFORMED
C        FOR PAIRWISE COMPARISONS OF UP TO 8 TREATMENT
C        GROUPS AT THE .05 AND .10 LEVELS OF SIGNIFICANCE.
C        IF INPUT FROM A DISK FILE IS DESIRED, THE FILE
C        SHOULD BE DESIGNATED AS DEVICE NUMBER 20.
         DIMENSION X(1000,3),N(10),SUMRK(10),AVRNK(10),DIFF(10)
         DIMENSION Z(2,8)
         DATA Z(1,1),Z(1,2),Z(1,3),Z(1,4)/1.960,2.241,2.394,2.498/
         DATA Z(1,5),Z(1,6),Z(1,7),Z(1,8)/2.576,2.638,2.690,2.734/
         DATA Z(2,1),Z(2,2),Z(2,3),Z(2,4)/1.645,1.960,2.128,2.241/
         DATA Z(2,5),Z(2,6),Z(2,7),Z(2,8)/2.326,2.394,2.450,2.498/
         INTEGER SAMSW
         IOUT=2
5        DO 10 I=1,1000
         DO 10 J=1,3
10       X(I,J)=0.
         DO 20 I=1,10
         SUMRK(I)=0
         AVRNK(I)=0
         DIFF(I)=0
20       N(I)=0
         TLSRK=0
         K=0
         TDIFF=0
         U=0
         U3=0
         NTL=0
         L=0
         SAMSW=0
         TLRNK=0
         TDIFF=0
         SUMR=0
         WRITE(IOUT,50)
50       FORMAT('1THIS PROGRAM PERFORMS NONPARAMETRIC ONE-WAY ANALYSIS '/
        1' OF VARIANCE USING THE KRUSKAL-WALLIS "H" PROCEDURE IF ',
        2'THE NUMBER '/' OF TREATMENT GROUPS IS THREE OR MORE.')
         WRITE(IOUT,53)
53       FORMAT(' DUNN''S TEST OF MULTIPLE COMPARISONS IS ALSO ',
        1'PERFORMED.'/
        2' IF THERE ARE ONLY TWO TREATMENT GROUPS, THE MANN-WHITNEY-',
        3'WILCOXON '/' TEST IS PERFORMED INSTEAD.')
         WRITE(IOUT,55)
55       FORMAT('0IF INPUT FROM A DISK FILE IS DESIRED, TYPE A 1.'/
        1' IF DATA ARE TO BE DIRECTLY ENTERED, TYPE A 0.')
         READ, IDISK
         WRITE(IOUT,60)
```

```
60       FORMAT('0ENTER THE NUMBER OF TREATMENTS (GROUPS).')
         READ, M
         WRITE(IOUT,70)
70       FORMAT('0IF ALL GROUPS ARE THE SAME SIZE, ENTER '/
         1' THAT SAMPLE SIZE.  OTHERWISE ENTER 999.')
         READ, NUM
         IF(NUM .NE. 999) GO TO 100
         DO 90 J=1,M
         WRITE(IOUT,80)J
80       FORMAT('0ENTER THE SAMPLE SIZE FOR GROUP ',I3)
         READ, N(J)
90       NTL=NTL+N(J)
         GO TO 120
100      DO 110 J=1,M
110      N(J)=NUM
         NTL=NUM*M
120      DO 170 J=1,M
         NUM=N(J)
         IF(IDISK .EQ. 0)WRITE(IOUT,130)J
130      FORMAT('0FOR GROUP ',I3,' ENTER OBSERVATIONS:')
         DO 170 I=1,NUM
         L=L+1
         IF(IDISK .EQ. 1)READ (20,*)X(L,1)
         IF(IDISK .EQ. 0)READ, X(L,1)
170      X(L,2)=L
         LM1=L-1
         DO 200 I1=1,LM1
         I2=I1+1
         DO 200 I3=I2,L
         IF(X(I1,1) .LE. X(I3,1))GO TO 200
         A=X(I1,1)
         X(I1,1)=X(I3,1)
         X(I3,1)=A
         B=X(I1,2)
         X(I1,2)=X(I3,2)
         X(I3,2)=B
200      CONTINUE
         DO 210 I=1,L
210      X(I,3)=I
         DO 250 I=1,LM1
         I1=I+1
         IF(X(I,1) .NE. X(I1,1))GO TO 240
         IF(SAMSW .EQ. 0)GO TO 230
         SAMSW=SAMSW+1
         TLRNK=TLRNK+X(I1,3)
         IF(I .NE. LM1) GO TO 250
         AVRK=TLRNK/SAMSW
         DO 220 I3=1,SAMSW
         M1=I-I3+2
220      X(M1,3)=AVRK
         U3=U3+SAMSW**3
         U=U+SAMSW
         GO TO 250
230      SAMSW=2
         TLRNK=X(I,3)+X(I1,3)
         IF(I .NE. LM1) GO TO 250
         AVRK=TLRNK/SAMSW
         X(LM1,3)=AVRK
         X(L,3)=AVRK
         U3=U3+SAMSW**3
         U=U+SAMSW
         GO TO 250
240      IF(SAMSW .EQ. 0)GO TO 250
         AVRK=TLRNK/SAMSW
         DO 245 I3=1,SAMSW
         M1=I-I3+1
245      X(M1,3)=AVRK
```

```
          U3=U3+SAMSW**3
          U=U+SAMSW
          SAMSW=0
          TLRNK=0
250       CONTINUE
          DO 280 I1=1,LM1
          I2=I1+1
          DO 280 I3=I2,L
          IF(X(I1,2) .LE. X(I3,2))GO TO 280
          A=X(I1,2)
          X(I1,2)=X(I3,2)
          X(I3,2)=A
          B=X(I1,1)
          X(I1,1)=X(I3,1)
          X(I3,1)=B
          C=X(I1,3)
          X(I1,3)=X(I3,3)
          X(I3,3)=C
280       CONTINUE
          DO 330 J=1,M
          NUM=N(J)
          DO 320 I=1,NUM
          K=K+1
320       SUMRK(J)=SUMRK(J)+X(K,3)
          TLSRK=TLSRK+SUMRK(J)
330       AVRNK(J)=SUMRK(J)/NUM
          TLSRK=TLSRK/L
          DO 340 J=1,M
          DIFF(J)=(SUMRK(J)-N(J)*TLSRK)**2/N(J)
340       TDIFF=TDIFF+DIFF(J)
          CONS=12./(L*(L+1))
          CORR=1.-(U3-U)/(L*(L**2-1))
          H=CONS*TDIFF/CORR
          IDF=M-1
          WRITE(IOUT,505)
505       FORMAT('0DO YOU WANT TO SEE THE RAW DATA?  IF YES, ',
         1'TYPE A 1.  IF NO, TYPE A 0.')
          READ, IDATA
          IF(IDATA .EQ. 0)GO TO 518
          WRITE(IOUT,500)
500       FORMAT('1 GROUP   OBSERVATION',13X,'INITIAL',
         14X,'FINAL')
          WRITE(IOUT,501)
501       FORMAT(' NUMBER',4X,'NUMBER',7X,'VALUE',6X,
         1'RANK',6X,'RANK')
          K=0
          DO 510 J=1,M
          NUM=N(J)
          DO 510 I=1,NUM
          K=K+1
          IF(I .EQ. 1)ISPACE=0
          IF(I .NE. 1)ISPACE=2
510       IF(IDATA .EQ. 1)WRITE(IOUT,515)ISPACE,J,I,X(K,1),X(K,2),X(K,3)
515       FORMAT(I1,1X,I3,7X,I4,5X,F7.2,5X,F6.2,4X,F6.2)
518       IF(M .EQ. 2) GO TO 560
          WRITE(IOUT,520)
520       FORMAT('0 GROUP',4X,'MEAN',2X,'SAMPLE',5X,'WEIGHTED'/
         1' NUMBER     RANK   SIZE',5X,'DIFFERENCE')
          DO 530 J=1,M
530       WRITE(IOUT,535)J,AVRNK(J),N(J),DIFF(J)
535       FORMAT(2X,I5,2X,F6.2,3X,I4,5X,F9.2)
          WRITE(IOUT,540)CORR
540       FORMAT('0THE TIE CORRECTION FACTOR=',F10.4)
C         THIS SECTION FOR KRUSKAL-WALLIS H TEST
          WRITE(IOUT,360)H,IDF
360       FORMAT('0H=',F8.2/'0IF THE SAMPLE SIZES ARE >5, ',
```

```
         1'THIS STATISTIC IS CHI-SQUARE DISTRIBUTED'/
         2' WITH ',I3,' DEGREES OF FREEDOM.')
C        THIS SECTION FOR DUNN'S TEST
         WRITE(IOUT,700)
700      FORMAT('0',10X,'RESULTS OF DUNN''S TEST')
         WRITE(IOUT,710)
710      FORMAT('0 GROUP   GROUP    ABS.DIFF.   CRITERION     AT'/
         132X,'.05      .10')
         DO 720 I=2,M
         IM1=I-1
         DO 720 J=1,IM1
         DR=ABS(AVRNK(I)-AVRNK(J))
         CR1=(NTL*(NTL**2-1))-(U3-U)
         CR2=1./N(I)+1./N(J)
         CR3=12.*(NTL-1)
         CR5=Z(1,M)*SQRT(CR1*CR2/CR3)
         CR10=Z(2,M)*SQRT(CR1*CR2/CR3)
720      WRITE(IOUT,730)J,I,DR,CR5,CR10
730      FORMAT(I5,3X,I5,5X,F6.3,7X,F6.3,3X,F6.3)
         GO TO 599
C        THIS SECTION FOR MANN-WHITNEY-WILCOXON TEST
560      NTL=N(1)+N(2)
         XNUMER=SUMRK(1)-N(1)*(NTL+1)/2.
         IF(XNUMER .LT. 0)XNUMER=XNUMER+0.5
         IF(XNUMER .GT. 0)XNUMER=XNUMER-0.5
         DENOM1=N(1)*N(2)/(12.*NTL*(NTL-1))
         DENOM2=NTL*(NTL**2-1)-(U3-U)
         DENOM=SQRT(DENOM1*DENOM2)
         ZCORR=XNUMER/DENOM
         ZUCORR=XNUMER/SQRT(N(1)*N(2)*(NTL+1)/12.)
         WRITE(IOUT,565)
565      FORMAT('0USING THE FIRST SAMPLE AS A BASE,')
         WRITE(IOUT,570)ZCORR,ZUCORR
570      FORMAT(' THE Z VALUE CORRECTED FOR CONTINUITY AND TIES IS ',F8.3/
         1' THE Z VALUE CORRECTED FOR CONTINUITY ONLY IS ',F8.3)
         WRITE(IOUT,580)
580      FORMAT('0THE NORMAL APPROXIMATION IS VALID FOR SAMPLE SIZES '/
         1' OF TEN OR MORE IN EACH SAMPLE.')
599      WRITE(IOUT,600)
600      FORMAT('0DO YOU WANT TO SOLVE ANOTHER PROBLEM?'/
         1' IF YES, TYPE 1.  IF NO, TYPE 0.')
         READ, IPROB
         IF(IPROB .EQ. 1) GO TO 5
         STOP
         END
```

INDEX